Undergraduate Lecture Notes in Physics

Undergraduate Lecture Notes in Physics (ULNP) publishes authoritative texts covering topics throughout pure and applied physics. Each title in the series is suitable as a basis for undergraduate instruction, typically containing practice problems, worked examples, chapter summaries, and suggestions for further reading.

ULNP titles must provide at least one of the following:

- An exceptionally clear and concise treatment of a standard undergraduate subject.
- A solid undergraduate-level introduction to a graduate, advanced, or non-standard subject.
- A novel perspective or an unusual approach to teaching a subject.

ULNP especially encourages new, original, and idiosyncratic approaches to physics teaching at the undergraduate level.

The purpose of ULNP is to provide intriguing, absorbing books that will continue to be the reader's preferred reference throughout their academic career.

Series Editors

Neil Ashby
University of Colorado, Boulder, CO, USA

William Brantley
Department of Physics, Furman University, Greenville, SC, USA

Matthew Deady
Physics Program, Bard College, Annandale-on-Hudson, NY, USA

Michael Fowler
Department of Physics, University of Virginia, Charlottesville, VA, USA

Morten Hjorth-Jensen
Department of Physics, University of Oslo, Oslo, Norway

Michael Inglis
Department of Physical Sciences, SUNY Suffolk County Community College, Selden, NY, USA

More information about this series at http://www.springer.com/series/8917

Ilya L. Shapiro

A Primer in Tensor Analysis and Relativity

 Springer

Ilya L. Shapiro
Departamento de Física
Universidade Federal de Juiz de Fora
Juiz de Fora, Minas Gerais, Brazil

ISSN 2192-4791   ISSN 2192-4805   (electronic)
Undergraduate Lecture Notes in Physics
ISBN 978-3-030-26894-7   ISBN 978-3-030-26895-4   (eBook)
https://doi.org/10.1007/978-3-030-26895-4

This Springer imprint is published by the registered company Springer Nature Switzerland AG
The registered company address is: Gewerbestrasse 11, 6330 Cham, Switzerland

To Elena

Preface

This book is based on the three sets of lecture notes that were prepared in the Federal University of Juiz de Fora starting from the late 1996. The three parts of the manuscript belong to different courses.

The contents of the first part of the book are designed for the second- or third-year undergraduate student with majoring in Physics. The knowledge of tensor calculus is necessary for successfully learning all parts of theoretical physics, and in my opinion it is better to learn tensors in the second year of an undergraduate program, e.g., after learning linear algebra and calculus of several variables. When I was a student at Tomsk State University in Russia, we had such a course taught by Veniamin Alexeevich Kuchin. I always thought that this course better prepared us for learning further levels of Physics. The first part of the present book is an essentially extended version of the tensor course, but the basic idea is the same. The main point of this part is that the geometric (and consequently) physical laws should be formulated in such a way that the dependence on the choice of coordinates can be kept under control.

The first part can also be used for independent learning tensors by an interested student. Typically, a motivated student can do it (solving all exercises) in 3 months. Our experience has shown that these lectures can benefit not only physicists, but also students majoring in Mathematics and related areas, such as computer graphics.

The second part of the present book is intermediate, linking the very introductory first part and the relatively advanced third part. Certainly, a physicist has to know much more about electrodynamics than it is possible to learn from Part II of this book. However, our aim is not to replace the Electrodynamics courses, but rather to show important intermediate-level applications of tensor analysis. From the practical side, this part of the book illustrates that (i) tensor methods can be extended to the pseudo-Euclidean space, and (ii) Maxwell equations can be made compatible with Mechanics by modifying the classical notions of space and time—that means introducing special relativity. The full version of my notes on Electrodynamics is essentially based on the book of Landau and Lifshits [1], and there is not much sense repeating it here.

The third part of the book contains an extended version of the lecture notes on general relativity, which were used in our research group to teach students in the last 20 years. The organization and contents of this part are a little bit different from the standard courses, mainly because they include material most useful for the students who are going to work in quantum field theory and quantum gravity. This part of the book was intended to be compact and very technical, and to have some sections that can be hardly found in other textbooks. For instance, it covers conformal transformations, Riemann normal coordinates, and the perturbative expansions of the gravitational actions. I suppose that this part of the book can be useful not only for students but also for the researchers who successfully work in Particle Physics or elsewhere, and intend to learn quickly the necessary information from general relativity.

Juiz de Fora, Minas Gerais, Brazil   Ilya L. Shapiro
December 2018

Reference

1. L.D. Landau, E.M. Lifshits, *Mechanics—Course of Theoretical Physics*, vol. 1 (Butterworth-Heinemann, 1976)

Acknowledgements

I first would like to thank my teacher Prof. Iosif Buchbinder, who taught us general relativity when I was undergraduate student in Tomsk State University—first by answering my questions about the book by S. Weinberg [1] and then giving us a very well-organized course on relativity. In the decades that passed after that, I benefited a lot from the discussion of many issues related to relativity with many colleagues, especially Manual Asorey, Roberto Balbinot, Alessandro Fabbri, Julio Fabris, Valeri Pavlovich Frolov, Alexei Alexandrovich Starobinsky, and Igor Tkachev. Let me also mention fruitful interactions with our former students Flavia Sobreira, Davi Rodrigues, Tibério de Paula Netto, and Breno Giacchini on various subjects treated in this book. I am very grateful to all of them and only wish to stress that all possible mistakes and faults in my knowledge are completely on my own responsibility.

The work on this book took about 20 years. Throughout this time, the work of the author has been partially supported by Conselho Nacional de Desenvolvimento Científico e Tecnológico—CNPq and Fundação de Amparo à Pesquisa de Minas Gerais—FAPEMIG. Let me mention that the project APQ-01205-16 from FAPEMIG provided specific technical aid for this book. Another essential source for support of this project came from International Centre for Theoretical Physics— ICTP. In particular, an essential part of the second and third parts of the book was prepared during the last visit of the author to ICTP in 2017 as Associate Member. I am really grateful to ICTP in general and especially to the high energy group for their kind hospitality and nice working conditions during all visits. The preliminary version of the first part on tensors has been published as "Lecture Notes" in 2003 in CBPF (Rio de Janeiro), and I am grateful to José Abdalla Helayël-Neto and Flavio Takakura for their generous assistance in organizing and preparing this edition.

A very special thanks is to Hannah Kaufman from Springer Nature, who was very kind and generous to improve my English to the level that readers can see.

Last but not least, I have to acknowledge the contribution of Alexandre Chapiro, who made useful, critical observations leading to essential improvements in the contents of Part I of the book. And finally, I am really grateful to my wife E. Konstantinova for her enormous patience and support for my work in general and especially this book.

Reference

1. S. Weinberg, *Gravitation and Cosmology* (Wiley, 1972)

Preliminary Observations and Notations

Let us briefly review the notations system and abbreviations used in this book.

(i) It is assumed that the student has a background in Calculus, Analytic Geometry, and Linear Algebra. Sometimes, the corresponding information will be repeated in the text of these notes. We do not try to substitute for the corresponding courses here, but only to supplement them.

(ii) Some objects with indices will be used below. Latin indices run the values

$$(a, b, c, \ldots, i, j, k, l, m, n, \ldots) = (1, 2, 3)$$

in $3D$ and

$$(a, b, c, \ldots, i, j, k, l, m, n, \ldots) = (1, 2, \ldots, D)$$

for an arbitrary D.

In the first part of the book, the indices $(a, b, c, \ldots)$ correspond to the orthonormal basis and to the Cartesian coordinates. The indices $(i, j, k, \ldots)$ correspond to the an arbitrary (generally nondegenerate) basis and to arbitrary, possibly curvilinear coordinates.

(iii) In the first part of the book, we consider, by default, that the space has dimension $D = 3$. However, in some cases we shall refer to an arbitrary dimension of space D, and sometimes consider $D = 2$, because this is the simplest nontrivial case. The indication of dimension is performed in shown as $3D$, meaning $D = 3$.

Starting from the second part of the book, we mainly work in $4D$. Then, the Greek indices $\alpha, \beta, \mu, \nu, etc$ run the values 0, 1, 2, 3, while the Latin indices i, j, k, etc run the values 1, 2, 3. In a few occasions, we use the first Latin letters a, b, c, d to denote the spacetime indices 0, 1, 2, 3 in the special Minkowski reference frame, when the metric is $\eta_{\mu\nu} = \eta_{ab} = diag$ $(1, -1, -1, -1)$. The signature of the space–time metric is therefore -2.

(iv) When using mathematical notations (mainly in Part I), we denote the set of the elements f_i as $\{f_i\}$. The properties of the elements are indicated after the vertical line. For example,

$$E = \{e \mid e = 2n, n \in N\}$$

means the set of even natural numbers. The comment may follow after the comma. For example,

$$\{e \mid e = 2n, n \in N, n \leq 3\} = \{2, 4, 6\}.$$

(v) The repeated upper and lower indices imply summation (Einstein convention). For example,

$$a^i b_i = \sum_{i=1}^{D} a^i b_i = a^1 b_1 + a^2 b_2 + \ldots + a^D b_D$$

for the D-dimensional case. It is important that the summation (umbral) index i here can be renamed in an arbitrary way, e.g.,

$$C_i^i = C_j^j = C_k^k = \ldots.$$

This is completely similar to the change of the notation for the variable of integration in a definite integral

$$\int_a^b f(x)dx = \int_a^b f(y)dy.$$

where, also, the name of the variable is irrelevant.

(vi) We use the notations like a'^i for the components of the vectors corresponding to the coordinates x'^i. In general, the same objects may be also denoted as $a^{i'}$ and $x^{i'}$. In the text we do not distinguish between these two notations, and one has to assume, e.g., that $a'^i = a^{i'}$. The same holds for any tensor indices, of course.

(vii) Throughout the book, we use the abbreviation

Def. $\equiv$ Definition

(viii) The exercises are dispersed in the text—an interested reader is advised to solve them in order of appearance. Some exercises are in fact simple theorems that will consequently be used; some of them may be important statements. Most of the exercises are quite simple and will consume only a small time to solve them. At the same time, in the first part, there are a few exercises marked by *. Those are presumably more difficult and are mainly recommended only for those students who are going to specialize in Mathematics or Theoretical Physics.

(ix) The vectors in $3D$ are denoted as bold Latin letters, such as **a** or **B**.

(x) When we refer to another part of the book, it can be done explicitly, or more generally as, e.g., "Part I" for the first part of the book.

Contents

Part I
Tensor Algebra and Analysis

The initial part of the book is intended for learning tensors at the level which is required for any student majoring in Physics or perhaps other related areas.

For the independent study, it is very important for the student to remember that the reading must be done in order and that it is highly recommended to complete all calculations and make all exercises before moving to the next section.

For the teachers who would like to use this part of the book in the class, let me in the classroom, let me spill my secret and say that some lengthy calculations (such as d'Alembertian operator in spherical coordinates, for instance) can be done in a much more economical way. The corresponding simple calculations are left as exercises in the text, but it is recommended to make an inversion and use a simpler version in the class, leaving lengthy calculation with all connections as an exercise. These were organized in this way for the expected benefit of the student in the third part of the book, where it is very useful to have some experience in deriving and using the coefficients of connections.

The main recommendation for further reading on differential geometry is the comprehensive monograph by Dubrovin, Novikov, and Fomenko [1]. This book contains all of what most physicists need to know about geometry, group theory, and relativity. I hope that after reading the present book the reader will find it easier to go through this complete work. Other editions that are at (more or less) the same level as Part 1 of the present book are [2–5].

References

1. A.T. Fomenko, S.P. Novikov, and B.A. Dubrovin, Modern Geometry-Methods and Applications, Part I: The Geometry of Surfaces, Transformation Groups, and Fields, (Springer, 1992).
2. A.J. McConnell, Applications of Tensor Analysis, (Dover, 1957).
3. I.S. Sokolnikoff, Tensor Analysis: Theory and Applications to Geometry and Mechanics of Continua, (John Wiley and Sons, N.Y. 1968).
4. A.I. Borisenko, I.E. Tarapov, Vector and Tensor Analysis with Applications, (Dover, N.Y. 1968-Translation from III Russian edition).
5. M.A. Akivis, V.V. Goldberg, Tensor Calculus With Applications (World Scientific, 2003); see also An Introduction to Linear Algebra and Tensors. (Dover Publications, 2010)

Chapter 1
Linear Spaces, Vectors, and Tensors

In this chapter, we shall introduce and discuss some necessary basic notions, which mainly belong to Analytic Geometry and Linear Algebra. We assume that the reader is already familiar with these disciplines, so our introduction will be brief.

1.1 Notion of Linear Space

Let us start this section from the main definition.

Def. 1 The linear space, say L, consists of the elements (vectors) that permit linear operations with the properties described below.[1] We will denote them here by bold letters, like **a**. The notion of linear space assumes that the linear operations over vectors are permitted. These operations include the following:

(i) Summing up the vectors. This means that if **a** and **b** are elements of L, then their sum **a** + **b** is also an element of L.

(ii) Multiplication by a number. This means that if **a** is an element of L, and k is a number, then the product $k\mathbf{a}$ is also an element of L. One can define that the number k must necessarily be real or permit it to be complex. In the last case, we meet a complex vector space and in the first case a real vector space. In the present book, we will almost exclusively deal with real spaces.

(iii) The linear operations defined above should satisfy the following requirements:

- $\mathbf{a} + \mathbf{b} = \mathbf{b} + \mathbf{a}$.
- $\mathbf{a} + (\mathbf{b} + \mathbf{c}) = (\mathbf{a} + \mathbf{b}) + \mathbf{c}$.
- There is a zero vector **0**, such that for any **a** we meet $\mathbf{a} + \mathbf{0} = \mathbf{a}$.

---

[1] The elements of linear spaces are conventionally called "vectors". In the subsequent sections of this chapter, we will introduce another definition of vectors and also definition of tensors. It is important to remember that in the sense of this definition, all vectors and tensors are elements of some linear spaces and, therefore, are all "vectors". For this reason, starting from the next section, we will avoid using the word "vector" when it means an element of arbitrary linear space.

© Springer Nature Switzerland AG 2019
I. L. Shapiro, *A Primer in Tensor Analysis and Relativity*, Undergraduate
Lecture Notes in Physics, https://doi.org/10.1007/978-3-030-26895-4_1

- For any $\mathbf{a}$, there is an opposite vector $-\mathbf{a}$, such that $\mathbf{a} + (-\mathbf{a}) = \mathbf{0}$.
- For any $\mathbf{a}$ we have $1 \cdot \mathbf{a} = \mathbf{a}$.
- For any $\mathbf{a}$ and for any numbers k, l we have $l(k\mathbf{a}) = (kl)\mathbf{a}$ and $(k + l)\mathbf{a} = k\mathbf{a} + l\mathbf{a}$.
- $k(\mathbf{a} + \mathbf{b}) = k\mathbf{a} + k\mathbf{b}$.

There is an infinite variety of vector spaces. It is useful to mention a few examples.
(1) The set of real numbers and the set of complex numbers form linear spaces. The set of real numbers is equivalent to the set of vectors on the straight line; this set is also a linear space.
(2) The set of vectors on the plane forms a vector space; the same is true for the set of vectors in the three-dimensional space. Both cases are considered in the standard courses of analytic geometry.
(3) The set of positive real numbers does not form a linear space, because the number 1 has no positive inverse. As usual, it is sufficient to have one counterexample to prove a negative statement.

Exercise 1 (a) Find another proof of that the set of positive real numbers does not form a linear space. (b) Prove that the set of complex numbers with negative real parts does not form a linear space. (c) Prove, by carefully inspecting all points in the part (iii), that the set of purely imaginary numbers ix forms a linear space. (d) Find another subset of the set of complex numbers that represents a linear space. Prove it by inspecting all points in (iii).

Def. 2 Consider the set of vectors, elements of the linear space L

$$\{\mathbf{a}_1, \mathbf{a}_2, ..., \mathbf{a}_n\} = \{\mathbf{a}_i | i = 1, ..., n\} \tag{1.1}$$

and the set of real numbers k_1, k_2, ... k_n. Vector

$$\mathbf{k} = \sum_{i=1}^{n} k^i \mathbf{a_i} \tag{1.2}$$

is called *linear combination* of the vectors (1.1).

Def. 3 Vectors $\{\mathbf{a_i}, i = 1, ..., n\}$ are called linearly independent if

$$\mathbf{k} = \sum_{i=1}^{n} k^i \mathbf{a}_i = k^i \mathbf{a}_i = 0 \quad \Longrightarrow \quad \sum_{i=1}^{n} (k^i)^2 = 0. \tag{1.3}$$

The last condition means that no one of the coefficients k_i can be different from zero.

Def. 4 If vectors $\{\mathbf{a}_i\}$ are not linearly independent, they are called linearly dependent.

Comment: For example, in $2D$, two vectors are linearly dependent if and only if they are parallel; in $3D$, three vectors are linearly dependent if and only if they belong to the same plane, etc.

Exercise 2 (a) Prove that vectors $\{\mathbf{0}, \mathbf{a}_1, \mathbf{a}_2, \mathbf{a}_3, ..., \mathbf{a}_n\}$ are always linearly dependent.

(b) Prove that $\mathbf{a}_1 = \mathbf{i} + \mathbf{k}$, $\mathbf{a}_2 = \mathbf{i} - \mathbf{k}$, $\mathbf{a}_3 = \mathbf{i} - 3\mathbf{k}$ are linearly dependent for $\forall \{\mathbf{i}, \mathbf{k}\}$.

Exercise 3 Prove that $\mathbf{a}_1 = \mathbf{i} + \mathbf{k}$ and $\mathbf{a}_2 = \mathbf{i} - \mathbf{k}$ are linearly independent if and only if $\mathbf{i}$ and $\mathbf{k}$ are linearly independent.

It is clear that in some cases the linear space L can only admit a restricted amount of linearly independent elements. For example, in case of the set of real numbers, there may be only one such element, because any two elements are linearly dependent. In this case, we say that the linear space is one-dimensional. In the general linear space, the maximal number of linearly independent elements is used as a definition of the dimension of a space.

Def. 5 The maximal number of linearly independent elements of the linear space L is called its *dimension*. It proves useful to denote L_D a linear space of dimension D.

For example, the usual plane studied in analytic geometry admits two linearly independent vectors, and so we call it two-dimensional space, or $2D$ space, or $2D$ plane. Furthermore, the space in which we live admits three linearly independent vectors, and so we can call it three-dimensional space, or $3D$ space. In Albert Einstein's theory of relativity, it is useful to consider the space–time, or Minkowski space, which includes space and time together. This is an example of a four-dimensional ($4D$) linear space, which will be considered in the second part of the book. The difference with the $2D$ plane is that, in the case of Minkowski space, certain types of transformations that mix space and time directions are forbidden. For this reason, the Minkowski space is sometimes called $3 + 1$ dimensional, instead of $4D$. An additional example of a similar sort is a configuration space of classical mechanics. In the simplest case of a single point-like particle, this space is composed of the radius vectors (or position vectors) $\mathbf{r}$ and velocities $\mathbf{v}$ of the particle. The configuration space has six dimensions—this is due to the fact that each of these two vectors has three components. However, in this space, there are some obvious restrictions on the transformations that mix different vectors. For instance, linear combinations such as $a\mathbf{r} + b\mathbf{v}$ are usually avoided.

Theorem Consider a D-dimensional linear space L_D and the set of linearly independent vectors $\mathbf{e}_i = (\mathbf{e}_1, \mathbf{e}_2, ..., \mathbf{e}_D)$. Then, for any vector $\mathbf{a}$ one can write

$$\mathbf{a} = \sum_{i=1}^{D} a^i \mathbf{e}_i = a^i \mathbf{e}_i, \tag{1.4}$$

where the coefficients a^i are defined in a unique way.

Proof The set of vectors $\left(\mathbf{a}, \mathbf{e}_1, \mathbf{e}_2, \ldots, \mathbf{e}_D\right)$ has $D + 1$ element and, according to Def. 4, these vectors are linearly dependent. Also, the set $\mathbf{e}_i$, without $\mathbf{a}$, is linearly independent. Therefore, we can have a zero linear combination

$$k \cdot \mathbf{a} + \tilde{a}^i \mathbf{e}_i = 0, \quad \text{with} \quad k \neq 0.$$

Dividing this equation by k and denoting $a_i = -\tilde{a}^i / k$, we arrive at (1.4).

Let us assume that there is, along with (1.4), another expansion

$$\mathbf{a} = a_*^i \mathbf{e}_i. \tag{1.5}$$

Subtracting (1.4) from (1.5), we arrive at

$$\mathbf{0} = (a^i - a_*^i) \mathbf{e}_i.$$

As far as the vectors $\mathbf{e}_i$ are linearly independent, this means $a^i - a_*^i = 0$, $\forall i$. Therefore, we have proven the uniqueness of the expansion (1.4) for a given basis $\mathbf{e}_i$.

Observation about terminology The coefficients a^i are called *components* or *contravariant components* of the vector $\mathbf{a}$. The set of vectors $\mathbf{e}_i$ is called a *basis* in the linear space L_D. The word "contravariant" here means that the components a^i have upper indices. Later on, we will also introduce components that have lower indices and give them a different name.

Def. 6 The coordinate system in the linear space L_D consists of the *initial point O* and the basis $\mathbf{e}_i$. The position of a point P can be characterized by its *position vector* or *radius vector*

$$\mathbf{r} = \overrightarrow{OP} = x^i \mathbf{e}_i. \tag{1.6}$$

The components x^i of the radius vector are called coordinates or contravariant coordinates of the point P.

When we change the basis $\mathbf{e}_i$, the vector $\mathbf{a}$ does not need to change. This is definitely true for the radius vector of the point P. However, the components of the vector can be different on a new basis. This change of components and many related issues will be a subject of discussion in the subsequent parts of the book. Let us now discuss a few particular cases of linear spaces, which have a special relevance for our further study.

Let us start from a physical example, which we already mentioned above. Consider a configuration space for one particle in classical mechanics. This space is composed of the radius vectors $\mathbf{r}$ and velocities $\mathbf{v}$ of the particle.

The velocity is nothing but the derivative of the radius vector, $\mathbf{v} = \dot{\mathbf{r}}$. But, independent on that, one can introduce two basis sets for the linear spaces of radius vectors and their velocities,

$$\mathbf{r} = x^i \mathbf{e}_i, \qquad \mathbf{v} = v^j \mathbf{f}_j. \tag{1.7}$$

The two bases $\{\mathbf{e}_i\}$ and $\{\mathbf{f}_j\}$ can be related or independent. The last means one can indeed choose two independent basis sets of vectors for the radius vectors and for their derivatives.

What is important is that the state of the system is characterized by the two sets of coefficients, namely, x^i and v^j. Let us assume that we are not allowed to perform such transformations of the basis vectors, which mix the two kinds of coefficients. At the same time, it is easy to see that the space of the states of the system—that is, the state that includes both $\mathbf{r}$ and $\mathbf{v}$ vectors—is a linear space. The coordinates that characterize the state of the system are x^i and v^j; therefore, it is a six-dimensional space.

Exercise 4 Verify that all requirements of the linear space are indeed satisfied for the space with the points given by a couple $(\mathbf{r}, \mathbf{v})$.

This example is a particular case of the linear space which is called a *direct product of the two linear spaces*. The element of the direct product of the two linear spaces L_{D_1} and L_{D_2} (they can have different dimensions) is the ordered set of the elements of each of the two spaces L_{D_1} and L_{D_2}. The basis of the direct product is given by the ordered set of the vectors from two bases. The notation for the basis in the case of configuration space is $\mathbf{e}_i \otimes \mathbf{f}_j$. Hence, the state of the point-like particle in the configuration space is characterized by the element of this linear space, which can be presented as

$$x^i v^j \, \mathbf{e}_i \otimes \mathbf{f}_j. \tag{1.8}$$

Of course, we can perfectly use the same basis (e.g., the Cartesian one, $\hat{\mathbf{i}}, \hat{\mathbf{j}}, \hat{\mathbf{k}}$) for both radius vector and velocity; then the element (1.8) becomes

$$x^i v^j \, \mathbf{e}_i \otimes \mathbf{e}_j. \tag{1.9}$$

The space of the elements (1.9) is six-dimensional, but it is a very special linear space, because not all transformations of its elements are allowed. For example, one cannot sum up the coordinates of the point and the components of its velocities.

In general, one can define the space that is a direct product of several linear spaces with different individual basis sets $\mathbf{e}_i$ each. In this case, we will have $\mathbf{e}_i^{(1)}, \mathbf{e}_i^{(2)}, ...,$ $\mathbf{e}_i^{(N)}$. The basis in the direct product space will be

$$\mathbf{e}_{i_1}^{(1)} \otimes \mathbf{e}_{i_2}^{(2)} \otimes \ ... \ \otimes \mathbf{e}_{i_N}^{(N)}. \tag{1.10}$$

The linear combinations of the basis vectors (1.10),

$$T^{i_1 i_2 \dots i_N} \, \mathbf{e}^{(1)}_{i_1} \otimes \mathbf{e}^{(2)}_{i_2} \otimes \dots \otimes \mathbf{e}^{(N)}_{i_N}, \tag{1.11}$$

form a linear space. We shall consider the features of elements of such linear space in the next sections.

1.2 Vector Basis and Its Transformation

The possibility to change the coordinate system is very important for several reasons. From the practical side, in many cases, one can use the coordinate system in which the given problem is more simple to describe or to solve. More generally, we describe different geometric and physical quantities by means of their components in a given coordinate system. For this reason, it is very important to know and to control how the change of coordinate system affects these components. Such a knowledge helps us to achieve better understanding of the geometric nature of the quantities and hence plays a fundamental role in many applications, e.g., to physics.

Let us start from the components of the vector, which were defined in (1.4). Consider, along with the original basis $\mathbf{e}_i$, another basis $\mathbf{e}'_i$. Since each vector of the new basis belongs to the same space, it can be expanded using the original basis as

$$\mathbf{e}'_i = \wedge^j_{i'} \mathbf{e}_j. \tag{1.12}$$

Then, the uniqueness of the expansion of the vector $\mathbf{a}$ leads to the following relation:

$$\mathbf{a} = a^i \mathbf{e}_i = a^{j'} \mathbf{e}_{j'} = a^{j'} \wedge^i_{j'} \mathbf{e}_i \,,$$

and hence

$$a^i = a^{j'} \wedge^i_{j'} \,. \tag{1.13}$$

This relation shows how the contravariant components of the vector transform from one basis to another.

Similarly, we can make the transformation inverse to (1.12). It is easy to see that the relation between the contravariant components can be also written as

$$a^{k'} = (\wedge^{-1})^{k'}_l a^l \,,$$

where the matrix $(\wedge^{-1})$ is inverse to $\wedge$:

$$(\wedge^{-1})^{k'}_l \cdot \wedge^l_{i'} = \delta^{k'}_{i'} \qquad \text{and} \qquad \wedge^l_{i'} \cdot (\wedge^{-1})^{i'}_k = \delta^l_k.$$

In the last formula, we introduced *Kronecker symbol*, which is defined as

$$\delta_l^i = \begin{cases} 1 \text{ if } i = j \\ 0 \text{ if } i \neq j \end{cases}. \tag{1.14}$$

Exercise 5 Verify and discuss the relations between the formulas

$$\mathbf{e}_{i'} = \wedge_{i'}^j \mathbf{e}_j, \qquad \mathbf{e}_i = (\wedge^{-1})_i^{k'} \mathbf{e}_{k'}$$

and

$$a^{i'} = (\wedge^{-1})_k^{i'} a^k, \qquad a^j = \wedge_{i'}^j a^{i'}.$$

Observation We can interpret the relations (1.13) in a different way. Suppose we have a set of three numbers that characterize some physical, geometrical, or other quantity. We can identify whether these numbers are components of a vector, by looking at their transformation rule. They form a vector if and only if they transform according to (1.13).

Consider in more details the coordinates of a point in a three-dimensional (3D) space. Later on, we will see that the generalization to the general case of an arbitrary D is straightforward. In what follows we will mainly perform our considerations for the 3D and 2D cases, but this is just for the sake of simplicity.

According to Def. 6, the coordinates of a point in space are contravariant components of the radius vector $\mathbf{r}$. The coordinates x^i depend on the choice of the basis $\{\mathbf{e}_i\}$, and also on the choice of the initial point O. For a while we do not consider the change of initial point and concentrate on the change from one basis to another one, assuming that both are *global*. This means that $\{\mathbf{e}_i\}$ is the same in all points of the space.

Similarly to Eq. (1.6), we can expand the same vector $\mathbf{r}$ using another basis

$$\mathbf{r} = x^i \mathbf{e}_i = x^{j'} \mathbf{e}_{j'} = x^{j'}. \wedge_{j'}^i \mathbf{e}_i$$

Then, according to (1.13),

$$x^i = \wedge_{j'}^i x^{j'}. \tag{1.15}$$

The last formula has an important consequence. Taking the partial derivatives, we arrive at the relations

$$\wedge_{j'}^i = \frac{\partial x^i}{\partial x^{j'}} \qquad \text{and} \qquad (\wedge^{-1})_l^{k'} = \frac{\partial x^{k'}}{\partial x^l}.$$

Now we can understand better the sense of the matrix $\wedge$. In particular, the relation

$$(\wedge^{-1})_l^{k'} \cdot \wedge_{k'}^i = \frac{\partial x^{k'}}{\partial x^l} \frac{\partial x^i}{\partial x^{k'}} = \frac{\partial x^i}{\partial x^l} = \delta_l^i$$

is nothing else but the chain rule for partial derivatives.

1.3 Scalar, Vector, and Tensor Fields

Def. 7 Function $\varphi(x)$ is called scalar field or simply scalar if it does not transform under the change of coordinates,

$$\varphi(x) = \varphi'(x'). \tag{1.16}$$

Observation 1 Geometrically, this means that when we change the coordinates, the value of the function $\varphi(x^i)$ remains the same in a given geometric point.

Let us give a clarifying example in $1D$. Consider some function, e.g., $y = x^2$. The plot is a parabola, as we all know. Now, let us change the variables $x' = x + 1$. The function $y = (x')^2$, obviously, represents another parabola. In order to preserve the plot intact, we need to modify the *form of the function*, that is, to go from φ to φ'. Then, the new function $y = (x' - 1)^2$ will represent the original parabola. Two formulas $y = x^2$ and $y = (x' - 1)^2$ represent the same parabola, because the change of the variable is completely compensated by the change of the form of the function.

Exercise 6 The reader has probably noticed that the example given above has a transformation rule for coordinate x that is different from Eq. (1.15). (a) Consider the transformation rule $x' = 5x$, which belongs to the class (1.15) and find the corresponding modification of the function $y'(x')$. (b) Formulate the most general global (this means point-independent) $1D$ transformation rule $x' = x'(x)$, which belongs to the class (1.15), and find the corresponding modification of the function $y'(x')$ for $y = x^2$. (c) Try to generalize the output of the previous point to an arbitrary function $y = f(x)$.

Observation 2 An important physical example of scalar is the temperature $T(x)$ in the given point of space at a given instant of time. Other examples are pressure $p(x)$ and air density $\rho(x)$.

Exercise 7 Discuss whether the three numbers T, p, ρ form a contravariant vector or not.

Observation 3 From Def. 7 it follows that for the generic change of coordinates and scalar field $\varphi'(x) \neq \varphi(x)$ and $\varphi(x') \neq \varphi(x)$. Let us calculate these quantities explicitly for the special case of infinitesimal transformation $x'^i = x^i + \xi^i$ where ξ are constant coefficients. We shall take into account only first order in ξ^i. Then

$$\varphi(x') = \varphi(x) + \frac{\partial \varphi}{\partial x^i} \xi^i \tag{1.17}$$

and, in the first-order approximation,

$$\varphi'(x^i) = \varphi'(x'^i - \xi^i) = \varphi'(x') - \frac{\partial \varphi'}{\partial x'^i} \cdot \xi^i.$$

Let us take into account that the partial derivative in the last term can be evaluated for $\xi = 0$. Then

$$\frac{\partial \varphi'}{\partial x'^i} = \frac{\partial \varphi(x)}{\partial x^i} + \mathcal{O}(\xi)$$

and therefore

$$\varphi'(x) = \varphi(x) - \xi^i \partial_i \varphi, \tag{1.18}$$

where we have introduced a useful notation $\partial_i = \partial/\partial x^i$.

Exercise 8a Using (1.17) and (1.18), verify relation (1.16).

Exercise 8b* Continue the expansions (1.17) and (1.18) until the second order in ξ, and consequently consider $\xi^i = \xi^i(x)$. Use (1.16) to verify your calculus.

Def. 8 The set of three functions $\{a^i(x)\} = \{a^1(x), a^2(x), a^3(x)\}$ forms contravariant vector field (or simply vector field) if they transform, under the change of coordinates $\{x^i\} \to \{x'^j\}$, as

$$a^{j'}(x') = \frac{\partial x^{j'}}{\partial x^i} \cdot a^i(x) \tag{1.19}$$

at any point of space.

Observation 4 One can easily see that this transformation rule is different from the one for scalar field (1.16). In particular, the components of the vector in a given geometrical point of space modify under the coordinate transformation, while the scalar field does not. The transformation rule (1.19) corresponds to the geometric object $\mathbf{a}$ in $3D$. Contrary to the scalar φ, the vector $\mathbf{a}$ has direction—exactly the origin of the difference between the transformation laws.

Observation 5 Some physical examples of quantities (fields) that have a vector transformation rule are perhaps familiar to the reader: electric field, magnetic field, instantaneous velocities of particles in a stationary flux of a fluid, etc. Later on, we shall consider particular examples of the transformation of vectors under rotations and inversions.

Observation 6 The scalar and vector fields can be considered as examples of the more general objects called tensors. Tensors are also defined through their transformation rules.

Def. 9 The set of 3^n functions $\{a^{i_1 \cdots i_n}(x)\}$ is called a contravariant tensor of rank n, if these functions transform, under $x^i \to x'^i$, as

$$a^{i'_1 \cdots i'_n}(x') = \frac{\partial x^{i'_1}}{\partial x^{j_1}} \cdots \frac{\partial x^{i'_n}}{\partial x^{j_n}} a^{j_1 \cdots j_n}(x). \tag{1.20}$$

Observation 7 According to the above definitions, the scalar and contravariant vector fields are nothing but the particular cases of the contravariant tensor field. Scalar is a tensor of rank 0, and vector is a tensor of rank 1.

Exercise 9 Show that the product $a^i b^j$ is a contravariant second-rank tensor, if both a^i and b^i are contravariant vectors.

1.4   Orthonormal Basis and Cartesian Coordinates

Def. 10 The scalar product of two vectors **a** and **b** in $3D$ is defined in a usual way,

$$(\mathbf{a}, \mathbf{b}) = |\mathbf{a}| \cdot |\mathbf{b}| \cdot \cos\theta,$$

where $|\mathbf{a}|$, $|\mathbf{b}|$ are absolute values of the vectors **a**, **b**, and θ is the angle between these two vectors.

Of course, the scalar product has the properties familiar from the Analytic Geometry, such as linearity

$$(\mathbf{a}, \alpha_1 \mathbf{b}_1 + \alpha_2 \mathbf{b}_2) = \alpha_1 (\mathbf{a}, \mathbf{b}_1) + \alpha_2 (\mathbf{a}, \mathbf{b}_2)$$

and symmetry

$$(\mathbf{a}, \mathbf{b}) = (\mathbf{b}, \mathbf{a}).$$

Observation Sometimes, we shall use a different notation for the scalar product

$$(\mathbf{a}, \mathbf{b}) \equiv \mathbf{a} \cdot \mathbf{b}.$$

Until now, we considered an arbitrary basis $\{\mathbf{e}_i\}$. Indeed, we requested the vectors $\{\mathbf{e}_i\}$ to be point-independent, but the length of each of these vectors and the angles between them were restricted only by the requirement of their linear independence. It proves useful to define a special basis, which corresponds to the conventional Cartesian coordinates.

Def. 11 Special orthonormal basis $\{\hat{\mathbf{n}}_a\}$ is the one with

$$(\hat{\mathbf{n}}_a, \hat{\mathbf{n}}_b) = \delta_{ab} = \begin{cases} 1 \text{ if } a = b \\ 0 \text{ if } a \neq b \end{cases}. \tag{1.21}$$

Here δ_{ab} is the same Kronecker symbol, which was defined earlier in (1.14).

Exercise 10 Making transformations of the basis vectors, verify that the change of coordinates

$$x' = \frac{x+y}{\sqrt{2}} + 3, \qquad y' = \frac{x-y}{\sqrt{2}} - 5$$

does not modify the type of coordinates x', y', which remains Cartesian.

Exercise 11 Using geometric intuition, discuss the form of coordinate transformations that do not violate the orthonormality of the basis.

Notations We shall denote the coordinates corresponding to the $\{\hat{\mathbf{n}}_a\}$ basis as X^a. With a few exceptions, we reserve the indices $a, b, c, d, \dots$ for these (Cartesian) coordinates, and use indices $i, j, k, l, m, \dots$ for arbitrary coordinates. For example, the radius vector $\mathbf{r}$ in $3D$ can be presented using different coordinates:

$$\mathbf{r} = x^i \mathbf{e}_i = X^a \hat{\mathbf{n}}_a = x\hat{\mathbf{n}}_x + y\hat{\mathbf{n}}_y + z\hat{\mathbf{n}}_z. \tag{1.22}$$

Indeed, the last equality is identity, because

$$X^1 = x, \qquad X^2 = y, \qquad X^3 = z.$$

Sometimes, we shall also use notations

$$\hat{\mathbf{n}}_x = \hat{\mathbf{n}}_1 = \hat{\mathbf{i}}, \qquad \hat{\mathbf{n}}_y = \hat{\mathbf{n}}_2 = \hat{\mathbf{j}}, \qquad \hat{\mathbf{n}}_z = \hat{\mathbf{n}}_3 = \hat{\mathbf{k}}.$$

Observation 1 One can express the vectors of an arbitrary basis $\{\mathbf{e}_i\}$ as linear combinations of the components of the orthonormal basis

$$\mathbf{e}_i = \frac{\partial X^a}{\partial x^i} \hat{\mathbf{n}}_a.$$

An inverse relation has the form

$$\hat{\mathbf{n}}_a = \frac{\partial x^j}{\partial X^a} \mathbf{e}_j.$$

Exercise 12 Prove the last formulas using the uniqueness of the components of the vector $\mathbf{r}$ in a given basis.

Observation 2 It is easy to see that

$$X^a \hat{\mathbf{n}}_a = x^i \mathbf{e}_i = x^i \cdot \frac{\partial X^a}{\partial x^i} \hat{\mathbf{n}}_a \implies X^a = \frac{\partial X^a}{\partial x^i} \cdot x^i. \tag{1.23}$$

Similarly,

$$x^i = \frac{\partial x^i}{\partial X^a} \cdot X^a. \tag{1.24}$$

Indeed, these relations hold only because the basis vectors $\mathbf{e}_i$ do not depend on the space–time points, and the matrices

$$\frac{\partial X^a}{\partial x^i} \qquad \text{and} \qquad \frac{\partial x^i}{\partial X^a}$$

have constant elements. Later on, we shall consider more general, nonlinear (or local) coordinate transformations where the relations (1.23) and (1.24) do not hold.

Now we can introduce a conjugated covariant basis.

Def. 12 Consider basis $\{\mathbf{e}_i\}$. The conjugated basis is defined as a set of vectors $\{\mathbf{e}^j\}$, which satisfy the relations

$$\mathbf{e}_i \cdot \mathbf{e}^j = \delta_i^j. \tag{1.25}$$

Remark The special property of the orthonormal basis is that $\hat{\mathbf{n}}^a = \hat{\mathbf{n}}_a$; therefore in this special case, the conjugated basis coincides with the original one.

Exercise 13 Prove that only an orthonormal basis may have this property.

Def. 13 Any vector $\mathbf{a}$ can be expanded using the conjugated basis $\mathbf{a} = a_i \mathbf{e}^i$. The coefficients a_i are called *covariant* components of the vector $\mathbf{a}$. Sometimes, we use this to say that these coefficients form the covariant vector.

Exercise 14 Consider, in $2D$, $\mathbf{e}_1 = \hat{\mathbf{i}} + \hat{\mathbf{j}}$ and $\mathbf{e}_2 = \hat{\mathbf{i}}$. Find the basis that is conjugated to $\{\mathbf{e}_1, \ \mathbf{e}_2\}$. Make a $2D$ plot with two basis sets.

The next question is: how do the coefficients a_i transform under the change of coordinates?

Theorem If we change the basis of the coordinate system from $\mathbf{e}_i$ to $\mathbf{e}'_i$, then the covariant components of the vector $\mathbf{a}$ transform as

$$a'_i = \frac{\partial x^j}{\partial x'^i} a_j. \tag{1.26}$$

Observation It is easy to see that in the case of covariant vector components, the transformation is done by means of the matrix inverse to the one for the contravariant components (1.19).

Proof First of all, we need to learn how do the vectors of the conjugated basis transform. Let us use the definition of the conjugated basis in both coordinate systems

$$(\mathbf{e}_i, \mathbf{e}^j) = (\mathbf{e}'_i, \ \mathbf{e}'^j) = \delta_i^j.$$

Since

$$\mathbf{e}'_i = \frac{\partial x^k}{\partial x'^i} \mathbf{e}_k ,$$

and, due to the linearity of a scalar product of the two vectors, we obtain

$$\frac{\partial x^k}{\partial x'^i} (\mathbf{e}_k , \mathbf{e}'^j) = \delta^j_i .$$

Therefore, the matrix $(\mathbf{e}_k, \mathbf{e}'^j)$ is inverse to the matrix $\left(\frac{\partial x^k}{\partial x'^i}\right)$. Using the uniqueness of the inverse matrix, we arrive at the relation

$$(\mathbf{e}_k, \mathbf{e}'^j) = \frac{\partial x^{j'}}{\partial x^k} = \frac{\partial x^{j'}}{\partial x^l} \delta^l_k = \frac{\partial x^{j'}}{\partial x^l}(\mathbf{e}_k, \mathbf{e}^l) = (\mathbf{e}_k, \frac{\partial x^{j'}}{\partial x^l} \mathbf{e}^l), \qquad \forall \mathbf{e}_k .$$

Thus, $\mathbf{e}'^j = \frac{\partial x^{j'}}{\partial x^l} \mathbf{e}^l$. Using the standard way of dealing with vector transformations, we obtain

$$a_{j'} \mathbf{e}'^j = a_{j'} \frac{\partial x^{j'}}{\partial x^l} \mathbf{e}^l \quad \Longrightarrow \quad a_l = \frac{\partial x'^j}{\partial x^l} a'_j .$$

1.5 Covariant and Mixed Vectors and Tensors

Now we are in a position to define covariant vectors, tensors, and mixed tensors.

Def. 14a The set of three functions $\{A_i(x)\}$ forms a covariant vector field, if they transform from one coordinate system to another one as

$$A'_i(x') = \frac{\partial x^j}{\partial x'^i} A_j(x). \tag{1.27}$$

Def. 14b The set of 3^n functions $\{A_{i_1 i_2 \ldots i_n}(x)\}$ form a covariant tensor of rank n if they transform from one coordinate system to another as

$$A'_{i_1 i_2 \ldots i_n}(x') = \frac{\partial x^{j_1}}{\partial x'^{i_1}} \frac{\partial x^{j_2}}{\partial x'^{i_2}} \cdots \frac{\partial x^{j_n}}{\partial x'^{i_n}} A_{j_1 j_2 \ldots j_n}(x).$$

Exercise 15 Show that the product of two covariant vectors $A_i(x)B_j(x)$ transforms as a second-rank covariant tensor. Discuss whether *any* second-rank covariant tensor can be presented as such product.

Hint. Try to evaluate the number of independent functions in both cases.

Def. 15 The set of 3^{n+m} functions $\{B_{i_1 \ldots i_n}{}^{j_1 \ldots j_m}(x)\}$ forms the tensor of the type (m, n), if these functions transform, under the change of coordinate basis, as

$$B_{i'_1 \ldots i'_n}{}^{j'_1 \ldots j'_m}(x') = \frac{\partial x^{j'_1}}{\partial x^{l_1}} \cdots \frac{\partial x^{j'_m}}{\partial x^{l_m}} \frac{\partial x^{k_1}}{\partial x'^{i_1}} \cdots \frac{\partial x^{k_n}}{\partial x'^{i_n}} B_{k_1 \ldots k_n}{}^{l_1 \ldots l_m}(x).$$

Other possible names are the mixed tensor of covariant rank n and contravariant rank m, or simply (m, n)-tensor.

Exercise 16 Verify that the scalar, co-, and contravariant vectors are, correspondingly, $(0, 0)$-tensor, $(0, 1)$-tensor, and $(1, 0)$-tensor.

Exercise 17 Prove that if the Kronecker symbol transforms as a mixed $(1, 1)$ tensor, then in any coordinates x^i it has the same form

$$\delta^i_j = \begin{cases} 1 \text{ if } i = j \\ 0 \text{ if } i \neq j \end{cases}.$$

Observation This property is very important, as it enables us to use the Kronecker symbol in any coordinates.

Exercise 18 Show that the product $A^i(x) B_j(x)$ of covariant and contravariant vectors transforms as a $(1, 1)$-type mixed tensor.

Observation Tensors are important due to the fact that they offer the *coordinate-independent* description of geometrical and physical laws. Any tensor can be viewed as a geometric object, independent of coordinates. Let us consider, as an example, the mixed $(1, 1)$-type tensor with components A^j_i. The tensor (as a geometric object) can be presented as a contraction of A^j_i with the basis vectors

$$\mathbf{A} = A^j_i \, \mathbf{e}^i \otimes \mathbf{e}_j . \tag{1.28}$$

The operation $\otimes$ is called a *direct product* and it indicates that the basis for the tensor $\mathbf{A}$ is composed of the products of the type $\mathbf{e}^i \otimes \mathbf{e}_j$. In other words, the tensor $\mathbf{A}$ is a linear combination of such "direct products". The most important observation is that the *l.h.s.* of Eq. (1.28) transforms as a scalar. Hence, despite the *components* of a tensor being dependent on the choice of the basis, the tensor is indeed a coordinate-independent geometric object.

Exercise 19 (a) Discuss the last observation for the case of a vector. (b) Discuss why in Eq. (1.28) there are, in general, nine independent components of the tensor A^j_i, while in Eq. (1.8) similar coefficient has only six independent components x^i and v^j. Is it true that any tensor is a product of some vectors or not?

1.6 Orthogonal Transformations

Let us consider important particular cases of the global coordinate transformations, which are called orthogonal. The orthogonal transformations may be classified rotations, inversions (parity transformations), and their combinations.

Fig. 1.1 Illustration of the
$2D$ rotation to the angle α
given by Eq. (1.29)

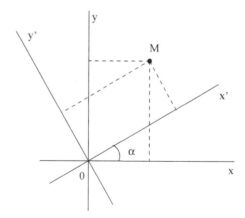

First, we consider the rotations in the $2D$ space with the initial Cartesian coordinates (x, y). Suppose other Cartesian coordinates (x', y') have the same origin as (x, y) and the difference is the rotation angle α. Then, the same point M (see Fig. 1.1) has coordinates (x,y) and (x', y'), and the relation between the coordinates is

$$x = x' \cos \alpha - y' \sin \alpha$$
$$y = x' \sin \alpha + y' \cos \alpha \, .$$
(1.29)

Exercise 20 Check that the inverse transformation has the form or rotation to the same angle but in the opposite direction.

$$x' = x \cos \alpha + y \sin \alpha$$
$$y' = -x \sin \alpha + y \cos \alpha \, .$$
(1.30)

The above transformations can be seen from the $3D$ viewpoint and also may be presented in a matrix form, e.g.,

$$\begin{pmatrix} x \\ y \\ z \end{pmatrix} = \hat{\wedge}_z \begin{pmatrix} x' \\ y' \\ z' \end{pmatrix}, \quad \text{where} \quad \hat{\wedge}_z = \hat{\wedge}_z(\alpha) = \begin{pmatrix} \cos \alpha & -\sin \alpha & 0 \\ \sin \alpha & \cos \alpha & 0 \\ 0 & 0 & 1 \end{pmatrix}. \quad (1.31)$$

Here, the index z indicates that the matrix $\hat{\wedge}_z$ describes rotation around the $\hat{\mathbf{z}}$-axis.

Exercise 21 Write the corresponding rotation matrices around $\hat{\mathbf{x}}$- and $\hat{\mathbf{y}}$-axes with the angles γ and β.

Exercise 22 Check the following property of the matrix $\hat{\wedge}_z$:

$$\hat{\wedge}_z^T = \hat{\wedge}_z^{-1} \, .$$
(1.32)

Def. 16 The matrix $\hat{\wedge}_z$ which satisfies $\hat{\wedge}_z^{-1} = \hat{\wedge}_z^T$ and the corresponding coordinate transformation is called orthogonal.[2]

In $3D$, any rotation of the rigid body may be represented as a combination of the rotations around the axes $\hat{z}$, $\hat{y}$, and $\hat{x}$ to the angles[3] α, β, and γ. Hence, the general $3D$ rotation matrix $\hat{\wedge}$ may be represented as

$$\hat{\wedge} = \hat{\wedge}_z(\alpha)\,\hat{\wedge}_y(\beta)\,\hat{\wedge}_x(\gamma). \tag{1.33}$$

If we use the known properties of the invertible square matrices

$$(A \cdot B)^T = B^T \cdot A^T \qquad \text{and} \qquad (A \cdot B)^{-1} = B^{-1} \cdot A^{-1}, \tag{1.34}$$

we can easily obtain that the general $3D$ rotation matrix satisfies the relation $\hat{\wedge}^T = \hat{\wedge}^{-1}$. Thus, we have proven that an arbitrary rotation matrix is orthogonal.

Exercise 23 Investigate the following relations and properties of the matrices $\hat{\wedge}$:

(i) Check explicitly that $\hat{\wedge}_z^T(\alpha) = \hat{\wedge}_z^{-1}(\alpha)$.
(ii) Check that $\hat{\wedge}_z(\alpha) \cdot \hat{\wedge}_z(\beta) = \hat{\wedge}_z(\alpha + \beta)$ (group property).
(iii) Verify that the determinant $\det \hat{\wedge}^z(\alpha) = 1$. Prove that the determinant equals to one for any rotation matrix.

Observation Since for the orthogonal matrix $\hat{\wedge}$ we have the equality

$$\hat{\wedge}^{-1} = \hat{\wedge}^T,$$

one can take the determinant and arrive at $\det \hat{\wedge} = \det \hat{\wedge}^{-1}$. Therefore, $\det \hat{\wedge} = \pm 1$. As far as any rotation matrix has a determinant equal to one, there must be some other orthogonal matrices with the determinant equal to -1. The following are examples of such matrices:

$$\hat{\pi}_z = \begin{pmatrix} 1 & 0 & 0 \\ 0 & 1 & 0 \\ 0 & 0 & -1 \end{pmatrix}, \qquad \hat{\pi} = \begin{pmatrix} -1 & 0 & 0 \\ 0 & -1 & 0 \\ 0 & 0 & -1 \end{pmatrix}. \tag{1.35}$$

The last matrix corresponds to the transformation called space inversion, or parity transformation. This transformation means that we simultaneously invert the direction of all axes. One can prove that an arbitrary orthogonal matrix can be presented as a product of inversion and rotation matrices.

---

[2]In case where the matrix elements are allowed to be complex, the matrix that satisfies the property $U^\dagger = U^{-1}$ is called unitary. The operation $U^\dagger$ is called Hermitian conjugation and consists of complex conjugation plus transposition $U^\dagger = (U^*)^T$.

[3]One can prove that it is always possible to replace the rotation of the rigid body around an arbitrary axis by performing the sequence of particular rotations $\hat{\wedge}_z(\alpha)\,\hat{\wedge}_y(\beta)\,\hat{\wedge}_z(-\alpha)$. The proof can be done using the Euler angles (see, e.g., [1, 2]).

Exercise 24 (i) Following the pattern of Eq. (1.35), construct $\hat{\pi}_y$ and $\hat{\pi}_x$.
(ii) Find rotation matrices that transform $\hat{\pi}_z$ into $\hat{\pi}_x$ and $\hat{\pi}$ into $\hat{\pi}_x$.
(iii) Suggest geometric interpretation of the matrices $\hat{\pi}_x$, $\hat{\pi}_y$, and $\hat{\pi}_z$.

Observation The issue of rotations and other orthogonal transformations is extremely important, and we refer the reader to the books on group theory where it is discussed in more detail. We do not go deeper into this subject here, as the target of the present book is restricted to tensor calculus and does not include group theory.

Exercise 25 Consider the transformation from some orthonormal basis $\hat{\mathbf{n}}_i$ to an arbitrary basis $\mathbf{e}_k$, $\mathbf{e}_k = \Lambda_k^i \hat{\mathbf{n}}_i$. Prove that the lines of the matrix Λ_k^i correspond to the components of the vectors $\mathbf{e}_k$ in the basis $\hat{\mathbf{n}}_i$.

Exercise 26 Consider the transformation from one orthonormal basis $\hat{\mathbf{n}}_i$ to another such basis $\hat{\mathbf{n}}'_k$, namely, $\hat{\mathbf{n}}'_k = R_k^i \hat{\mathbf{n}}_i$. Prove that the matrix $\| R_k^i \|$ is orthogonal.

Exercise 27 Consider the rotation (1.29), (1.30) on the XOY plane in the $3D$ space.

(i) Write the same transformation in the $3D$ notations.
(ii) Calculate all partial derivatives $\frac{\partial x_i}{\partial x'_j}$ and $\frac{\partial x'_k}{\partial x_l}$.
(iii) Vector $\mathbf{a}$ has components $a_x = 1$, $a_y = 2$, and $a_z = 3$. Find its components $a_{x'}$, $a_{y'}$, and $a_{z'}$.
(iv) Tensor b_{ij} has nonzero components $b_{xy} = 1$, $b_{xz} = 2$, $b_{yy} = 3$, $b_{yz} = 4$, and $b_{zz} = 5$. Find all components $b_{i'j'}$.

Exercise 28 Consider the so-called pseudo-Euclidean plane[4] with coordinates t and x. Consider the coordinate transformation

$$t = t' \cosh \psi + x' \sinh \psi$$
$$x = t' \sinh \psi + x' \cosh \alpha \quad (1.36)$$

(i) Show, by direct inspection, that $t'^2 - x'^2 = t^2 - x^2$.
(ii) Find an inverse transformation to (1.36).
(iii) Calculate all partial derivatives $\partial x_i / \partial x'_j$ and $\partial x'_k / \partial x_l$, where $x^i = (t, x)$.
(iv) Vector $\mathbf{a}$ has components $a_x = 1$ and $a_t = 2$. Find its components $a_{x'}$ and $a_{t'}$.
(v) Tensor b_{ij} has components $b_{xt} = 1$, $b_{xx} = 2$, $b_{tx} = 0$, and $b_{tt} = 3$. Find all components $b_{i'j'}$.
(vi) Repeat the same program for the c_i^j with components $c_x^t = 1$, $c_x^x = 2$, $c_t^x = 0$ and $c_t^t = 3$. This means one has to calculate all the components $c_{i'}^{j'}$.

---

[4]In the second part of the book, we shall see that this is the Lorentz transformation in special relativity in the units $c = 1$.

References

1. L.D. Landau, E.M. Lifshits, *Mechanics—Course of Theoretical Physics*, vol. 1 (Butterworth-Heinemann, 1976)
2. J.J. Sakurai, *Modern Quantum Mechanics* (Addison-Wesley Pub Co., 1994)

Chapter 2
Operations over Tensors, Metric Tensor

In this chapter, we shall define operations over tensors. These operations may involve one or two tensors. The most important point is that the result is always a tensor.

Operation 1 Summation is defined only for the tensors of the same type. The sum of two two tensors of the type (m, n) is also a tensor type (m, n), which has components equal to the sums of the corresponding components of the summands,

$$(A + B)_{i_1...i_n}{}^{j_1...j_m} = A_{i_1...i_n}{}^{j_1...j_m} + B_{i_1...i_n}{}^{j_1...j_m}. \tag{2.1}$$

Exercise 1 Prove that the components of $(A + B)$ in (2.1) form a tensor, if $A_{i_1...i_n}{}^{j_1...j_m}$ and $B_{i_1...i_n}{}^{j_1...j_m}$ form tensors.

Hint. Try first to prove this statement for scalar and vector fields.

Exercise 2 Prove the commutativity and associativity of the tensor summation.

Hint. Use the definition of tensor Def. 1.15.

Operation 2 Multiplication of a tensor by a number produces a tensor of the same type. This operation is equivalent to the multiplication of all tensor components to the same number α, namely,

$$(\alpha A)_{i_1...i_n}{}^{j_1...j_m} = \alpha \cdot A_{i_1...i_n}{}^{j_1...j_m}. \tag{2.2}$$

Exercise 3 Prove that the components of (αA) in (2.2) form a tensor, in the case when $A_{i_1...i_n}{}^{j_1...j_m}$ are components of a tensor.

Operation 3 Multiplication of two tensors is defined for a couple of tensors of any type. The product of a (m, n)-tensor and a (t, s)-tensor results in the $(m + t, n + s)$-tensor, e.g.,

$$A_{i_1...i_n}{}^{j_1...j_m} \cdot C_{l_1...l_s}{}^{k_1...k_t} = D_{i_1...i_n}{}^{j_1...j_m}{}_{l_1...l_s}{}^{k_1...k_t}. \tag{2.3}$$

The order of indices is important here, because a_{ij} may be different from a_{ji}.

© Springer Nature Switzerland AG 2019
I. L. Shapiro, *A Primer in Tensor Analysis and Relativity*, Undergraduate
Lecture Notes in Physics, https://doi.org/10.1007/978-3-030-26895-4_2

Exercise 4 Prove, by checking the transformation law, that the product of the contravariant vector a^i and mixed tensor b^k_j is a mixed (2, 1)-type tensor.

Exercise 5 Prove the following theorem: the product of an arbitrary (m, n)-type tensor and a scalar is a (m, n)-type tensor. Formulate and prove similar statements concerning multiplication by co- and contravariant vectors.

Operation 4 Contraction reduces the (n, m)-tensor to the $(n - 1, m - 1)$-tensor through the summation over two (always upper and lower, of course) indices.

Example

$$A_{ijk}{}^{ln} \longrightarrow A_{ijk}{}^{lk} = \sum_{k=1}^{3} A_{ijk}{}^{lk}. \tag{2.4}$$

Operation 5 The internal product of the two tensors consists in their multiplication with the consequent contraction over some couple of indices. Internal product of (m, n) and (r, s)-type tensors results in the $(m + r - 1, n + s - 1)$-type tensor.

Example

$$A_{ijk} \cdot B^{lj} = \sum_{j=1}^{3} A_{ijk} B^{lj}. \tag{2.5}$$

Observation Contracting another couple of indices may give another internal product.

Exercise 6 Prove that the internal product $a_i \cdot b^i$ is a scalar if $a_i(x)$ and $b^i(x)$ are co- and contravariant vectors.

Now we can introduce one of the central notions of this book.

Def. 1 Consider a basis $\{\mathbf{e}_i\}$. The scalar product of the two basis vectors,

$$g_{ij} = (\mathbf{e}_i, \mathbf{e}_j), \tag{2.6}$$

is called *metric*.

Properties:

1. Symmetry of the metric $g_{ij} = g_{ji}$ follows from the symmetry of a scalar product $(\mathbf{e}_i, \mathbf{e}_j) = (\mathbf{e}_j, \mathbf{e}_i)$.
2. For the orthonormal basis $\hat{\mathbf{n}}_a$, the metric is nothing but the Kronecker symbol

$$g_{ab} = (\hat{\mathbf{n}}_a, \hat{\mathbf{n}}_b) = \delta_{ab}.$$

3. Metric is a (0, 2) - tensor. The proof of this statement is very simple,

$$g_{i'j'} = (\mathbf{e}'_i, \mathbf{e}'_j) = \left(\frac{\partial x^l}{\partial x'^i}\mathbf{e}_l, \frac{\partial x^k}{\partial x'^j}\mathbf{e}_k\right)$$

$$= \frac{\partial x^l}{\partial x'^i}\frac{\partial x^k}{\partial x'^j}(\mathbf{e}_l, \mathbf{e}_k) = \frac{\partial x^l}{\partial x'^i}\frac{\partial x^k}{\partial x'^j}\cdot g_{kl}.$$

Sometimes, we shall call metric the metric tensor.

4. The distance between two points: $M_1(x^i)$ and $M_2(y^i)$ is defined by

$$S_{12}^2 = g_{ij}(x^i - y^i)(x^j - y^j). \tag{2.7}$$

Proof Since g_{ij} is (0, 2) - tensor and $(x^i - y^i)$ is (1, 0)-tensor (contravariant vector), S_{12}^2 is a scalar. Therefore, S_{12}^2 is the same in any coordinate system. In particular, we can use the $\hat{\mathbf{n}}_a$ basis, where the (Cartesian) coordinates of the two points are $M_1(X^a)$ and $M_2(Y^a)$. Making the transformation, we obtain

$$S_{12}^2 = g_{ij}(x^i - y^i)(x^j - y^j) = g_{ab}(X^a - Y^a)(X^b - Y^b) =$$

$$= \delta_{ab}(X^a - Y^a)(X^b - Y^b) = (X^1 - Y^1)^2 + (X^2 - Y^2)^2 + (X^3 - Y^3)^2,$$

that is, the standard expression for the square of the distance between the two points in Cartesian coordinates.

Exercise 7 Check, by direct inspection of the transformation law, that the double internal product $g_{ij}z^iz^j$ is a scalar. In the proof of Property 4, we used this for $z^i = x^i - y^i$.

Exercise 8 Use the definition of the metric and the relations

$$x^i = \frac{\partial x^i}{\partial X^a}X^a \qquad \text{and} \qquad \mathbf{e}_i = \frac{\partial X^a}{\partial x^i}\hat{\mathbf{n}}_a$$

to prove the Property 4, starting from $S_{12}^2 = g_{ab}(X^a - Y^a)(X^b - Y^b)$.

Def. 2 The conjugated metric is defined as

$$g^{ij} = (\mathbf{e}^i, \mathbf{e}^j) \qquad \text{where} \qquad (\mathbf{e}^i, \mathbf{e}_k) = \delta_k^i.$$

Indeed, $\mathbf{e}^i$ are the vectors of the basis conjugated to $\mathbf{e}_k$ (see Def. 1.11).

Exercise 9 Prove that g^{ij} is a contravariant second-rank tensor.

Theorem 1

$$g^{ik}g_{kj} = \delta_j^i. \tag{2.8}$$

Due to this theorem, the conjugated metric g^{ij} is conventionally called inverse metric.

Proof For the special orthonormal basis $\hat{\mathbf{n}}^a = \hat{\mathbf{n}}_a$, we have $g^{ab} = \delta^{ab}$ and $g_{bc} = \delta_{bc}$. Of course, in this case the two metrics are inverse matrices,

$$g^{ab} \cdot g_{bc} = \delta^a_c .$$

Now we can use the result of Exercise 9. The last equality can be transformed to an arbitrary basis $\mathbf{e}_i$, by multiplying both sides by $\frac{\partial x^i}{\partial X^a}$ and $\frac{\partial X^c}{\partial x^j}$ and inserting the identity matrix (in the parenthesis) as follows:

$$\frac{\partial x^i}{\partial X^a} \left(g^{ab} \cdot g_{bc} \right) \frac{\partial X^c}{\partial x^j} = \frac{\partial x^i}{\partial X^a} g^{ab} \left(\frac{\partial x^k}{\partial X^b} \frac{\partial X^d}{\partial x^k} \right) g_{dc} \frac{\partial X^c}{\partial x^j} . \qquad (2.9)$$

The last expression can be rewritten as

$$\frac{\partial x^i}{\partial X^a} g^{ab} \frac{\partial x^k}{\partial X^b} \cdot \frac{\partial X^d}{\partial x^k} g_{dc} \frac{\partial X^c}{\partial x^j} = g^{ik} g_{kj} .$$

But, at the same time, the same expression (2.9) can be presented in another form

$$\frac{\partial x^i}{\partial X^a} g^{ab} \cdot g_{bc} \frac{\partial X^c}{\partial x^j} = \frac{\partial x^i}{\partial X^a} \delta^a_c \frac{\partial X^c}{\partial x^j} = \frac{\partial x^i}{\partial X^a} \frac{\partial X^a}{\partial x^j} = \delta^i_j ,$$

which completes the proof.

Operation 6 Raising and lowering indices of a tensor. This operation consists of taking an appropriate internal product of a given tensor and the corresponding metric tensor.

Example 1

Lowering the index, $A_i(x) = g_{ij} A^j(x)$, $B_{ik}(x) = g_{ij} B^j_{\cdot k}(x)$.
Raising the index, $C^l(x) = g^{lj} C_j(x)$, $D^{ik}(x) = g^{ij} D_j^{\cdot k}(x)$.

Exercise 10 Prove the relations:

$$\mathbf{e}_i \, g^{ij} = \mathbf{e}^j , \qquad \mathbf{e}^k \, g_{kl} = \mathbf{e}_l . \qquad (2.10)$$

Exercise 11* Try to solve the previous exercise in several different ways.

The following theorem clarifies the geometric sense of the metric and its relation with the change of basis and coordinate systems. Furthermore, it has serious practical importance and will be extensively used below.

Theorem 2 *Let the metric* g_{ij} *correspond to the basis* $\mathbf{e}_k$ *and to the coordinates* x^k. *The determinants of the metric tensors and of the matrices of the transformations to Cartesian coordinates* X^a *satisfy the following relations:*

$$g = det(g_{ij}) = det\left(\frac{\partial X^a}{\partial x^k}\right)^2, \qquad g^{-1} = det(g^{kl}) = det\left(\frac{\partial x^l}{\partial X^b}\right)^2. \quad (2.11)$$

Proof The first observation is that in the Cartesian coordinates, the metric is an identity matrix $g_{ab} = \delta_{ab}$ and therefore $det(g_{ab}) = 1$. Next, since the metric is a tensor,

$$g_{ij} = \frac{\partial X^a}{\partial x^i}\frac{\partial X^b}{\partial x^j}g_{ab}.$$

Using the known property of determinants (you are going to prove this relation in the next chapter as an exercise)

$$det(A \cdot B) = det(A) \cdot det(B),$$

we arrive at the first relation in (2.11). The proof of the second relation is quite similar and we leave it to the reader as an exercise.

Observation 1 The possibility of lowering and raising the indices may help us in contracting two contravariant or two covariant indices of a tensor. To this end, we simply need to raise or lower one of two indices and then perform contraction.

Observation 2 Looking at Theorem 2, one may note that solving Eq. (2.11) with respect to the determinants $\left|\frac{\partial X^a}{\partial x^k}\right|$ yields solutions with both signs: positive and negative. The general rule is that the positive sign should be taken for the transformations that do not involve change or orientation (parity-even), while the negative sign corresponds to the parity-odd change of coordinates. One can observe the effect of this sign issue in Exercise 9 of the next section.

Example 2

$$A^{ij} \quad \longrightarrow \quad A^{j}_{.k} = A^{ij}g_{ik} \quad \longrightarrow \quad A^{k}_{.k} = A^{ik}g_{ik}.$$

Here the point indicates the position of the raised index. Sometimes, this is an important information that helps to avoid an ambiguity. Let us illustrate this using another simple example.

Example 3 Suppose we need to contract the two first indices of the $(3, 0)$-tensor B^{ijk}. After we lower the second index, we arrive at the tensor

$$B^{i\ k}_{.l.} = B^{ijk}g_{jl} \quad\quad\quad\quad (2.12)$$

and now can contract the indices i and l. But, if we forget to indicate the order of indices, we obtain, instead of (2.12), the formula B^{ik}_{l}, and it is not immediately clear which index was lowered and in which couple of indices one has to perform the contraction.

Exercise 12 Transform the $2D$ vector $\mathbf{a} = a_x\hat{\mathbf{i}} + a_y\hat{\mathbf{j}}$ to another basis, composed of the two vectors

$$\mathbf{e}_1 = 3\hat{\mathbf{i}} + \hat{\mathbf{j}}, \qquad \mathbf{e}_2 = \hat{\mathbf{i}} - \hat{\mathbf{j}}, \qquad \text{using}$$

by means of (i) direct substitution of the basis vectors and (ii) matrix elements of the corresponding transformation $\frac{\partial x^i}{\partial X^a}$. Calculate both contravariant a^1, a^2 and covariant a_1, a_2 components of the vector $\mathbf{a}$. (iii) Derive the two metric tensors g_{ij} and g^{ij}. Verify the relations $a_i = g_{ij}a^j$ and $a^i = g^{ik}a_k$.

Chapter 3
Symmetric, Skew(Anti) Symmetric Tensors, and Determinants

Symmetric and antisymmetric tensors play important roles in Mathematics and applications. Here, we consider some relevant aspects of these special tensors.

3.1 Definitions and General Considerations

Def. 1 Tensor $A^{ijk\cdots}$ is called symmetric in the indices i and j, if

$$A^{ijk\cdots} = A^{jik\cdots}. \tag{3.1}$$

Example As we already know, the metric tensor is symmetric, because

$$g_{ij} = (\mathbf{e}_i, \mathbf{e}_j) = (\mathbf{e}_j, \mathbf{e}_i) = g_{ji}. \tag{3.2}$$

Exercise 1 Prove that the inverse metric is also a symmetric tensor, $g^{ij} = g^{ji}$.

Exercise 2 Let a^i be the components of a contravariant vector. Prove that the product $a^i a^j$ is a symmetric contravariant $(2, 0)$-tensor.

Exercise 3 Let a^i and b^j be components of two (generally different) contravariant vectors. As we already know, the products $a^i b^j$ and $a^j b^i$ are components of a contravariant tensor. Construct a linear combination of $a^i b^j$ and $a^j b^i$ that would be a symmetric tensor.

Exercise 4 Let c^{ij} be components of a contravariant tensor. Construct a linear combination of c^{ij} and c^{ji} that would be a symmetric tensor.

Observation 1 In general, the symmetry reduces the number of independent components of a tensor. For example, an arbitrary $(2, 0)$-tensor has nine independent components, while a symmetric tensor of second rank has only six independent components.

© Springer Nature Switzerland AG 2019
I. L. Shapiro, *A Primer in Tensor Analysis and Relativity*, Undergraduate
Lecture Notes in Physics, https://doi.org/10.1007/978-3-030-26895-4_3

Exercise 5 Evaluate the number of independent components of a symmetric second-rank tensor B^{ij} in D-dimensional space.

Def. 2 Tensor $A^{i_1 i_2 \ldots i_n}$ is called completely (or absolutely) symmetric in the indices $(i_1, i_2, \ldots, i_n)$, if it is symmetric in any couple of these indices.

$$A^{i_1 \ldots i_{k-1} i_k i_{k+1} \ldots i_{l-1} i_l i_{l+1} \ldots i_n} = A^{i_1 \ldots i_{k-1} i_l i_{k+1} \ldots i_{l-1} i_k i_{l+1} \ldots i_n}. \tag{3.3}$$

Observation 2 Tensor can be symmetric in any couple of contravariant indices or in any couple of covariant indices, and not symmetric in other couples of indices.

Exercise 6* Evaluate the number of independent components of an absolutely symmetric third-rank tensor B^{ijk} in a D-dimensional space.

Def. 3 Tensor A^{ij} is called skew-symmetric or antisymmetric, if

$$A^{ij} = -A^{ji}. \tag{3.4}$$

Observation 3 We shall formulate the most of consequent definitions and theorems for the contravariant tensors. The statements for the covariant tensors are very similar and can be formulated by the reader as a simple exercise.

Observation 4 The above definitions and consequent properties and relations use only algebraic operations over tensors, and do not address the transformation from one basis to another. Hence, one can consider, instead of symmetric or antisymmetric tensors, rather some symmetric or antisymmetric objects provided by indices. The advantage of tensors is that their (anti)symmetry holds under the transformation from one basis to another. But the same property is shared by more general objects called tensor densities (see Def. 8 below).

Exercise 7 Let c_{ij} be components of a covariant tensor. Construct a linear combination of c_{ij} and c_{ji} that would be an antisymmetric tensor.

Exercise 8 Let c_{ij} be the components of a covariant tensor. Show that c_{ij} can always be presented as a sum of a symmetric tensor $c_{(ij)}$

$$c_{(ij)} = c_{(ji)} \tag{3.5}$$

and an antisymmetric tensor $c_{[ij]}$

$$c_{[ij]} = -c_{[ji]}. \tag{3.6}$$

Discuss the peculiarities of this representation for cases when the initial tensor c_{ij} is, by itself, symmetric or antisymmetric. Discuss whether there are any restrictions on the dimension D for this representation.

Hint 1. One can always represent the tensor c_{ij} as

$$c_{ij} = \frac{1}{2} \left(c_{ij} + c_{ji} \right) + \frac{1}{2} \left(c_{ij} - c_{ji} \right). \tag{3.7}$$

Hint 2. Consider particular cases $D = 2$, $D = 3$, and $D = 1$.

Exercise 9 Consider the basis in $2D$,

$$\mathbf{e}_1 = \hat{\mathbf{i}} + \hat{\mathbf{j}}, \quad \mathbf{e}_2 = \hat{\mathbf{i}} - \hat{\mathbf{j}}.$$

(i) Derive all components of the absolutely antisymmetric tensor ε^{ij} in the new basis, taking into account that $\varepsilon^{ab} = \epsilon^{ab}$, where $\epsilon^{12} = 1$ in the orthonormal basis $\hat{\mathbf{n}}_1 = \hat{\mathbf{i}}$, $\hat{\mathbf{n}}_2 = \hat{\mathbf{j}}$. The calculation should be performed directly and also by using the formula for antisymmetric tensor. Explain the difference between the two results.
(ii) Repeat the calculation for ε_{ij}. Calculate the metric components and verify that $\varepsilon_{ij} = g_{ik}g_{jl} \varepsilon^{kl}$ and that $\varepsilon^{ij} = g^{ik}g^{jl}\varepsilon_{kl}$.

Answers. $\varepsilon_{12} = -2$ and $\varepsilon^{12} = -1/2$.

3.2 Completely Antisymmetric Tensors

Def. 4 $(n, 0)$-tensor $A^{i_1 \ldots i_n}$ is called completely (or absolutely) antisymmetric, if it is antisymmetric in any couple of its indices

$$\forall (k, l) , \qquad A^{i_1 \ldots i_l \ldots i_k \ldots i_n} = -A^{i_1 \ldots i_k \ldots i_l \ldots i_n}.$$

In the case of an absolutely antisymmetric tensor, the sign changes when we perform permutation of any two indices.

Exercise 10 (i) Evaluate the number of independent components of an antisymmetric tensor A^{ij} in 2- and 3-dimensional space. (ii) Evaluate the number of independent components of absolutely antisymmetric tensors A^{ij} and A^{ijk} in a D-dimensional space, where $D > 3$.

Theorem 1 *In a D-dimensional space, all completely antisymmetric tensors of rank $n > D$ are equal to zero.*

Proof First, we notice that if two of its indices are equal, an absolutely skew-symmetric tensor is zero:

$$A^{\ldots i \ldots i \ldots} = -A^{\ldots i \ldots i \ldots} \qquad \Longrightarrow \qquad A^{\ldots i \ldots i \ldots} = 0.$$

Furthermore, in a D-dimensional space, any index can take only D distinct values, and thus among $n > D$ indices there are at least two equal ones. Therefore, all components of $A^{i_1 i_2 \ldots i_n}$ are equal to zero.

Theorem 2 *An absolutely antisymmetric tensor* $A^{i_1...i_D}$ *in a D-dimensional space has only one independent component.*

Proof All components with two equal indices are zeros. Consider an arbitrary component with all indices distinct from each other. Using permutations, it can be transformed into one single component $A^{12...D}$. Since each permutation changes the sign of the tensor, the nonzero components have the absolute value equal to $\left| A^{12...D} \right|$, and the sign dependent on whether the number of necessary permutations is even or odd.

Observation 5 We shall call an absolutely antisymmetric tensor with D indices in the D-dimensional space the *maximal absolutely antisymmetric tensor*. There is a special case of the maximal absolutely antisymmetric tensor $\varepsilon^{i_1 i_2...i_D}$ which is called *Levi-Civita tensor*. In the Cartesian coordinates, the unique relevant nonzero component of this tensor equals

$$\varepsilon^{123...D} = E^{123...D} = 1. \tag{3.8}$$

It is customary to consider a non-tensor absolutely antisymmetric object $E^{i_1 i_2 i_3...i_D}$, which has the same components (3.8) in *any* coordinates. The nonzero components of the tensor $\varepsilon^{i_1 i_2...i_D}$ in arbitrary coordinates will be derived below.

Observation 6 In general, the tensor $A^{ijk...}$ may be symmetric in some couples of indexes, antisymmetric in other couples of indices and have no symmetry in yet another couple of indices.

Exercise 11 Prove that for symmetric a^{ij} and antisymmetric b_{ij} tensors the scalar internal product is zero $a^{ij} \cdot b_{ij} = 0$.

Observation 7 This is an extremely important statement!

   Solution. Changing the names of the indices and using symmetries, we get

$$a^{ij} b_{ij} = a^{ji} b_{ji} = (+a^{ij})(-b_{ij}) = -a^{ij} b_{ij}.$$

The quantity that equals itself with an opposite sign is zero.

Exercise 12 Suppose that for some tensor a_{ij} and some numbers x and y, there is a relation:

$$x\, a_{ij} + y\, a_{ji} = 0.$$

Prove that either

$$x = y \quad \text{and} \quad a_{ij} = -a_{ji}$$

or

$$x = -y \quad \text{and} \quad a_{ij} = a_{ji}.$$

Exercise 13* (i) In $2D$, how many distinct components have tensor c^{ijk}, if

$$c^{ijk} = c^{jik} = -c^{ikj} ?$$

(ii) Investigate the same problem in $3D$. (iii) Prove that $c^{ijk} \equiv 0$ in $\forall D$.

Exercise 14 (i) Prove that if $b^{ij}(x)$ is a tensor and $a_l(x)$ and $c_k(x)$ are covariant vector fields, then $f(x) = b^{ij}(x) \cdot a_i(x) \cdot c_j(x)$ is a scalar field.
(ii) Try to generalize the situation described in the previous point for arbitrary tensors.
(iii) In case $a_i(x) \equiv c_i(x)$, formulate the sufficient condition for $b^{ij}(x)$, such that the product $f(x) \equiv 0$.

Exercise 15* (i) Prove that, in a D-dimensional space, with $D > 3$, the numbers of independent components of absolutely antisymmetric tensors $A^{k_1 k_2 \dots k_n}$ and $\tilde{A}_{i_1 i_2 i_3 \dots i_{D-n}}$ are equal, for any $n \le D$. (ii) Using the absolutely antisymmetric tensor $\varepsilon_{j_1 j_2 \dots j_D}$, try to find an explicit relation (which is called dual correspondence) between the tensor $A^{k_1 k_2 \dots k_n}$ and some $\tilde{A}_{i_1 i_2 i_3 \dots i_{D-n}}$ and then invert this relation.

Hint. The reader is advised to start from the $4D$ space and explore the relation between an absolutely antisymmetric tensor T_{ijk} and the dual vector $S^l = T_{ijk}\varepsilon^{ijkl}$. Show that the inverse relation holds for the Euclidean space, $T_{ijk} = \frac{1}{6}\varepsilon_{ijkl}S^l$. The word Euclidean means that in the pseudo-Euclidean, Minkowski space that we will discuss in the second part of the book, the sign of the inverse relation changes.

3.3 Determinants

There is a deep relation between maximal absolutely antisymmetric tensors and determinants. Let us formulate the main aspects of this relation as a set of theorems. Some of the theorems will first be formulated in $2D$ and then generalized for the case of arbitrary D.

Consider first $2D$ and special coordinates X^1 and X^2 corresponding to the orthonormal basis $\hat{n}_1$ and $\hat{n}_2$.

Def. 5 We define a special maximal absolutely antisymmetric object in $2D$ using the relations (see Observation 5)

$$E_{ab} = -E_{ba} \quad \text{and} \quad E_{12} = 1.$$

Theorem 3 For any matrix $\|C_b^a\|$ we can write its determinant as

$$\det \|C_b^a\| = E_{ab}C_1^a C_2^b. \tag{3.9}$$

Proof Using $E_{12} = -E_{21} = 1$ and $E_{11} = E_{22} = 0$, we get

$$E_{ab}C_1^a C_2^b = E_{12}C_1^1 C_2^2 + E_{21}C_1^2 C_2^1 + E_{11}C_1^1 C_2^1 + E_{22}C_1^2 C_2^2$$
$$= C_1^1 C_2^2 - C_1^2 C_2^1 + 0 + 0 = \begin{vmatrix} C_1^1 & C_2^1 \\ C_1^2 & C_2^2 \end{vmatrix}. \tag{3.10}$$

Remark One can note that this result is not an accident. The source of the coincidence is that the $\det(C_b^a)$ is the sum of the products of the elements of the matrix C_b^a, taken with the sign corresponding to the parity of the permutation. The internal product with E_{ab} provides the correct signs to all these products. In particular, the terms with equal indices a and b vanish, because $E_{aa} \equiv 0$. This exactly corresponds to the fact that the determinant with two equal lines vanishes. But the determinant with two equal rows also vanishes, and hence we can formulate a more general theorem.

Theorem 4

$$\det(C_b^a) = \frac{1}{2} E_{ab} E^{de} C_d^a C_e^b. \tag{3.11}$$

Proof The difference between (3.9) and (3.11) is that the latter admits two expressions $C_1^a C_2^b$ and $C_2^a C_1^b$, while the former admits only the first combination. But, since in both cases the indices run the same values $(a, b) = (1, 2)$, we have, in (3.11), two equal expressions compensated by the $1/2$ factor.

Exercise 16 Check the equality (3.11) by direct calculus.

Theorem 5

$$E_{ab}C_e^a C_d^b = -E_{ab}C_e^b C_d^a. \tag{3.12}$$

Proof

$$E_{ab}C_e^a C_d^b = E_{ba}C_e^b C_d^a = -E_{ab}C_d^a C_e^b.$$

Here in the first equality, we have exchanged the names of the umbral indices $a \leftrightarrow b$, and in the second one used antisymmetry $E_{ab} = -E_{ba}$.

Now we consider an arbitrary dimension of space D.

Theorem 6 *Consider* $a, b = 1, 2, 3, \ldots, D$. $\forall$ *matrix* $\|C_b^a\|$, *the determinant is*

$$\det \|C_b^a\| = E_{a_1 a_2 \ldots a_D} \cdot C_1^{a_1} C_2^{a_2} \ldots C_D^{a_D}, \tag{3.13}$$

where

$$E_{a_1 a_2 \ldots a_D} \text{ is absolutely antisymmetric and } E_{123 \ldots D} = 1.$$

Proof The determinant on the left-hand side (*l.h.s.*) of Eq. (3.13) is a sum of $D!$ terms, each of which is a product of elements from different lines and columns, taken with the positive sign for even and with the negative sign for odd parity of the permutations.

The expression on the right-hand side (*r.h.s.*) of (3.13) also has $D!$ terms. In order to see this, let us note that the terms with equal indices give zero. Hence, for the first index, we have D choices, for the second $D - 1$ choices, for the third $D - 2$ choices, etc. Each nonzero term has one element from each line and one element from each column, because two equal numbers of columns give zero when we make a contraction with $E_{a_1 a_2 \ldots a_D}$. Finally, the sign of each term depends on the permutations' parity and is identical to the sign of the corresponding term in the *l.h.s.* Finally, the terms on both sides are identical.

Theorem 7 *The expression*

$$\mathcal{A}_{a_1 \ldots a_D} = E_{b_1 \ldots b_D} A_{a_1}^{b_1} A_{a_2}^{b_2} \ldots A_{a_D}^{b_D} \tag{3.14}$$

is absolutely antisymmetric in the indices $\{a_1, \ldots, a_D\}$ and therefore is proportional to $E_{a_1 \ldots a_D}$ (see Theorem 2).

Proof For any couple b_l, b_k $(l \neq k)$, the proof performs exactly as in Theorem 5 and we obtain that the *r.h.s.* is antisymmetric in $\forall$ couple of indices. Due to Def. 4, this proves the theorem.

Theorem 8 *In an arbitrary dimension D*

$$\det \| C_b^a \| = \frac{1}{D!} E_{a_1 a_2 \ldots a_D} E^{b_1 b_2 \ldots b_D} \cdot C_{b_1}^{a_1} C_{b_2}^{a_2} \ldots C_{b_D}^{a_D}. \tag{3.15}$$

Proof The proof can be easily performed by analogy with Theorems 4 and 7.

Theorem 9 *The following relation is valid:*

$$E^{a_1 a_2 \ldots a_D} \cdot E_{b_1 b_2 \ldots b_D} = \begin{vmatrix} \delta_{b_1}^{a_1} & \delta_{b_2}^{a_1} & \ldots & \delta_{b_D}^{a_1} \\ \delta_{b_1}^{a_2} & \delta_{b_2}^{a_2} & \ldots & \delta_{b_D}^{a_2} \\ \ldots & \ldots & \ldots & \ldots \\ \delta_{b_1}^{a_D} & \ldots & \ldots & \delta_{b_D}^{a_D} \end{vmatrix}. \tag{3.16}$$

Proof First, we show that both *l.h.s.* and *r.h.s.* of Eq. (3.16) are completely antisymmetric in both $\{a_1, \ldots, a_D\}$ and $\{b_1, \ldots, b_D\}$. For the *l.h.s.* this is obvious. For the *r.h.s.* case we exchange $a_l \leftrightarrow a_k$ for any $l \neq k$. This is equivalent to the permutation of the two lines in the determinant, and therefore the sign changes. The same happens if we make a similar permutation of the columns. Making such permutations, we can reduce Eq. (3.16) to the obviously correct equality

$$1 = E^{12 \ldots D} \cdot E_{12 \ldots D} = \det \| \delta_j^i \| = 1.$$

Consequence: For the special dimension $3D$, we have

$$E^{abc} E_{def} = \begin{vmatrix} \delta_d^a & \delta_e^a & \delta_f^a \\ \delta_d^b & \delta_e^b & \delta_f^b \\ \delta_d^c & \delta_e^c & \delta_f^c \end{vmatrix}. \tag{3.17}$$

Let us make a contraction of the indices c and f in the last equation. Then (remember that $\delta_c^c = 3$ in $3D$)

$$E^{abc} E_{dec} = \begin{vmatrix} \delta_d^a & \delta_e^a & \delta_c^a \\ \delta_d^b & \delta_e^b & \delta_c^b \\ \delta_d^c & \delta_e^c & 3 \end{vmatrix} = \delta_d^a \delta_e^b - \delta_e^a \delta_d^b. \tag{3.18}$$

The last formula is sometimes called magic, as it is an extremely useful tool for the calculations in analytic geometry and vector calculus. We are going to extensively use this formula in what follows.

Let us proceed and contract indices b and e. We obtain

$$E^{abc} E_{dbc} = 3\delta_d^a - \delta_d^a = 2\delta_d^a. \tag{3.19}$$

Finally, contracting the last remaining couple of indices, we arrive at

$$E^{abc} E_{abc} = 6. \tag{3.20}$$

Exercise 17 (a) Verify the last equation by direct calculus. (b) Without making detailed intermediate calculations, even without using the formula (3.16), prove that for an arbitrary dimension D of Euclidean space the analogous formula looks like $E^{a_1 a_2 \ldots a_D} E_{a_1 a_2 \ldots a_D} = D!$. (c) Using the previous result and symmetry arguments, find the partial contraction $E^{a a_2 \ldots a_D} E_{b a_2 \ldots a_D}$. Try to continue the procedure and finally arrive at Eq. (3.16).

Exercise 18* Using definition of the maximal antisymmetric symbol $E^{a_1 a_2 \ldots a_D}$, prove the rule for the product of matrix determinants

$$\det (A \cdot B) = \det A \cdot \det B. \tag{3.21}$$

Here both $A = \|a_k^i\|$ and $B = \|b_k^i\|$ are $n \times n$ matrices.

Hint. It is recommended to consider the simplest case $D = 2$ first and only then proceed with an arbitrary dimension D. The proof may be performed in a most compact way by using the representation of the determinants from Theorem 8 and, of course, Theorem 9.

Solution. Consider

$$\det (A \cdot B) = E_{i_1 i_2 \ldots i_D} (A \cdot B)_1^{i_1} (A \cdot B)_2^{i_2} \ldots (A \cdot B)_D^{i_D}$$

$$= E_{i_1 i_2 \ldots i_D} \, a_{k_1}^{i_1} \, a_{k_2}^{i_2} \ldots a_{k_D}^{i_D} \, b_1^{k_1} \, b_2^{k_2} \ldots b_D^{k_D} \, .$$

Now we notice that

$$E_{i_1 i_2 \ldots i_D} \, a_{k_1}^{i_1} \, a_{k_2}^{i_2} \ldots a_{k_D}^{i_D} \, = \, E_{k_1 k_2 \ldots k_D} \cdot \det A .$$

Therefore

$$\det (A \cdot B) \, = \, \det A \; E_{k_1 k_2 \ldots k_D} \, b_1^{k_1} \, b_2^{k_2} \ldots b_D^{k_D} \, ,$$

which is nothing else but (3.21).

Finally, let us consider two useful, but a bit more complicated, statements concerning inverse matrix and derivative of the determinant.

Lemma *For any nondegenerate $D \times D$ matrix M_b^a with the determinant M, the elements of the inverse matrix $(M^{-1})_c^b$ are given by the expressions*

$$(M^{-1})_a^b \, = \, \frac{1}{M \, (D-1)!} \, E_{a \, a_2 \ldots a_D} \, E^{b \, b_2 \ldots b_D} \, M_{b_2}^{a_2} \, M_{b_3}^{a_3} \ldots M_{b_D}^{a_D} . \tag{3.22}$$

Proof In order to prove this statement, let us multiply the expression (3.22) by the element M_b^c. We want to prove that $(M^{-1})_a^b \, M_b^c \, = \, \delta_a^c$, that means exactly

$$\frac{1}{(D-1)!} \, E_{a \, a_2 \ldots a_D} \, E^{b \, b_2 \ldots b_D} \, M_b^c \, M_{b_2}^{a_2} \, M_{b_3}^{a_3} \ldots M_{b_D}^{a_D} \, = \, M \, \delta_a^c . \tag{3.23}$$

The first observation is that, due to Theorem 7, the expression

$$E^{b \, b_2 \ldots b_D} \, M_b^c \, M_{b_2}^{a_2} \, M_{b_3}^{a_3} \ldots M_{b_D}^{a_D} \tag{3.24}$$

is a maximal absolutely antisymmetric object. Thus, it may be different from zero only if the index c is different from all the indices a_2, a_3, ...a_D. The same is of course valid for the index a in (3.23). Therefore, the expression (3.23) is zero if $a \neq c$. Now we have to prove that the *l.h.s.* of this expression equals M for $a = c$. Consider, e.g., $a = c = 1$. It is easy to see that the coefficient of the term $\| M_b^1 \|$ in the *l.h.s* of (3.23) is the $(D-1) \times (D-1)$ determinant of the matrix $\| M \|$ without the first line and without the row b. The sign is positive for even b and negative for odd b. Summing up over the index b, we meet exactly $M = \det \| M \|$. It is easy to see that the same situation holds for all $a = c = 1, 2, ..., D$.

Theorem 10 *Consider the nondegenerate $D \times D$ matrix $\| M_b^a \|$ with the elements $M_b^a = M_b^a(\kappa)$ being functions of some parameter κ. In general, the determinant of this matrix $M = \det \| M_b^a \|$ also depends on κ. Suppose all functions $M_b^a(\kappa)$ are differentiable. Then the derivative of the determinant equals*

$$\dot{M} \, = \, \frac{dM}{d\kappa} \, = \, M \, (M^{-1})_a^b \, \frac{d \, M_b^a}{d\kappa} \, , \tag{3.25}$$

where $(M^{-1})^b_a$ are the elements of the matrix inverse to (M^a_b) and the dot over a function indicates its derivative with respect to κ.

Proof Let us use Theorem 8 as a starting point. Taking the derivative of

$$M = \frac{1}{D!} E_{a_1 a_2 \ldots a_D} E^{b_1 b_2 \ldots b_D} \cdot M^{a_1}_{b_1} M^{a_2}_{b_2} \ldots M^{a_D}_{b_D} \tag{3.26}$$

we arrive at the expression

$$
\begin{aligned}
\dot{M} = \frac{1}{D!} E_{a_1 a_2 \ldots a_D} E^{b_1 b_2 \ldots b_D} \big(\dot{M}^{a_1}_{b_1} M^{a_2}_{b_2} \ldots M^{a_D}_{b_D} \\
+ M^{a_1}_{b_1} \dot{M}^{a_2}_{b_2} \ldots M^{a_D}_{b_D} + \ldots + M^{a_1}_{b_1} M^{a_2}_{b_2} \ldots \dot{M}^{a_D}_{b_D} \big).
\end{aligned}
\tag{3.27}
$$

By making permutations of the indices, it is easy to see that the last expression is nothing but the sum of D equal terms, and therefore it can be rewritten as

$$\dot{M} = \dot{M}^a_b \left[\frac{1}{(D-1)!} E_{a\, a_2 \ldots a_D} \cdot E^{b\, b_2 \ldots b_D} \cdot M^{a_2}_{b_2} \ldots M^{a_D}_{b_D} \right]. \tag{3.28}$$

According to the previous lemma, the coefficient of the derivative $\dot{M}^a_b$ in (3.28) is $M\,(M^{-1})^b_a$, which completes the proof.

3.4 Applications to Vector Algebra

Consider applications of the formula (3.18) to vector algebra. For the sake of simplicity, we will be working in a special orthonormal basis $\hat{\mathbf{n}}_a = (\hat{\mathbf{i}}, \hat{\mathbf{j}}, \hat{\mathbf{k}})$, corresponding to the Cartesian coordinates $X^a = (X^1, X^2, X^3) = (x, y, z)$.

Def. 6 The vector product of the two vectors $\mathbf{a}$ and $\mathbf{b}$ in $3D$ is the third vector[1]

$$\mathbf{c} = \mathbf{a} \times \mathbf{b} = [\mathbf{a}, \mathbf{b}], \tag{3.29}$$

which satisfies the following conditions: (i) $\mathbf{c} \perp \mathbf{a}$ and $\mathbf{c} \perp \mathbf{b}$; (ii) $c = |\mathbf{c}| = |\mathbf{a}| \cdot |\mathbf{b}| \cdot \sin \varphi$, where φ is the angle between the two vectors $\mathbf{a}$ and $\mathbf{b}$; (iii) the direction of the vector product $\mathbf{c}$ is defined by the right-hand rule. That is, looking along $\mathbf{c}$ (from beginning to end), one observes the smallest rotation angle from $\mathbf{a}$ to $\mathbf{b}$ performed clockwise (see Fig. 3.1).

The properties of the vector product are discussed in the courses of analytic geometry and thus will not be repeated here. The property that is the most important for us is linearity

---

[1] Exactly as in the case of the scalar product of two vectors, it proves useful to introduce two different notations for the vector product of two vectors.

Fig. 3.1 Positively
(right-handed) oriented basis
in $3D$

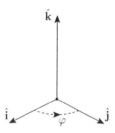

$$\mathbf{a} \times (\mathbf{b}_1 + \mathbf{b}_2) = \mathbf{a} \times \mathbf{b}_1 + \mathbf{a} \times \mathbf{b}_2 .$$

Vector products of the elements of orthonormal basis satisfy the cyclic rule,

$$\hat{\mathbf{i}} \times \hat{\mathbf{j}} = \hat{\mathbf{k}}, \qquad \hat{\mathbf{j}} \times \hat{\mathbf{k}} = \hat{\mathbf{i}}, \qquad \hat{\mathbf{k}} \times \hat{\mathbf{i}} = \hat{\mathbf{j}}$$

or, in other notations

$$\left[\hat{\mathbf{n}}_1 , \hat{\mathbf{n}}_2\right] = \hat{\mathbf{n}}_1 \times \hat{\mathbf{n}}_2 = \hat{\mathbf{n}}_3,$$
$$\left[\hat{\mathbf{n}}_2 , \hat{\mathbf{n}}_3\right] = \hat{\mathbf{n}}_2 \times \hat{\mathbf{n}}_3 = \hat{\mathbf{n}}_1,$$
$$\left[\hat{\mathbf{n}}_3 , \hat{\mathbf{n}}_1\right] = \hat{\mathbf{n}}_3 \times \hat{\mathbf{n}}_1 = \hat{\mathbf{n}}_2. \tag{3.30}$$

It is easy to see that all the three products (3.30) can be written in a compact form using index notations and the maximal antisymmetric symbol E^{abc},

$$\left[\hat{\mathbf{n}}_a , \hat{\mathbf{n}}_b\right] = E_{abc} \cdot \hat{\mathbf{n}}^c . \tag{3.31}$$

Let us remember that in orthonormal basis $\hat{\mathbf{n}}_a = \hat{\mathbf{n}}^a$, and hence we do not need to distinguish between covariant and contravariant indices. Until we are considering only this type of basis, symbol E^{abc} is a tensor and $E^{abc} = E_{abc}$.

Theorem 11 *Consider two vectors* $\mathbf{V} = V^a \hat{\mathbf{n}}_a$ *and* $\mathbf{W} = W^a \hat{\mathbf{n}}_a$. *Then*

$$[\mathbf{V}, \mathbf{W}] = E_{abc} \cdot V^a \cdot W^b \cdot \hat{\mathbf{n}}^c .$$

Proof Due to the linearity of the vector product $[\mathbf{V}, \mathbf{W}]$, we can write

$$[\mathbf{V}, \mathbf{W}] = \left[V^a \hat{\mathbf{n}}_a , W^b \hat{\mathbf{n}}_b\right] = V^a W^b \left[\hat{\mathbf{n}}_a , \hat{\mathbf{n}}_b\right] = V^a W^b E_{abc} \cdot \hat{\mathbf{n}}^c .$$

Observation 8 Together with the contraction formula for E_{abc}, Theorem 11 provides a simple way to solve many problems of vector algebra.

Example 1 Consider the mixed product of the three vectors

$$(\mathbf{U}, \mathbf{V}, \mathbf{W}) = (\mathbf{U}, [\mathbf{V}, \mathbf{W}]) = U^a \cdot [\mathbf{V}, \mathbf{W}]_a$$

$$= U^a \cdot E_{abc} V^b W^c = E_{abc} U^a V^b W^c = \begin{vmatrix} U^1 & U^2 & U^3 \\ V^1 & V^2 & V^3 \\ W^1 & W^2 & W^3 \end{vmatrix}. \tag{3.32}$$

Exercise 19 Using the formulas above, prove the following properties of the mixed product:

(i) Cyclic identity: $(\mathbf{U}, \mathbf{V}, \mathbf{W}) = (\mathbf{W}, \mathbf{U}, \mathbf{V}) = (\mathbf{V}, \mathbf{W}, \mathbf{U})$.
(ii) Antisymmetry: $(\mathbf{U}, \mathbf{V}, \mathbf{W}) = -(\mathbf{V}, \mathbf{U}, \mathbf{W}) = -(\mathbf{U}, \mathbf{W}, \mathbf{V})$.

Example 2 Let us calculate the double vector product $[\mathbf{U}, \mathbf{V}] \times [\mathbf{W}, \mathbf{Y}]$. It proves useful to work with the components, e.g., with the component a. We make this calculation in full detail,

$$([\mathbf{U}, \mathbf{V}] \times [\mathbf{W}, \mathbf{Y}])_a = E_{abc} [\mathbf{U}, \mathbf{V}]^b [\mathbf{W}, \mathbf{Y}]^c = E_{abc} E^{bde} U_d V_e E^{cfg} W_f Y_g$$

$$= -E_{bac} E^{bde} E^{cfg} U_d V_e W_f Y_g = -\left(\delta_a^d \delta_c^e - \delta_a^e \delta_c^d\right) E^{cfg} U_d V_e W_f Y_g$$

$$= E^{cfg} \left(U_c V_a W_f Y_g - U_a V_c W_f Y_g\right) = V_a (\mathbf{U}, \mathbf{W}, \mathbf{Y}) - U_a (\mathbf{V}, \mathbf{W}, \mathbf{Y}).$$

In the geometric vector notations, this result can be written as

$$[\mathbf{U}, \mathbf{V}] \times [\mathbf{W}, \mathbf{Y}] = \mathbf{V} \cdot (\mathbf{U}, \mathbf{W}, \mathbf{Y}) - \mathbf{U} \cdot (\mathbf{V}, \mathbf{W}, \mathbf{Y}). \tag{3.33}$$

Warning: Do not try to reproduce this formula in a "usual" way without using the index notations, for it may take too much time.

Exercise 20

(i) Prove the following dual formula:

$$[[\mathbf{U}, \mathbf{V}] , [\mathbf{W}, \mathbf{Y}]] = \mathbf{W} \cdot (\mathbf{Y}, \mathbf{U}, \mathbf{V}) - \mathbf{Y} \cdot (\mathbf{W}, \mathbf{U}, \mathbf{V}).$$

(ii) Using (i), prove the following relation:

$$\mathbf{V} (\mathbf{W}, \mathbf{Y}, \mathbf{U}) - \mathbf{U} (\mathbf{W}, \mathbf{Y}, \mathbf{V}) = \mathbf{W} (\mathbf{U}, \mathbf{V}, \mathbf{Y}) - \mathbf{Y} (\mathbf{U}, \mathbf{V}, \mathbf{W}).$$

(iii) Using index notations and properties of the symbol E_{abc}, prove the identity

$$[\mathbf{U} \times \mathbf{V}] \cdot [\mathbf{W} \times \mathbf{Y}] = (\mathbf{U} \cdot \mathbf{W}) (\mathbf{V} \cdot \mathbf{Y}) - (\mathbf{U} \cdot \mathbf{Y}) (\mathbf{V} \cdot \mathbf{W}).$$

(iv) Prove the identity $[\mathbf{A}, [\mathbf{B}, \mathbf{C}]] = \mathbf{B} (\mathbf{A}, \mathbf{C}) - \mathbf{C} (\mathbf{A}, \mathbf{B})$.

Exercise 21 Prove the Jacobi identify for the vector product

$$[U, [V, W]] + [W, [U, V]] + [V, [W, U]] = 0.$$

Until this moment, we were considering the maximal antisymmetric symbol E_{abc} only in the special Cartesian coordinates $\{X^a\}$ associated with the orthonormal basis $\{\hat{n}_a\}$. Let us now construct a tensor version of this symbol.

Def. 7 An absolutely antisymmetric Levi-Civita symbol ε_{ijk} is a tensor that coincides with E_{abc} in the special coordinates $\{X^a\}$.

The recipe for constructing this tensor in arbitrary coordinates is obvious. One has to use the transformation rule for the covariant tensor of the third rank, starting from the special Cartesian coordinates X^a. In this way, for arbitrary coordinates x^i we obtain the following components:

$$\varepsilon_{ijk} = \frac{\partial X^a}{\partial x^i} \frac{\partial X^b}{\partial x^j} \frac{\partial X^c}{\partial x^k} E_{abc}. \tag{3.34}$$

Exercise 22 Using the definition (3.34), prove that
(i) Levi-Civita symbol satisfies the tensor transformation rule

$$\varepsilon'_{lmn} = \frac{\partial x^i}{\partial x'^l} \frac{\partial x^j}{\partial x'^m} \frac{\partial x^k}{\partial x'^n} \varepsilon_{ijk};$$

(ii) ε_{ijk} is absolutely antisymmetric

$$\varepsilon_{ijk} = -\varepsilon_{jik} = -\varepsilon_{ikj}.$$

Hint. Compare this exercise with Theorem 7.

According to the last exercise, ε_{ijk} is a maximal absolutely antisymmetric tensor, and then Theorem 2 tells us it has only one relevant component. For example, if we derive the component ε_{123}, we will be able to obtain *all* other components that are either zeros (if any two indices coincide) or can be obtained from ε_{123} via the permutations of the indices

$$\varepsilon_{123} = \varepsilon_{312} = \varepsilon_{231} = -\varepsilon_{213} = -\varepsilon_{132} = -\varepsilon_{321}.$$

Observation 9 A similar property holds in any dimension D.

Theorem 12 *The component* ε_{123} *is a square root of the metric determinant,*

$$\varepsilon_{123} = g^{1/2}, \quad where \quad g = \det \|g_{ij}\|. \tag{3.35}$$

Proof Let us use the relation (3.34) for ε_{123},

$$\varepsilon_{123} = \frac{\partial X^a}{\partial x^1} \frac{\partial X^b}{\partial x^2} \frac{\partial X^c}{\partial x^3} E_{abc}. \tag{3.36}$$

According to Theorem 6, this is nothing but

$$\varepsilon_{123} = \det \left\| \frac{\partial X^a}{\partial x^i} \right\|. \tag{3.37}$$

On the other hand, we can use Theorem 2 from Chap. 2 and immediately arrive at (3.35).

Exercise 23 Define ε^{ijk} as a contravariant tensor that coincides with $E^{abc} = E_{abc}$ in Cartesian coordinates, and prove that
(i) ε^{ijk} is absolutely antisymmetric; (ii) $\varepsilon'^{lmn} = \frac{\partial x'^l}{\partial x^i} \frac{\partial x'^m}{\partial x^j} \frac{\partial x'^n}{\partial x^k} \varepsilon^{ijk}$;
(iii) the unique nontrivial component is

$$\varepsilon^{123} = \frac{1}{\sqrt{g}}, \quad \text{where} \quad g = \det \|g_{ij}\|. \tag{3.38}$$

Observation 10 Let us note that using Eq. (3.37) does not require too much care, but using Eq. (3.38) does. The reason is that the coordinate transformations that break parity can change the sign in the r.h.s. in Eq. (3.38). The sign written in this equation and in the rest of this chapter corresponds to continuous transformations only, when parity is preserved.

Observation 11 E_{ijk} and E^{ijk} are examples of quantities that transform in such a way that their value, in a given coordinate system, is related to a certain power of $g = \det \|g_{ij}\|$, where g_{ij} is a metric corresponding to these coordinates. Let us remember that

$$E_{ijk} = \frac{1}{\sqrt{g}} \varepsilon_{ijk} ,$$

where ε_{ijk} is a tensor. Of course, E_{ijk} is not a tensor. The values of a unique nontrivial component for both symbols are in general different:

$$E_{123} = 1 \quad \text{and} \quad \varepsilon_{123} = \sqrt{g}.$$

Despite E_{ijk} is not being a tensor, we can easily control its transformation from one coordinate system to another. E_{ijk} is a particular example of objects which are called tensor densities.

Def. 8 The quantity $A_{i_1...i_n}^{\quad\ \ j_1...j_m}$ is a tensor density of the (m, n)-type with the weight r, if the quantity

$$g^{-r/2} \cdot A_{i_1...i_n}^{\quad\ \ j_1...j_m} \quad \text{is a tensor.} \tag{3.39}$$

It is easy to see that the tensor density transforms as

$$\left(g'\right)^{-r/2} A_{l'_1 \dots l'_n}{}^{k'_1 \dots k'_m}(x')$$

$$= g^{-r/2} \frac{\partial x^{i_1}}{\partial x'^{l_1}} \cdots \frac{\partial x^{i_n}}{\partial x'^{l_n}} \frac{\partial x^{k'_1}}{\partial x^{j_1}} \cdots \frac{\partial x^{k'_m}}{\partial x^{j_m}} A_{i_1 \dots i_n}{}^{j_1 \dots j_m}(x),$$

or, more explicitly,

$$A_{l'_1 \dots l'_n}{}^{k'_1 \dots k'_m}(x') \tag{3.40}$$

$$= \left(\det \left\| \frac{\partial x^i}{\partial x'^l} \right\| \right)^r \cdot \frac{\partial x^{i_1}}{\partial x'^{l_1}} \cdots \frac{\partial x^{i_n}}{\partial x'^{l_n}} \frac{\partial x^{k'_1}}{\partial x^{j_1}} \cdots \frac{\partial x^{k'_m}}{\partial x^{j_m}} A_{i_1 \dots i_n}{}^{j_1 \dots j_m}(x).$$

Remark All operations over densities can be defined in an obvious way. We shall leave the construction of these definitions as exercises for the reader.

Exercise 24 Prove that a product of the tensor density of the weight r and another tensor density of the weight s is a tensor density of the weight $r + s$.

Exercise 25 (i) Construct an example of the covariant tensor of the fifth rank, which is symmetric in the first two indices and absolutely antisymmetric in the other three. (ii) Complete the same task using, as building blocks, only the maximally antisymmetric tensor ε_{ijk} and the metric tensor g_{ij}.

Exercise 26 Generalize Def. 7 and Theorem 12 for an arbitrary dimension D.

3.5 Symmetric Tensor of Second Rank and Its Reduction to Principal Axes

In the last section of this chapter, we review an operation over symmetric tensors of the second rank, which has very special relevance in physics and other applications of tensors and beyond. The idea is that, for the symmetric tensor, one can make such coordinate changes that the matrix representation becomes diagonal. In fact, one can diagonalize two symmetric tensors (also called bilinear forms in this case) by a single transformation, if at least one of these bilinear forms is positively defined, as explained below. The purpose of this section is to show how such a diagonalization can be done.

Let us start with an important example. The total energy of the multidimensional harmonic oscillator can be presented in the form (see, e.g., [1])

$$E = K + U, \quad \text{where} \quad K = \frac{1}{2} m_{ij} \dot{q}^i \dot{q}^j \quad \text{and} \quad U = \frac{1}{2} k_{ij} q^i q^j. \tag{3.41}$$

Here, $q_i(t)$ are generalized coordinates that depend on time t, $i = 1, 2, \dots, N$. Furthermore, dots indicate time derivatives; m_{ij} is a positively defined mass matrix and

k_{ij} is a positively defined bilinear form of the interaction term. In the harmonic approximation, both m_{ij} and k_{ij} can be regarded as constants.

The expression "positively defined" means that $k_{ij} q^i q^j \geq 0$ for any choice of q^i, while zero value is possible only if all $q^i = 0$. From the viewpoint of physics, it is necessary that both m_{ij} and k_{ij} be positively defined. In the case of m_{ij}, this guarantees that the system cannot have negative kinetic energy K and, in the case of k_{ij}, the positive sign guarantees that the system has potential, bound from below, such that it can only perform oscillations near a stable equilibrium point at $q^i = 0$.

The Lagrange function of the system is defined as $L = K - U$, and the Lagrange equations

$$m_{ij} \ddot{q}^j = k_{ij} q^j$$

represent a set of interacting second-order differential equations. Needless to say, it is very much desirable to find another set of variables $Q_a(t)$, in which these equations separate from each other, becoming a set of individual oscillator equations

$$\ddot{Q}_a + \omega_a^2 Q_a = 0, \quad a = 1, 2, ..., N. \tag{3.42}$$

The variables $Q_a(t)$ are called normal modes of the system and ω_a are corresponding frequencies. The problem is how to find the proper change of variables between $q^i(t)$ and $Q_a(t)$. In the case when k_{ij} is not positively defined, some of the normal modes will be anti-oscillators, meaning the corresponding frequency is imaginary, $\omega_a^2 < 0$ and the solutions are exponentials instead of linear combinations of sin and cos.

The first observation is that the time dependence does not play a significant role here and that the problem of our interest is purely algebraic. As we have mentioned above, the two bilinear forms can be diagonalized even if only one of the two tensors m_{ij} and k_{ij} is positively defined. In what follows, we assume that m_{ij} is positively defined and will not assume any restrictions of this sort for k_{ij}. Our consideration will be mathematically general and is applicable not only to the oscillator problem, but also to many others, e.g., to the moment of inertia of rigid bodies and a lot of other cases. The problem is traditionally characterized as finding principal axes of inertia, so the general procedure of diagonalization for a symmetric positively defined second-rank tensor is often identified as reducing the symmetric tensor to the principle axes.

The more complete procedure than the one presented here can be found in many textbooks on Linear Algebra, e.g., [2]. Many important references to original papers and useful explanations can be found in the book by Petrov [3]. We will present a slightly reduced version of the proof, which has the advantage of being also must simpler. Our strategy will be, as always, to consider the $2D$ version first and then generalize it to the general D-dimensional case.

In the $2D$ case, consider generalized coordinates $q^i = q^{1,2}$. For the positively defined mass term $m_{ij} q^i q^j$, perform the rotation

$$\begin{aligned} q^1 &= x_1 \cos \alpha - x_2 \sin \alpha \\ q^2 &= x_1 \sin \alpha + x_2 \cos \alpha \end{aligned}, \tag{3.43}$$

where $x_{1,2}$ are new variables. The rotation angle α can always be chosen in such a way that in the new variables, the bilinear expression

$$\tilde{m}_{ij}\,\dot{x}_i\,\dot{x}_j \;=\; m_{ij}\,\dot{q}_i\,\dot{q}_j \tag{3.44}$$

becomes diagonal, with $\tilde{m}_{12} = 0$. The direct replacement gives

$$\tilde{m}_{12} \;=\; m_{12}(\cos^2\alpha - \sin^2\alpha) - (m_{11} - m_{22})\cos\alpha\sin\alpha.$$

Thus, we have two solutions

$$
\begin{aligned}
\text{(i) } & m_{11} = m_{22} && \Longrightarrow && \cos 2\alpha = 0, \\
\text{(ii) } & m_{11} \neq m_{22} && \Longrightarrow && \tan 2\alpha = \frac{2\,m_{12}}{m_{11} - m_{22}}.
\end{aligned} \tag{3.45}
$$

The remaining coefficients $\tilde{m}_{11}$ and $\tilde{m}_{22}$ can be easily found in both cases. For instance, in the first case of $m_{11} = m_{22}$ we can take $\alpha = \frac{\pi}{4}$ and then

$$\tilde{m}_{11} = m_{11} + m_{12} \quad \text{and} \quad \tilde{m}_{22} = m_{11} - m_{12}.$$

One can note that both quantities $\tilde{m}_{11}$ and $\tilde{m}_{22}$ are positive, as otherwise the original bilinear form would not be positively defined due to degeneracy. The same is true for case 2) in (3.45). Thus, we achieved the diagonalization of the first bilinear form m_{ij}.

Consider the situation with the second form k_{ij}. As a result of rotation (3.43) with the angle (3.45), we arrive at the new matrix $\tilde{k}_{ij}$, which is not necessary diagonal, of course. Our intention is to perform second rotation and diagonalize $\tilde{k}_{ij}$, but before that, we have to make one more change of variables, which guarantees that the diagonalization of the first bilinear form m_{ij} does not break down.

To this end, one has to perform the rescalings of the variables

$$x_1 = y_1 \cdot \sqrt{\tilde{m}_{11}} \quad \text{and} \quad x_2 = y_2 \cdot \sqrt{\tilde{m}_{22}}, \tag{3.46}$$

such that the new expression for the mass term becomes simply

$$\frac{1}{2}m_{ij}q^i q^j \;=\; \frac{1}{2}\left(y_1^2 + y_2^2\right). \tag{3.47}$$

The remarkable feature of expression (3.47) is that it does not change under rotation. Therefore, if using rotation to diagonalize the second bilinear form $\tilde{k}_{ij}$, the first one will still remain in the same form (3.47). All in all, we have shown that one can diagonalize the two $2D$ matrices m_{ij} and k_{ij} at the same time. Moreover, it is easy to note that we use the positive definiteness only for the first bilinear form m_{ij} and not for k_{ij}.

The last part of our consideration concerns the case of an arbitrary dimension D. Let us first perform the operations described above in the sector of the variables q^1 and q^2. As a result, we arrive at the new m and k matrices with zero (12)-elements. After that, we can start the same cycle of rotations and rescalings for the next pair of variables, e.g., q^1 and q^3. It is important to observe that when we perform rotations and rescalings in the (13) plane, the (12)-elements of the m and k matrices remain zero. The reason is that the rotations and rescalings in the (13) plane do not involve the q^2 variable. In fact, the $3D$ case is sufficiently general. Further increase of dimension of the bilinear forms does not bring qualitatively new difficulties.

References

1. L.D. Landau, E.M. Lifshits, *Mechanics - Course of Theoretical Physics*, vol. 1 (Butterworth-Heinemann, Oxford, 1976)
2. D.K. Faddeev, *Lectures on Algebra* (Nauka, Moscow, 1984)
3. A.Z. Petrov, *Einstein Spaces* (Pergamon Press, Oxford, 1969)

Chapter 4
Curvilinear Coordinates, Local Coordinate Transformations

The reader probably has experience in using polar coordinates on the $2D$ plane, or spherical coordinates in the $3D$ space. Below, we consider a general treatment of curvilinear coordinate systems, which include these and many other examples.

4.1 Curvilinear Coordinates and Change of Basis

The purpose of this chapter is to consider an arbitrary change of coordinates

$$x'^{\alpha} = x'^{\alpha}(x^{\mu}), \tag{4.1}$$

where $x'^{\alpha}(x)$ are not necessary linear functions as before. It is supposed that $x'^{\alpha}(x)$ are smooth functions (this means that all partial derivatives $\partial x'^{\alpha}/\partial x^{\mu}$ are continuous functions), possibly except finite number of isolated points.

It is easy to see that all the definitions we have introduced before can be easily generalized for the case of arbitrary change of coordinates. Say, the scalar field may be defined as

$$\varphi'(x') = \varphi(x). \tag{4.2}$$

Also, vectors (contra- and covariant) are defined as quantities that transform according to the rules

$$a'^{i}(x') = \frac{\partial x'^{i}}{\partial x^{j}} a^{j}(x) \tag{4.3}$$

for the contravariant and

$$b'_{l}(x') = \frac{\partial x^{k}}{\partial x'^{l}} b_{k}(x) \tag{4.4}$$

© Springer Nature Switzerland AG 2019
I. L. Shapiro, *A Primer in Tensor Analysis and Relativity*, Undergraduate
Lecture Notes in Physics, https://doi.org/10.1007/978-3-030-26895-4_4

for the covariant cases. The same scheme works for tensors, for example, we define the $(1, 1)$-type tensor as an object, whose components are transforming according to

$$A^{i'}_{j'}\left(x'\right) = \frac{\partial x^{i'}}{\partial x^k} \frac{\partial x^l}{\partial x'^j} A^k_l(x).$$ (4.5)

The metric tensor g_{ij} is defined as

$$g_{ij}(x) = \frac{\partial X^a}{\partial x^i} \frac{\partial X^b}{\partial x^j} g_{ab},$$ (4.6)

where $g_{ab} = \delta_{ab}$ is a metric in Cartesian coordinates. Also, the inverse metric is

$$g^{ij} = \frac{\partial x^i}{\partial X^a} \frac{\partial x^j}{\partial X^b} \delta^{ab},$$

where X^a (as before) are Cartesian coordinates corresponding to the orthonormal basis $\hat{\mathbf{n}}_a$. Of course, the basic vectors $\hat{\mathbf{n}}_a$ are the same in all points of the space.

It is so easy to generalize the notion of tensor and algebraic operations over tensors, because all these operations are defined in the same point of the space. Thus, the main difference between general coordinate transformation $x'^\alpha = x'^\alpha(x)$ and the special one $x'^\alpha = \wedge^\alpha_\beta x^\beta + B^{\alpha'}$ with $\wedge^{\alpha'}_\beta = $ const and $B^{\alpha'} = $ const is that, in the general case, the transition coefficients $\partial x^i / \partial x'^j$ are not necessary constants.

One of important consequences is that the metric tensor g_{ij} also depends on the point. Additionally, the antisymmetric tensor ε^{ijk} also depends on the coordinates

$$\varepsilon^{ijk} = \frac{\partial x^i}{\partial X^a} \frac{\partial x^j}{\partial X^b} \frac{\partial x^k}{\partial X^c} E^{abc}.$$

Using $E_{123} = 1$ and according to Theorem 3.10, we have

$$\varepsilon_{123} = \sqrt{g} \quad \text{and} \quad \varepsilon^{123} = \frac{1}{\sqrt{g}}, \quad \text{where} \quad g = g(x) = \det \|g_{ij}(x)\|.$$ (4.7)

Before turning to the applications and particular cases, let us consider the definition of the basis vectors corresponding to the curvilinear coordinates. It is intuitively clear that these vectors must be different in different points. Let us suppose that the vectors of the point-dependent basis $\mathbf{e}_i(x)$ are such that the vector $\mathbf{a} = a^i \mathbf{e}_i(x)$ is a coordinate-independent geometric object. As far as the rule for the vector transformation is (4.3), the transformation for the basis vectors must have the form

$$\mathbf{e}'_i = \frac{\partial x^l}{\partial x'^i} \mathbf{e}_l.$$ (4.8)

In particular, we obtain the following relation between the point-dependent and permanent orthonormal basis

$$\mathbf{e}_i = \frac{\partial X^a}{\partial x^i} \hat{\mathbf{n}}_a. \tag{4.9}$$

Exercise 1 Derive the relation (4.8) starting from (4.3).

Exercise 2 Verify that Eq. (4.9) is compatible with the expression (4.6) for g_{ij}.

4.2 Polar Coordinates on the Plane

As an example of local coordinate transformations, consider polar coordinates on the $2D$ plane. The polar coordinates r, φ are defined as follows:

$$x = r \cos \varphi, \qquad y = r \sin \varphi, \tag{4.10}$$

where x, y are Cartesian coordinates corresponding to the basis vectors $\hat{\mathbf{n}}_x = \hat{\mathbf{i}}$ and $\hat{\mathbf{n}}_y = \hat{\mathbf{j}}$. Our purpose is to learn how to transform an arbitrary tensor to polar coordinates.

Let us denote, using our standard notations, $X^a = (x, y)$ and $x^i = (r, \varphi)$. The first step is to derive the two mutually inverse matrices

$$\left(\frac{\partial X^a}{\partial x^i} \right) = \begin{pmatrix} \cos \varphi & \sin \varphi \\ -r \sin \varphi & r \cos \varphi \end{pmatrix}$$

$$\text{and} \quad \left(\frac{\partial x^j}{\partial X^b} \right) = \begin{pmatrix} \cos \varphi & -\frac{1}{r} \sin \varphi \\ \sin \varphi & \frac{1}{r} \cos \varphi \end{pmatrix}. \tag{4.11}$$

Exercise 3 (i) Derive the first matrix by taking derivatives of (4.10) and the second one by inverting the first. Interpret both transformations on the circumference $r = 1$ as rotations.

(ii) For an arbitrary value of r, write both matrices in (4.11) as products of rotation transformations and another matrix. Explain the geometric sense of the second matrix.

Now we are in a position to calculate the components of an arbitrary tensor in polar coordinates. Let us start from the basis vectors. Using the general formula (4.9) in the specific case of polar coordinates, we obtain

$$\mathbf{e}_r = \frac{\partial x}{\partial r} \hat{\mathbf{n}}_x + \frac{\partial y}{\partial r} \hat{\mathbf{n}}_y = \hat{\mathbf{i}} \cos \varphi + \hat{\mathbf{j}} \sin \varphi,$$

$$\mathbf{e}_\varphi = \frac{\partial x}{\partial \varphi} \hat{\mathbf{n}}_x + \frac{\partial y}{\partial \varphi} \hat{\mathbf{n}}_y = r \left(-\hat{\mathbf{i}} \sin \varphi + \hat{\mathbf{j}} \cos \varphi \right). \tag{4.12}$$

Let us analyze these formulas in detail. The first observation is that the basic vectors $\mathbf{e}_r$ and $\mathbf{e}_\varphi$ are orthogonal $\mathbf{e}_r \cdot \mathbf{e}_\varphi = 0$. As a result, the metric in polar coordinates is diagonal

$$g_{ij} = \begin{pmatrix} g_{rr} & g_{r\varphi} \\ g_{\varphi r} & g_{\varphi\varphi} \end{pmatrix} = \begin{pmatrix} \mathbf{e}_r \cdot \mathbf{e}_r & \mathbf{e}_r \cdot \mathbf{e}_\varphi \\ \mathbf{e}_\varphi \cdot \mathbf{e}_r & \mathbf{e}_\varphi \cdot \mathbf{e}_\varphi \end{pmatrix} = \begin{pmatrix} 1 & 0 \\ 0 & r^2 \end{pmatrix}. \tag{4.13}$$

Furthermore, we can see that the vector $\mathbf{e}_r$ is nothing but the normalized radius vector of the point

$$\mathbf{e}_r = \frac{\mathbf{r}}{r}. \tag{4.14}$$

Correspondingly, the basic vector $\mathbf{e}_\varphi$ is orthogonal to the radius vector $\mathbf{r}$. Another observation is that only the basic vector $\mathbf{e}_r$ is normalized $|\mathbf{e}_r| = 1$, while the basic vector $\hat{\mathbf{e}}_\varphi$ is (in general) not normalized, $|\hat{\mathbf{e}}_\varphi| = r$. One can introduce the normalized basis

$$\hat{\mathbf{n}}_r = \mathbf{e}_r, \qquad \hat{\mathbf{n}}_\varphi = \frac{1}{r} \mathbf{e}_\varphi. \tag{4.15}$$

One may note that the normalized basis $\hat{\mathbf{n}}_r$, $\hat{\mathbf{n}}_\varphi$ is not a usual *coordinate basis*, which we described and dealt with previously. This means that there are no coordinates that correspond to this basis in the sense of the formula (4.8). The basis corresponding to the polar coordinates (r, φ) is $(\mathbf{e}_r, \mathbf{e}_\varphi)$ and not $(\hat{\mathbf{n}}_r, \hat{\mathbf{n}}_\varphi)$. Therefore, some extra work will be necessary to transform vector components into a normalized basis. One can expand an arbitrary vector $\mathbf{A}$ using one or another basis,

$$\mathbf{A} = \tilde{A}^r \mathbf{e}_r + \tilde{A}^\varphi \mathbf{e}_\varphi = A^r \hat{\mathbf{n}}_r + A^\varphi \hat{\mathbf{n}}_\varphi, \tag{4.16}$$

where

$$A^r = \tilde{A}^r \qquad \text{and} \qquad A^\varphi = r \tilde{A}^\varphi. \tag{4.17}$$

In what follows we shall always observe this pattern: if the coordinate basis (like $\mathbf{e}_r, \mathbf{e}_\varphi$) is not normalized, we shall mark the components of the vector in this basis by tilde. The simpler notations without tilde are always reserved for the components of the vector in the normalized basis.

Why do we need to give this preference to the normalized basis? The reason is that the normalized basis is much more useful for the applications, e.g., in physics. Let us remember that the physical quantities usually have the specific dimensionality. Using the normalized basis means that all components of the vector, even if we are using curvilinear coordinates, have the same dimension. At the same time, the dimensions of the distinct vector components in a non-normalized basis may be different. For example, if we consider the vector of velocity $\mathbf{v}$, its magnitude is measured in centimeters per seconds. Using our notations, in the normalized basis, both v^r and v^φ are also measured in centimeters per seconds. But, in the original coordinate basis $(\mathbf{e}_r, \mathbf{e}_\varphi)$, the component $\tilde{v}^\varphi$ is measured in $1/sec$, while the unit for $\tilde{v}^r$ is cm/sec. As we see, in this coordinate Basis, the different components of the same vector have different dimensionalities, while in the normalized basis they are equal. It is not necessary to be a physicist to understand which basis is more useful. However, we cannot disregard completely the coordinate basis, because it enables us to perform practical transformations from one coordinate system to another in a general form. Also, for more complicated cases (say, for elliptic or hyperbolic coordinates), we are forced to use the coordinate basis. Below we shall organize all the calculations in such a way that one first has to perform the transformation between the coordinate basis systems and then make an extra transformation to the normalized basis.

Exercise 4 For the polar coordinates on the $2D$ plane, find the metric by performing tensor transformation of the metric in Cartesian coordinates.

Solution: In Cartesian coordinates

$$g_{ab} = \delta_{ab}, \quad \text{that is} \quad g_{xx} = 1 = g_{yy}, \quad g_{xy} = g_{yx} = 0.$$

Let us apply the tensor transformation rule:

$$g_{\varphi\varphi} = \frac{\partial x}{\partial \varphi}\frac{\partial x}{\partial \varphi}g_{xx} + \frac{\partial x}{\partial \varphi}\frac{\partial y}{\partial \varphi}g_{xy} + \frac{\partial y}{\partial \varphi}\frac{\partial x}{\partial \varphi}g_{yx} + \frac{\partial y}{\partial \varphi}\frac{\partial y}{\partial \varphi}g_{yy} =$$

$$= \left(\frac{\partial x}{\partial \varphi}\right)^2 + \left(\frac{\partial y}{\partial \varphi}\right)^2 = r^2\sin^2\varphi + r^2\cos^2\varphi = r^2.$$

Similarly we find, using $g_{xy} = g_{yx} = 0$,

$$g_{\varphi r} = \frac{\partial x}{\partial \varphi}\frac{\partial x}{\partial r}g_{xx} + \frac{\partial y}{\partial \varphi}\frac{\partial y}{\partial r}g_{yy} = (-r\sin\varphi)\cdot\cos\varphi + r\cos\varphi\cdot\sin\varphi = 0 = g_{r\varphi}$$

and

$$g_{rr} = \frac{\partial x}{\partial r}\frac{\partial x}{\partial r}g_{xx} + \frac{\partial y}{\partial r}\frac{\partial y}{\partial r}g_{yy} = \left(\frac{\partial x}{\partial r}\right)^2 + \left(\frac{\partial y}{\partial r}\right)^2 = \cos^2\varphi + \sin^2\varphi = 1.$$

Thus, the metric is

$$\begin{pmatrix} g_{rr} & g_{r\varphi} \\ g_{\varphi r} & g_{\varphi\varphi} \end{pmatrix} = \begin{pmatrix} 1 & 0 \\ 0 & r^2 \end{pmatrix},$$

exactly the same result that we obtained earlier in (4.13).

This derivation of the metric has a simple geometric interpretation. Consider two points that have infinitesimally close φ and r (see Fig. 4.1). The distance between these two points dl is defined by the relation

Fig. 4.1 Line element in
$2D$ polar coordinates,
corresponding to the
infinitesimal $d\phi$ and dr

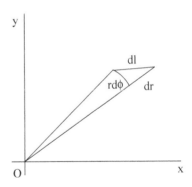

$$ds^2 = dx^2 + dy^2 = g_{ab}\, dX^a\, dX^b = g_{ij}\, dx^i\, dx^j = dr^2 + r^2 d\varphi^2$$
$$= g_{rr} dr dr + g_{\varphi\varphi} d\varphi d\varphi + 2g_{\varphi r} d\varphi dr.$$

Hence, the tensor form of the transformation of the metric corresponds to the coordinate-independent distance between two infinitesimally close points.

Let us consider the motion of the particle on the plane in polar coordinates. The position of the particle in the instant of time t is given by the radius vector $\mathbf{r} = \mathbf{r}(t)$, its velocity by $\mathbf{v} = \dot{\mathbf{r}}$ and acceleration by $\mathbf{a} = \dot{\mathbf{v}} = \ddot{\mathbf{r}}$. Our purpose is to find the components of these three vectors in polar coordinates. We can work directly in the normalized basis. Using (4.12), we arrive at

$$\hat{\mathbf{n}}_r = \hat{\mathbf{i}} \cos\varphi + \hat{\mathbf{j}} \sin\varphi, \qquad \hat{\mathbf{n}}_\varphi = -\hat{\mathbf{i}} \sin\varphi + \hat{\mathbf{j}} \cos\varphi. \qquad (4.18)$$

As we have already mentioned, the transformation from one orthonormal basis $(\hat{\mathbf{i}}, \hat{\mathbf{j}})$ to another one $(\hat{\mathbf{n}}_r, \hat{\mathbf{n}}_\varphi)$ is a rotation, and the angle of rotation φ depends on time. Taking the first and second derivatives, we obtain

$$\frac{d\hat{\mathbf{n}}_r}{dt} = \dot\varphi\,(-\hat{\mathbf{i}} \sin\varphi + \hat{\mathbf{j}} \cos\varphi) = \dot\varphi\, \hat{\mathbf{n}}_\varphi,$$
$$\frac{d\hat{\mathbf{n}}_\varphi}{dt} = \dot\varphi\,(-\hat{\mathbf{i}} \cos\varphi - \hat{\mathbf{j}} \sin\varphi) = -\dot\varphi\, \hat{\mathbf{n}}_r \qquad (4.19)$$

and

$$\frac{d^2\hat{\mathbf{n}}_r}{dt^2} = \ddot\varphi\, \hat{\mathbf{n}}_\varphi - \dot\varphi^2\, \hat{\mathbf{n}}_r, \qquad \frac{d^2\hat{\mathbf{n}}_\varphi}{dt^2} = = -\ddot\varphi\, \hat{\mathbf{n}}_r - \dot\varphi^2\, \hat{\mathbf{n}}_\varphi, \qquad (4.20)$$

where $\dot\varphi$ and $\ddot\varphi$ are angular velocity and acceleration.

The above formulas enable us to derive the velocity and acceleration of the particle. In the Cartesian coordinates, of course,

$$\mathbf{r} = x\hat{\mathbf{i}} + y\hat{\mathbf{j}}, \qquad \mathbf{v} = \dot{x}\hat{\mathbf{i}} + \dot{y}\hat{\mathbf{j}}, \qquad \mathbf{a} = \ddot{x}\hat{\mathbf{i}} + \ddot{y}\hat{\mathbf{j}}. \qquad (4.21)$$

In polar coordinates, using the formulas (4.20), after some algebra we get

$$\mathbf{r} = r\,\hat{\mathbf{n}}_r\,, \qquad \mathbf{v} = \dot{r}\,\hat{\mathbf{n}}_r + r\dot{\varphi}\,\hat{\mathbf{n}}_\varphi\,,$$
$$\mathbf{a} = (\ddot{r} - r\dot{\varphi}^2)\,\hat{\mathbf{n}}_r + (r\ddot{\varphi} + 2\dot{r}\,\dot{\varphi})\,\hat{\mathbf{n}}_\varphi\,. \tag{4.22}$$

Exercise 5 Suggest a physical interpretation of the last two formulas.

4.3 Cylindric and Spherical Coordinates

Def. 1 Cylindric coordinates (r, φ, z) in $3D$ are defined by the relations

$$x = r\cos\varphi\,, \qquad y = r\sin\varphi\,, \qquad z = z\,, \tag{4.23}$$
$$\text{where} \quad 0 \le r < \infty\,, \qquad 0 \le \varphi < 2\pi \quad \text{and} \quad -\infty < z < \infty.$$

Exercise 6 (a) Derive explicit expressions for the basic vectors for the case of cylindric coordinates in $3D$; (b) Using the results of the point (a), calculate all metric components. (c) Using the results of the point (a), construct the normalized non-coordinate basis. (d) Compare all previous results to the ones for the polar coordinates in $2D$.

Answers:

$$\mathbf{e}_r = \hat{\mathbf{i}}\cos\varphi + \hat{\mathbf{j}}\sin\varphi\,, \qquad \mathbf{e}_\varphi = -\hat{\mathbf{i}}r\sin\varphi + \hat{\mathbf{j}}r\cos\varphi\,, \qquad \mathbf{e}_z = \hat{\mathbf{k}}. \tag{4.24}$$

$$g_{ij} = \begin{pmatrix} g_{rr} & g_{r\varphi} & g_{rz} \\ g_{\varphi r} & g_{\varphi\varphi} & g_{\varphi z} \\ g_{zr} & g_{z\varphi} & g_{zz} \end{pmatrix} = \begin{pmatrix} 1 & 0 & 0 \\ 0 & r^2 & 0 \\ 0 & 0 & 1 \end{pmatrix}. \tag{4.25}$$

$$\hat{\mathbf{n}}_r = \hat{\mathbf{i}}\cos\varphi + \hat{\mathbf{j}}\sin\varphi\,, \qquad \hat{\mathbf{n}}_\varphi = -\hat{\mathbf{i}}\sin\varphi + \hat{\mathbf{j}}\cos\varphi\,, \qquad \hat{\mathbf{n}}_z = \hat{\mathbf{k}}. \tag{4.26}$$

Def. 2 Spherical coordinates are defined by the relations

$$x = r\cos\varphi\sin\chi\,, \qquad y = r\sin\varphi\sin\chi\,, \qquad z = r\cos\chi\,, \tag{4.27}$$
$$\text{where} \quad 0 \le r < \infty\,, \qquad 0 \le \varphi < 2\pi \quad \text{and} \quad 0 \le \chi \le \pi.$$

Exercise 7 Repeat the whole program of Exercise 4 for the case of spherical coordinates. Verify that the ordering (r, χ, φ) of these coordinates corresponds to the right-hand orientation of the basis $\hat{\mathbf{n}}_r, \hat{\mathbf{n}}_\chi, \hat{\mathbf{n}}_\varphi$.

Answers : $\mathbf{e}_r = \hat{\mathbf{i}} \cos \varphi \sin \chi + \hat{\mathbf{j}} \sin \varphi \sin \chi + \hat{\mathbf{k}} \cos \chi$,

$\quad\quad\quad \mathbf{e}_\chi = \hat{\mathbf{i}} r \cos \varphi \cos \chi + \hat{\mathbf{j}} r \sin \varphi \cos \chi - \hat{\mathbf{k}} r \sin \chi$,

$\quad\quad\quad \mathbf{e}_\varphi = -\hat{\mathbf{i}} r \sin \varphi \sin \chi + \hat{\mathbf{j}} r \cos \varphi \sin \chi ;$ \hfill (4.28)

$$ g_{ij} = \begin{pmatrix} g_{rr} & g_{r\chi} & g_{r\varphi} \\ g_{\chi r} & g_{\chi\chi} & g_{\chi\varphi} \\ g_{\varphi r} & g_{\varphi\chi} & g_{\varphi\varphi} \end{pmatrix} = \begin{pmatrix} 1 & 0 & 0 \\ 0 & r^2 & 0 \\ 0 & 0 & r^2 \sin^2 \chi \end{pmatrix} ; \quad (4.29) $$

$\hat{\mathbf{n}}_r = \hat{\mathbf{i}} \cos \varphi \sin \chi + \hat{\mathbf{j}} \sin \varphi \sin \chi + \hat{\mathbf{k}} \cos \chi$,

$\hat{\mathbf{n}}_\varphi = -\hat{\mathbf{i}} \sin \varphi + \hat{\mathbf{j}} \cos \varphi$,

$\hat{\mathbf{n}}_\chi = \hat{\mathbf{i}} \cos \varphi \cos \chi + \hat{\mathbf{j}} \sin \varphi \cos \chi - \hat{\mathbf{k}} \sin \chi$. \hfill (4.30)

Exercise 8 Repeat the calculations of Exercise 2 for the case of cylindric and spherical coordinates.

Let us derive the expressions for the velocity and acceleration in $3D$ (analogs of Eqs. (4.19), (4.20), and (4.22)), for the case of the spherical coordinates. The starting point is Eq. (4.30). Now we need to derive the first and second time derivatives for the vectors of this orthonormal basis. A simple calculus gives the following result for the first derivatives:

$$
\begin{aligned}
\dot{\hat{\mathbf{n}}}_r &= \dot{\varphi} \sin \chi \left(-\hat{\mathbf{i}} \sin \varphi + \hat{\mathbf{j}} \cos \varphi \right) \\
&\quad + \dot{\chi} \left(\hat{\mathbf{i}} \cos \varphi \cos \chi + \hat{\mathbf{j}} \sin \varphi \cos \chi - \hat{\mathbf{k}} \sin \chi \right), \\
\dot{\hat{\mathbf{n}}}_\varphi &= -\dot{\varphi} \left(\hat{\mathbf{i}} \cos \varphi + \hat{\mathbf{j}} \sin \varphi \right), \\
\dot{\hat{\mathbf{n}}}_\chi &= \dot{\varphi} \cos \chi \left(-\hat{\mathbf{i}} \sin \varphi + \hat{\mathbf{j}} \cos \varphi \right) \\
&\quad - \dot{\chi} \left(\hat{\mathbf{i}} \sin \chi \cos \varphi + \hat{\mathbf{j}} \sin \chi \sin \varphi + \hat{\mathbf{k}} \cos \chi \right),
\end{aligned}
\quad (4.31)
$$

where, as before,

$$ \dot{\hat{\mathbf{n}}}_r = \frac{d\,\hat{\mathbf{n}}_r}{dt}, \qquad \dot{\hat{\mathbf{n}}}_\varphi = \frac{d\,\hat{\mathbf{n}}_\varphi}{dt}, \qquad \dot{\hat{\mathbf{n}}}_\chi = \frac{d\,\hat{\mathbf{n}}_\chi}{dt}. $$

It is easy to understand that this is not what we really need, because the derivatives (4.31) are given in the constant Cartesian basis and not in the $\hat{\mathbf{n}}_r$, $\hat{\mathbf{n}}_\varphi$, $\hat{\mathbf{n}}_\chi$ basis. Therefore, we need to perform an inverse transformation from the basis $\hat{\mathbf{i}}$, $\hat{\mathbf{j}}$, $\hat{\mathbf{k}}$ to the basis $\hat{\mathbf{n}}_r$, $\hat{\mathbf{n}}_\chi$, $\hat{\mathbf{n}}_\varphi$. The general transformation formula for the coordinate basis looks like

$$ \mathbf{e}'_i = \frac{\partial x^k}{\partial x'^i} \mathbf{e}_k $$

and we get

$$\hat{\mathbf{i}} = \frac{\partial r}{\partial x}\mathbf{e}_r + \frac{\partial \varphi}{\partial x}\mathbf{e}_\varphi + \frac{\partial \chi}{\partial x}\mathbf{e}_\chi,$$

$$\hat{\mathbf{j}} = \frac{\partial r}{\partial y}\mathbf{e}_r + \frac{\partial \varphi}{\partial y}\mathbf{e}_\varphi + \frac{\partial \chi}{\partial y}\mathbf{e}_\chi,$$

$$\mathbf{k} = \frac{\partial r}{\partial z}\mathbf{e}_r + \frac{\partial \varphi}{\partial z}\mathbf{e}_\varphi + \frac{\partial \chi}{\partial z}\mathbf{e}_\chi. \tag{4.32}$$

The nine partial derivatives that appear in the last relations can be found from the observation that

$$\begin{pmatrix} x'_r & x'_\varphi & x'_\chi \\ y'_r & y'_\varphi & y'_\chi \\ z'_r & z'_\varphi & z'_\chi \end{pmatrix} \cdot \begin{pmatrix} r'_x & r'_y & r'_z \\ \varphi'_x & \varphi'_y & \varphi'_z \\ \chi'_x & \chi'_y & \chi'_z \end{pmatrix} = \begin{pmatrix} 1 & 0 & 0 \\ 0 & 1 & 0 \\ 0 & 0 & 1 \end{pmatrix}. \tag{4.33}$$

Inverting the first matrix, we obtain

$$r'_x = \sin\chi\cos\varphi, \qquad \varphi'_x = -\frac{\sin\varphi}{r\sin\chi}, \qquad \chi'_x = \frac{\cos\varphi\cos\chi}{r}$$

$$r'_y = \sin\chi\sin\varphi, \qquad \varphi'_y = \frac{\cos\varphi}{r\sin\chi}, \qquad \chi'_y = \frac{\sin\varphi\cos\chi}{r},$$

$$r'_z = \cos\chi, \qquad \varphi'_z = 0 \qquad \chi'_z = -\frac{\sin\chi}{r}. \tag{4.34}$$

After replacing (4.34) into (4.32), we arrive at the relations

$$\hat{\mathbf{i}} = \hat{\mathbf{n}}_r \cos\varphi\sin\chi - \hat{\mathbf{n}}_\varphi \sin\varphi + \hat{\mathbf{n}}_\chi \cos\varphi\cos\chi,$$

$$\hat{\mathbf{j}} = \hat{\mathbf{n}}_r \sin\varphi\sin\chi + \hat{\mathbf{n}}_\varphi \cos\varphi + \hat{\mathbf{n}}_\chi \sin\varphi\cos\chi,$$

$$\hat{\mathbf{k}} = \hat{\mathbf{n}}_r \cos\chi - \hat{\mathbf{n}}_\chi \sin\chi, \tag{4.35}$$

which are inverse to (4.30). Using these relations, one can derive the first derivatives of the vectors $\hat{\mathbf{n}}_r$, $\hat{\mathbf{n}}_\varphi$, $\hat{\mathbf{n}}_\chi$ in the final form

$$\dot{\hat{\mathbf{n}}}_r = \dot\varphi \sin\chi\,\hat{\mathbf{n}}_\varphi + \dot\chi\,\hat{\mathbf{n}}_\chi,$$

$$\dot{\hat{\mathbf{n}}}_\varphi = -\dot\varphi\left(\sin\chi\,\hat{\mathbf{n}}_r + \cos\chi\,\hat{\mathbf{n}}_\chi\right),$$

$$\dot{\hat{\mathbf{n}}}_\chi = \dot\varphi \cos\chi\,\hat{\mathbf{n}}_\varphi - \dot\chi\,\hat{\mathbf{n}}_r. \tag{4.36}$$

Now we are in a position to derive the particle's velocity

$$\mathbf{v} = \dot{\mathbf{r}} = \frac{d}{dt}\left(r\hat{\mathbf{n}}_r\right) = \dot{r}\,\hat{\mathbf{n}}_r + r\dot{\chi}\,\hat{\mathbf{n}}_\chi + r\dot{\varphi}\sin\chi\,\hat{\mathbf{n}}_\varphi \qquad (4.37)$$

and acceleration

$$\mathbf{a} = \ddot{\mathbf{r}} = \left(\ddot{r} - r\dot{\chi}^2 - r\dot{\varphi}^2\sin^2\chi\right)\hat{\mathbf{n}}_r + \left(2r\dot{\varphi}\dot{\chi}\cos\chi + 2\dot{\varphi}\dot{r}\sin\chi + r\ddot{\varphi}\sin\chi\right)\hat{\mathbf{n}}_\varphi$$
$$+ \left(2\dot{r}\dot{\chi} + r\ddot{\chi} - r\dot{\varphi}^2\sin\chi\cos\chi\right)\hat{\mathbf{n}}_\chi. \qquad (4.38)$$

Exercise 9 Find the basic vectors and metric for the case of the hyperbolic coordinates in the right half-plane in $2D$,

$$x = r\cdot\cosh\varphi, \qquad y = r\cdot\sinh\varphi. \qquad (4.39)$$

Reminder. Hyperbolic functions are defined as follows:

$$\cosh\varphi = \frac{1}{2}\left(e^\varphi + e^{-\varphi}\right), \qquad \sinh\varphi = \frac{1}{2}\left(e^\varphi - e^{-\varphi}\right)$$

and have the following properties:

$$\cosh^2\varphi - \sinh^2\varphi = 1\,; \qquad \cosh'\varphi = \sinh\varphi\,; \qquad \sinh'\varphi = \cosh\varphi.$$

Answer:

$$g_{rr} = \cosh 2\varphi, \quad g_{\varphi\varphi} = r^2\cosh 2\varphi, \quad g_{r\varphi} = r\sinh 2\varphi.$$

We can observe that in this case the basis is not orthogonal.

Exercise 10 Find the components of Levi-Civita tensors ε_{123} and ε^{123} for cylindric and spherical coordinates.

Hint. It is an easy exercise. One has to derive the metric determinants

$$g = \det\left\|g_{ij}\right\| \quad \text{and} \quad \det\left\|g^{ij}\right\| = \frac{1}{g}$$

for both cases, be careful with the ordering of variables and use Eqs. (3.35) and (3.38).

Chapter 5
Derivatives of Tensors, Covariant Derivatives

In the previous chapters, we learned how to use the tensor transformation rule. Consider a tensor in some (maybe curvilinear) coordinates $\{x^i\}$. If one knows the components of the tensor in these coordinates and the relation between original and new coordinates, $x'^i = x'^i\left(x^j\right)$, it is possible to derive the components of the tensor in new coordinates. As we already know, the tensor transformation rule means that the tensor components transform while the tensor itself corresponds to a (geometric or physical) coordinate-independent quantity. When we change the coordinate system, our point of view changes, and hence we may see something different despite the object being the same in both cases.

Why do we need to have a control over the change of coordinates? One of the reasons is that physical laws must be formulated in such a way that they remain valid in any coordinates. However, at this point we meet a problem. Many physical laws include not only the functions themselves but also, quite often, their derivatives. How can we deal with derivatives in the tensor framework?

The two important questions are

(1) Is the partial derivative of a tensor always producing a tensor?

(2) If not, how one can define a "correct" derivative of a tensor, so that differentiation produces a new tensor?

Let us start from a scalar field φ. Consider its partial derivative

$$\partial_i \varphi = \varphi_{,i} = \frac{\partial \varphi}{\partial x^i}. \tag{5.1}$$

Notice that here we used three different notations for the same partial derivative. In the transformed coordinates $x'^i = x'^i\left(x^j\right)$ we obtain, using the chain rule and $\varphi'(x') = \varphi(x)$,

$$\partial_{i'} \varphi' = \frac{\partial \varphi'}{\partial x'^i} = \frac{\partial x^j}{\partial x'^i} \frac{\partial \varphi}{\partial x^j} = \frac{\partial x^j}{\partial x'^i} \partial_j \varphi.$$

© Springer Nature Switzerland AG 2019
I. L. Shapiro, *A Primer in Tensor Analysis and Relativity*, Undergraduate
Lecture Notes in Physics, https://doi.org/10.1007/978-3-030-26895-4_5

The last formula shows that the partial derivative of a scalar field $\varphi_{,i} = \partial_i \varphi$ transforms as a covariant vector field. Certainly, there is no need to modify a partial derivative in this case.

Def. 1 The vector (5.1) is called the *gradient* of a scalar field $\varphi(x)$, and is denoted by

$$\text{grad } \varphi = \mathbf{e}^i \, \varphi_{,i} \quad \text{or} \quad \nabla \varphi = \text{grad } \varphi. \tag{5.2}$$

In this last formula, we have introduced a new object: vector differential operator ∇

$$\nabla = \mathbf{e}^i \frac{\partial}{\partial x^i},$$

which is also called the Hamilton (or nabla) operator. When acting on a scalar, ∇ produces a covariant vector which is called gradient. However, as we shall see later on, the operator ∇ may also act on vectors and tensors.

The next step is to consider a partial derivative of a covariant vector $b_i(x)$. This partial derivative $\partial_j b_i$ looks like a second rank $(0, 2)$-type tensor. But it is easy to see that it is not a tensor! Let us make a corresponding transformation

$$\partial_{j'} b_{i'} = \frac{\partial}{\partial x'^j} b_{i'} = \frac{\partial}{\partial x'^j} \left(\frac{\partial x^k}{\partial x'^i} b_k \right) = \frac{\partial x^k}{\partial x'^i} \frac{\partial x^l}{\partial x'^j} \partial_l b_k + \frac{\partial^2 x^k}{\partial x'^j \partial x'^i} b_k. \tag{5.3}$$

Let us note that in this and subsequent formulas we do not write arguments, assuming that $b_i = b_i(x)$ and $b_{i'} = b_{i'}(x')$. Furthermore, it is worth commenting on a technical subtlety in the calculation (5.3). The parenthesis contains a product of the two expressions that are functions of different coordinates x'^j and x^i. When the partial derivative $\partial / \partial x'^j$ acts on the function of the coordinates x^i (like $b_k = b_k(x)$), it must be applied following the chain rule

$$\frac{\partial}{\partial x'^j} = \frac{\partial x^l}{\partial x'^j} \frac{\partial}{\partial x^l}.$$

The relation (5.3) provides important information. Since the last term in the *r.h.s.* of this equation is nonzero, in general, partial derivatives of a covariant vector do not form a tensor. Indeed, the last term in (5.3) may be equal to zero and then (for the special transformation of coordinates) the partial derivative of a vector transforms as a tensor. This is the case for the global coordinate transformation, when the matrix $\partial x^k / \partial x'^i$ is constant. Then its derivatives are zeros and the tensor rule in (5.3) really holds. But, for a general case of curvilinear coordinates (e.g., polar in $2D$ or spherical in $3D$), the formula (5.3) shows the non-tensor nature of the transformation.

Exercise 1   Derive formulas, similar to (5.3), for the cases of (i) Contravariant vector $a^i(x)$; (ii) Mixed tensor $T_i^j(x)$.

We need to construct a derivative of a tensor which should be also tensor. To this end, we shall use the following strategy. First of all, we choose a name for this new derivative: it will be called *covariant*. After that, the problem is solved through the following definition:

Def. 2 The covariant derivative ∇_i satisfies the following two conditions:
(i) Tensor transformation rule when the new derivative is applied to a tensor;
(ii) In the Cartesian coordinates, $\{X^a\}$ the covariant derivative coincides with the usual partial derivative

$$\nabla_a = \partial_a = \frac{\partial}{\partial X^a}.$$

As we shall see in a moment, these two conditions fix the form of the covariant derivative of a tensor in a unique way.

The simplest way of obtaining an explicit expression for the covariant derivative is to apply what may be called a *covariant continuation*. Let us consider an arbitrary tensor, say mixed $(1, 1)$-type one W_i^j. In order to define its covariant derivative, we perform the following steps:
(1) Transform it to Cartesian coordinates X^a

$$W_b^a = \frac{\partial X^a}{\partial x^i} \frac{\partial x^j}{\partial X^b} W_j^i.$$

(2) Take the partial derivative with respect to the Cartesian coordinates X^c

$$\partial_c W_b^a = \frac{\partial}{\partial X^c} \left(\frac{\partial X^a}{\partial x^i} \frac{\partial x^j}{\partial X^b} W_j^i \right).$$

(3) Transform this derivative back to the original coordinates x^i using the tensor rule.

$$\nabla_k W_m^l = \frac{\partial X^c}{\partial x^k} \frac{\partial x^l}{\partial X^a} \frac{\partial X^b}{\partial x^m} \partial_c W_b^a.$$

According to the procedure described above, the conditions (i) and (ii) are satisfied automatically. Also, by construction, the covariant derivative follows the Leibnitz rule for the product of two tensors A and B

$$\nabla_i (A \cdot B) = \nabla_i A \cdot B + A \cdot \nabla_i B. \tag{5.4}$$

It is easy to see that this procedure works well for any tensor. As an example, let us make an explicit calculation for the covariant vector T_i. Following the scheme of covariant continuation described above and also the chain rule, we obtain

$$\nabla_i T_j = \frac{\partial X^a}{\partial x^i} \frac{\partial X^b}{\partial x^j} (\partial_a T_b) = \frac{\partial X^a}{\partial x^i} \frac{\partial X^b}{\partial x^j} \cdot \frac{\partial}{\partial X^a} \left(\frac{\partial x^k}{\partial X^b} T_k \right)$$

$$= \frac{\partial X^a}{\partial x^i} \frac{\partial X^b}{\partial x^j} \frac{\partial x^k}{\partial X^b} \frac{\partial x^l}{\partial X^a} \frac{\partial T_k}{\partial x^l} + \frac{\partial X^a}{\partial x^i} \frac{\partial X^b}{\partial x^j} T_k \frac{\partial^2 x^k}{\partial X^a \partial X^b}$$

$$= \partial_i T_j - \Gamma_{ji}^k T_k, \tag{5.5}$$

where

$$\Gamma_{ji}^k = -\frac{\partial X^a}{\partial x^i} \frac{\partial X^b}{\partial x^j} \frac{\partial^2 x^k}{\partial X^a \partial X^b}. \tag{5.6}$$

Here we meet, for the first time, the symbol Γ_{ki}^l, which will play a prominent role in what follows. One can obtain another representation for this symbol. Let us start from the general statement, formulated as the following theorem.

Theorem 1 *Suppose the elements of the matrix $\wedge$ depend on the parameter κ and $\wedge(\kappa)$ is a differentiable and invertible matrix for $\kappa \in (a, b)$. Within the region (a, b), the inverse matrix $\wedge^{-1}(\kappa)$ is also differentiable and its derivative is*

$$\frac{d \wedge^{-1}}{d\kappa} = - \wedge^{-1} \frac{\partial \wedge}{\partial \kappa} \wedge^{-1}. \tag{5.7}$$

Proof The differentiability of the inverse matrix can be proved in the same way as for the inverse function in analysis. Hence, we skip this part of the proof. Taking the derivative of $\wedge \cdot \wedge^{-1} = I$ we obtain

$$\frac{d \wedge}{d\kappa} \wedge^{-1} + \wedge \frac{\partial \wedge^{-1}}{\partial \kappa} = 0.$$

After multiplying this equation by $\wedge^{-1}$ from the left, we arrive at (5.7).

Consider

$$\wedge_k^b = \frac{\partial X^b}{\partial x^k}, \qquad \text{then the inverse matrix is} \qquad \left(\wedge^{-1}\right)_a^k = \frac{\partial x^k}{\partial X^a}.$$

Using (5.7) with x^i playing the role of parameter κ, we arrive at

$$\frac{\partial^2 X^b}{\partial x^i \partial x^k} = \frac{\partial}{\partial x^i} \wedge_k^b = - \wedge_l^b \frac{\partial \left(\wedge^{-1}\right)_a^l}{\partial x^i} \wedge_k^a$$

$$= -\frac{\partial X^b}{\partial x^l} \left(\frac{\partial X^c}{\partial x^i} \frac{\partial}{\partial X^c} \frac{\partial x^l}{\partial X^a} \right) \frac{\partial X^a}{\partial x^k} = -\frac{\partial X^b}{\partial x^l} \frac{\partial X^a}{\partial x^k} \frac{\partial X^c}{\partial x^i} \frac{\partial^2 x^l}{\partial X^c \partial X^a}. \tag{5.8}$$

Applying this equality to (5.5), the second equivalent form of the symbol Γ_{ki}^i emerges

$$\Gamma^{j}_{ki} = \frac{\partial x^{j}}{\partial X^{b}} \frac{\partial^{2} X^{b}}{\partial x^{i} \partial x^{k}} = -\frac{\partial^{2} x^{j}}{\partial X^{b} \partial X^{a}} \frac{\partial X^{a}}{\partial x^{k}} \frac{\partial X^{b}}{\partial x^{i}}. \tag{5.9}$$

Exercise 2 Check, by making direct calculation and using (5.9) that

$$\nabla_{i} S^{j} = \partial_{i} S^{j} + \Gamma^{j}_{ki} S^{k}. \tag{5.10}$$

Exercise 3 Repeat the same calculation for a mixed tensor and show that

$$\nabla_{i} W^{j}_{k} = \partial_{i} W^{j}_{k} + \Gamma^{j}_{li} W^{l}_{k} - \Gamma^{l}_{ki} W^{j}_{l}. \tag{5.11}$$

Observation It is important that the expressions (5.10) and (5.11) are consistent with (5.9). In order to explore the origin of this consistency, one can note that the contraction $T_{i} S^{i}$ is a scalar, since T_{i} and S^{j} are co- and contravariant vectors. Using (5.4), we obtain

$$\nabla_{j} \left(T_{i} S^{i} \right) = T_{i} \cdot \nabla_{j} S^{i} + \nabla_{j} T_{i} \cdot S^{i}.$$

Let us assume that $\nabla_{j} S^{i} = \partial_{j} S^{i} + \tilde{\Gamma}^{i}_{kj} S^{k}$. For a while we will treat $\tilde{\Gamma}^{i}_{kj}$ as an independent quantity, but later on we shall see that it actually coincides with the Γ^{i}_{kj}. Then

$$\partial_{j} \left(T_{i} S^{i} \right) = \nabla_{j} \left(T_{i} S^{i} \right) = T_{i} \nabla_{j} S^{i} + \nabla_{j} T_{i} \cdot S^{i} =$$

$$= T_{i} \partial_{j} S^{i} + T_{i} \tilde{\Gamma}^{i}_{kj} S^{k} + \partial_{j} T_{i} \cdot S^{i} - T_{k} \Gamma^{k}_{ij} S^{i}.$$

Changing the names of the umbral indices, we rewrite this equality in the form

$$\partial_{j} \left(T_{i} S^{i} \right) = S^{i} \cdot \partial_{j} T_{i} + T_{i} \cdot \partial_{j} S^{i} = T_{i} \partial_{j} S^{i} + \partial_{j} T_{i} \cdot S^{i} + T_{i} S^{k} \left(\tilde{\Gamma}^{i}_{kj} - \Gamma^{i}_{kj} \right).$$

It is easy to see that the unique, consistent option is to take $\tilde{\Gamma}^{i}_{kj} = \Gamma^{i}_{kj}$. Similar consideration may be used for any tensor. For example, in order to obtain the covariant derivative for the mixed $(1, 1)$-tensor, one can use the fact that $T_{i} S^{k} W^{i}_{k}$ is a scalar. Then Eq. (5.11) follows and so on. This method has a great advantage over direct calculation, not simply because it is more economic. The main point is that using this approach, we can formulate the general rule for constructing a covariant derivative of an arbitrary tensor,

$$\nabla_{i} T^{j_{1} j_{2} \cdots}_{\ k_{1} k_{2} \cdots} = \partial_{i} T^{j_{1} j_{2} \cdots}_{\ k_{1} k_{2} \cdots} + \Gamma^{j_{1}}_{li} T^{l j_{2} \cdots}_{\ k_{1} k_{2} \cdots} + \Gamma^{j_{2}}_{li} T^{j_{1} l \cdots}_{\ k_{1} k_{2} \cdots}$$
$$- \Gamma^{l}_{k_{1} i} T^{j_{1} j_{2} \cdots}_{\ k_{1} l \cdots} - \Gamma^{l}_{k_{2} i} T^{j_{1} j_{2} \cdots}_{\ l k_{2} \cdots} - \cdots . \tag{5.12}$$

As we already know, the covariant derivative differs from the partial derivative by the terms that are linear in the coefficients Γ^{i}_{jk}. These coefficients thus play an important role and deserve a special attention. The first observation is that the transformation rule for Γ^{i}_{jk} must be non-tensor. In Eq. (5.12), the non-tensor transformation for Γ^i_{jk} cancels with the non-tensor transformation for the partial derivative and, in sum, they give a tensor transformation for the covariant derivative.

Exercise 4 Starting from each of the two representations (5.9), derive the transformation rule for the symbol $\Gamma^i_{jk}(x)$ between coordinates x^i and x'^j.

Solution.

$$\Gamma^{i'}_{j'k'} = \frac{\partial x^{i'}}{\partial x^l} \frac{\partial x^m}{\partial x^{j'}} \frac{\partial x^n}{\partial x^{k'}} \Gamma^l_{mn} + \frac{\partial x^{i'}}{\partial x^r} \frac{\partial^2 x^r}{\partial x^{j'} \partial x^{k'}}. \tag{5.13}$$

Observation (i) We shall come back to this transformation rule many times, especially in the third part of the book devoted to general relativity. (ii) The symbol Γ^i_{jk} is zero in the Cartesian coordinates X^a, where covariant derivative is nothing but the partial derivative. Taking X^a as a second coordinates x^i, we arrive back at (5.9).

Indeed, both representations for the symbol Γ^i_{jk} that we have formulated by now are not the most useful ones for practical calculations, so we will have to construct one more. At present, we shall give a special name to it.

Def. 3 Γ^i_{jk} is called the Christoffel symbol, or affine connection.

For a while, we are considering only the flat space—that is, the space that admits global Cartesian coordinates and (equivalently) global orthonormal basis. In this special case, Christoffel symbol and affine connection are the same thing. But there are other, more general geometries which we do not consider in this chapter of the book. For these geometries, Christoffel symbol may differ from the affine connection. For this reason, it is customary to denote the Christoffel symbol as $\left\{ {i \atop jk} \right\}$ and keep the notation Γ^i_{jk} for the affine connection.

The most useful form of the affine connection Γ^i_{jk} is expressed via the metric tensor. In order to obtain this form of Γ^i_{jk}, we remember that in the flat space there is a special orthonormal basis X^a where metric has the form

$$g_{ab} \equiv \delta_{ab}, \quad \text{and hence} \quad \partial_c g_{ab} \equiv \nabla_c g_{ab} \equiv 0.$$

Therefore, in any other coordinate system

$$\nabla_i g_{jk} = \frac{\partial X^c}{\partial x^i} \frac{\partial X^b}{\partial x^j} \frac{\partial X^a}{\partial x^k} \partial_c g_{ab} = 0. \tag{5.14}$$

If we apply to (5.14), the explicit form of the covariant derivative (5.12), we arrive at the equation

$$\nabla_i g_{jk} = \partial_i g_{jk} - \Gamma^l_{ji} g_{lk} - \Gamma^l_{ki} g_{lj} = 0. \tag{5.15}$$

Making permutations of indices, we get

$$\partial_i g_{jk} = \Gamma^l_{ji}\, g_{lk} + \Gamma^l_{ki}\, g_{lj} \quad (i)$$
$$\partial_j g_{ik} = \Gamma^l_{ij}\, g_{lk} + \Gamma^l_{kj}\, g_{il} \quad (ii) \ .$$
$$\partial_k g_{ij} = \Gamma^l_{ik}\, g_{lj} + \Gamma^l_{jk}\, g_{li} \quad (iii)$$

Taking linear combination $(i) + (ii) - (iii)$, we arrive at the relation

$$2\,\Gamma^l_{ij}\, g_{lk} = \partial_i g_{jk} + \partial_j g_{ik} - \partial_k g_{ij}.$$

Contracting both parts with g^{km} (remember that $g^{km}\, g_{ml} = \delta^k_l$), we arrive at

$$\Gamma^k_{ij} = \frac{1}{2}\, g^{kl}\left(\partial_i\, g_{jl} + \partial_j\, g_{il} - \partial_l\, g_{ij}\right). \tag{5.16}$$

The last formula (5.16) is the most useful one for the practical calculation of Γ^i_{jk}. As an instructive example, let us calculate Γ^i_{jk} for the polar coordinates in $2D$,

$$x = r\cos\varphi, \qquad y = r\sin\varphi$$

in $2D$. We start with the very detailed calculation and consider

$$\Gamma^\varphi_{\varphi\varphi} = \frac{1}{2}g^{\varphi\varphi}\left(\partial_\varphi g_{\varphi\varphi} + \partial_\varphi g_{\varphi\varphi} - \partial_\varphi g_{\varphi\varphi}\right) + \frac{1}{2}\,g^{\varphi r}\left(\partial_\varphi g_{r\varphi} + \partial_\varphi g_{r\varphi} - \partial_r g_{\varphi\varphi}\right).$$

The second term here is zero, since $g^{\varphi r} = 0$. In what follows, one can safely disregard all terms with non-diagonal metric components. Furthermore, the inverse of the diagonal metric $g_{ij} = \mathrm{diag}\left(r^2, 1\right)$ is nothing else but $g^{ij} = \mathrm{diag}\left(1/r^2, 1\right)$, such that $g^{\varphi\varphi} = 1/r^2$, $g^{rr} = 1$. Then, since $g_{\varphi\varphi}$ does not depend on φ, we have $\Gamma^\varphi_{\varphi\varphi} = 0$.

Other coefficients can be obtained in a similar fashion,

$$\Gamma^\varphi_{rr} = \frac{1}{2}g^{\varphi\varphi}\left(2\partial_r g_{\varphi r} - \partial_\varphi g_{rr}\right) = 0,$$

$$\Gamma^r_{r\varphi} = \frac{1}{2}g^{rr}\left(\partial_r g_{\varphi r} + \partial_\varphi g_{rr} - \partial_r g_{\varphi r}\right) = 0,$$

$$\Gamma^\varphi_{r\varphi} = \frac{1}{2}g^{\varphi\varphi}\left(\partial_r g_{\varphi\varphi} + \partial_\varphi g_{r\varphi} - \partial_\varphi g_{r\varphi}\right) = \frac{1}{2r^2}\partial_r r^2 = \frac{1}{r},$$

$$\Gamma^r_{\varphi\varphi} = \frac{1}{2}g^{rr}\left(\partial_\varphi g_{r\varphi} + \partial_\varphi g_{r\varphi} - \partial_r g_{\varphi\varphi}\right) = \frac{1}{2}\left(-\partial_r r^2\right) = -r,$$

$$\Gamma^r_{rr} = \frac{1}{2}g^{rr}\left(\partial_r g_{rr}\right) = 0. \tag{5.17}$$

After all, one can see that only $\Gamma^\varphi_{r\varphi} = \frac{1}{r}$ and $\Gamma^r_{\varphi\varphi} = -r$ are nonzero.

As an application of these formulas, we can consider the derivation of the Laplace operator acting on scalar and vector fields. For any kind of field, the Laplace operator can be defined as

$$\Delta = g^{ij} \nabla_i \nabla_j. \tag{5.18}$$

In the Cartesian coordinates, this operator has the well-known universal form

$$\Delta = \frac{\partial^2}{\partial x^2} + \frac{\partial^2}{\partial y^2} = g^{ab} \nabla_a \nabla_b = \nabla^a \nabla_a = \partial^a \partial_a.$$

What we want to know is what the operator Δ looks like in arbitrary coordinates. This is why we have to construct it as a covariant scalar operator expressed in terms of covariant derivatives. In the case of a scalar field Ψ, we have

$$\Delta \Psi = g^{ij} \nabla_i \nabla_j \Psi = g^{ij} \nabla_i \partial_j \Psi. \tag{5.19}$$

In (5.19), the second covariant derivative acts on the vector $\partial_i \Psi$, hence

$$\Delta \Psi = g^{ij} \left(\partial_i \partial_j - \Gamma_{ij}^k \partial_k \right) \Psi = \left(g^{ij} \partial_i \partial_j - g^{ij} \Gamma_{ij}^k \partial_k \right) \Psi. \tag{5.20}$$

Similarly, for the vector field $A^i(x)$, we obtain

$$\begin{aligned}
\Delta A^i &= g^{jk} \nabla_j \nabla_k A^i = g^{jk} \left[\partial_j \left(\nabla_k A^i \right) - \Gamma_{kj}^l \nabla_l A^i + \Gamma_{lj}^i \nabla_k A^l \right] \\
&= g^{jk} \left[\partial_j \left(\partial_k A^i + \Gamma_{lk}^i A^l \right) - \Gamma_{kj}^l \left(\partial_l A^i + \Gamma_{ml}^i A^m \right) + \Gamma_{lj}^i \left(\partial_k A^l + \Gamma_{mk}^l A^m \right) \right] \\
&= g^{jk} \left(\partial_j \partial_k A^i + \Gamma_{lk}^i \partial_j A^l + A^l \partial_j \Gamma_{lk}^i - \Gamma_{kj}^l \partial_l A^i + \Gamma_{lj}^i \partial_k A^l \right. \\
&\quad \left. - \Gamma_{kj}^l \Gamma_{ml}^i A^m + \Gamma_{lj}^i \Gamma_{mk}^l A^m \right).
\end{aligned} \tag{5.21}$$

Observation 1 The Laplace operators acting on vectors and scalars look quite different in curvilinear coordinates. They have the same form only in the Cartesian coordinates.

Observation 2 The formulas (5.19) and (5.21) are quite general; they hold for any dimension D and for an arbitrary choice of coordinates.

Let us perform an explicit calculation for the Laplace operator acting on scalars in the polar coordinates in $2D$. The relevant nonzero contractions of Christoffel symbol are

$$g^{ij} \Gamma_{ij}^k = g^{\varphi\varphi} \Gamma_{\varphi\varphi}^k + g^{rr} \Gamma_{rr}^k. \tag{5.22}$$

Thus, using (5.17), we can see that only one trace

$$g^{\varphi\varphi} \Gamma_{\varphi\varphi}^r = -\frac{1}{r} \quad \text{is nonzero.}$$

Then,

$$\Delta\Psi = \left[g^{\varphi\varphi}\frac{\partial^2}{\partial\varphi^2} + g^{rr}\frac{\partial^2}{\partial r^2} - \left(-\frac{1}{r}\right)\frac{\partial}{\partial r}\right]\Psi = \left[\frac{1}{r^2}\frac{\partial^2}{\partial\varphi^2} + \frac{\partial^2}{\partial r^2} + \frac{1}{r}\frac{\partial}{\partial r}\right]\Psi$$

and finally we obtain the well-known formula

$$\Delta\Psi = \frac{1}{r}\frac{\partial}{\partial r}\left(r\frac{\partial\Psi}{\partial r}\right) + \frac{1}{r^2}\frac{\partial^2\Psi}{\partial\varphi^2}. \tag{5.23}$$

Def. 4 For vector field $\mathbf{A}(x) = A^i \mathbf{e}_i$ the scalar quantity $\nabla_i A^i$ is called divergence of $\mathbf{A}$. The operation of taking divergence of the vector can be seen as a contraction of the vector with the Hamilton operator (5.2), so we have (exactly as in the case of a gradient) two distinct notations for divergence

$$\div\mathbf{A} = \nabla\mathbf{A} = \nabla_i A^i = \frac{\partial A^i}{\partial x^i} + \Gamma^i_{ji} A^j.$$

Exercise 5* Complete the calculation of $\Delta A^i (x)$ from Eq. (5.21) in polar coordinates.

Observation This is a really cumbersome calculation. We recommend using some software to make it. The result can be found in the book [1].

Exercise 6 Write the general expression for the Laplace operator acting on the covariant vector $\Delta B_i (x)$, similar to (5.21). Discuss the relation between these two formulas and Eq. (5.19).

Exercise 7 Write the general expression for $\nabla_i A^i$ and derive it in the polar coordinates (the result can be checked through comparison with similar calculations in cylindric and spherical coordinates in $3D$ in Chap. 7).

Exercise 8 The commutator of two covariant derivatives is defined as

$$\left[\nabla_i, \nabla_j\right] = \nabla_i \nabla_j - \nabla_j \nabla_i . \tag{5.24}$$

Prove, without *any* calculations, that for $\forall$ tensor T the commutator of two covariant derivatives is zero

$$\left[\nabla_i, \nabla_j\right] T^{k_1...k_s}{}_{l_1...l_t} = 0.$$

Hint. Use the existence of global Cartesian coordinates in the flat space and the tensor form of the transformations from one coordinate to another.

Observation Of course, the same commutator is *nonzero* if we consider a more general geometry, which does not admit a global orthonormal basis. For example, after reading Chap. 8 one can check that this commutator is nonzero for the geometry on a curved $2D$ surface in a $3D$ space, e.g., for the sphere.

Exercise 9   Verify that for the scalar field Ψ, the following relation takes place:

$$\Delta\Psi = \text{div} \left(\text{grad}\, \Psi\right).$$

Hint. Use the fact that $\text{grad}\, \Psi = \nabla\Psi = \mathbf{e}^i \partial_i \Psi$ is a covariant vector. Before taking $\div$, which is defined as an operation over the contravariant vector, one has to *raise* the index: $\nabla^i \Psi = g^{ij} \nabla_j \Psi$. Also, one has to use the metricity condition $\nabla_i g_{jk} = 0$.

Exercise 10 Derive grad (divA) for a vector field $\mathbf{A}$ in polar coordinates.

Exercise 11 Prove the following relation between the contraction of the Christoffel symbol Γ^k_{ij} and the derivative of the metric determinant $g = \det \|g_{\mu\nu}\|$:

$$\Gamma^j_{ij} = \frac{1}{2g} \frac{\partial g}{\partial x^i} = \partial_i \ln \sqrt{g}. \tag{5.25}$$

Hint. Use formula (3.25).

Exercise 12 Use the relation (5.25) for the derivation of the divA in polar coordinates. Also, calculate divA for elliptic coordinates

$$x = ar \cos\varphi, \qquad y = br \sin\varphi \tag{5.26}$$

and hyperbolic coordinates

$$x = \rho \cosh\chi, \qquad y = \rho \sinh\chi. \tag{5.27}$$

Exercise 13 Prove the relation

$$g^{ij} \Gamma^k_{ij} = -\frac{1}{\sqrt{g}} \partial_i \left(\sqrt{g}\, g^{ik}\right) \tag{5.28}$$

and use it for the derivation of $\Delta\Psi$ in polar, elliptic, and hyperbolic coordinates.

Exercise 14 Invent the way to calculate $\Delta\Psi$ in polar and elliptic coordinates using Eq. (5.25). Verify that this calculation is extremely simple, especially for the case of polar coordinates.

Hint. Just raise the index after taking the first derivative, that is, grad Ψ, and then recall about metricity property of the covariant derivative and Eq. (5.25).

Observation The same approach can be used for derivation of the Laplace operator in spherical coordinates in $3D$, where it makes calculations relatively simple. In Chap. 7, we are going to follow a direct approach to this calculation, just to show how it works. However, the derivation based on Eq. (5.25) definitely adds much in the way of simplicity.

Reference

1. J.D. Jackson, *Classical Electrodynamics*, 3rd edn. (Wiley, New York, 1998)

Chapter 6
Grad, div, rot, and Relations Between Them

6.1 Basic Definitions and Relations

Here, we consider some important operations over vector and scalar fields, and relations between them. In this chapter, all the consideration will be restricted to the special case of Cartesian coordinates, but the results can be easily generalized to any other coordinates by using the tensor transformation rule. As far as we are dealing with vectors and tensors, we know how to transform them.

Caution: When transforming the relations into curvilinear coordinates, one has to take care to use only the Levi-Civita tensor ε_{ijk} and not the maximal antisymmetric symbol E_{ijk}. The two objects are indeed related as $\varepsilon_{ijk} = \sqrt{g}\, E_{ijk}$, where $E_{123} = 1$ in any coordinate system.

We already know three of four relevant differential operators

$$
\begin{aligned}
\mathrm{div}\mathbf{V} &= \partial_a V^a = \nabla \mathbf{V} & \text{divergence ;} \\
\mathrm{grad}\,\Psi &= \hat{\mathbf{n}}^a \partial_a \Psi = \nabla \Psi & \text{gradient ;} \\
\Delta &= g^{ab}\partial_a \partial_b & \text{Laplace operator}
\end{aligned}
\tag{6.1}
$$

where

$$
\nabla = \hat{\mathbf{n}}^a \partial_a = \hat{\mathbf{i}}\,\frac{\partial}{\partial x} + \hat{\mathbf{j}}\,\frac{\partial}{\partial y} + \hat{\mathbf{k}}\,\frac{\partial}{\partial z}
\tag{6.2}
$$

is the *Hamilton operator*, which is also called the *nabla* operator. Of course, the vector operator ∇ can be formulated in $\vee$ other coordinate systems using the covariant derivative

$$
\nabla = \mathbf{e}^i \nabla_i = g^{ij} \mathbf{e}_j \nabla_i
\tag{6.3}
$$

acting on the corresponding (scalar, vector or tensor) fields. There is the fourth relevant differential operator, called *rotor*, *curl*, or *rotation*. This operator acts on

© Springer Nature Switzerland AG 2019
I. L. Shapiro, *A Primer in Tensor Analysis and Relativity*, Undergraduate Lecture Notes in Physics, https://doi.org/10.1007/978-3-030-26895-4_6

vectors. Contrary to grad and div, the rot as a vector function of a vector argument can be defined only in $3D$.

Def. 1 The rotor of the vector $\mathbf{V}$ is defined as

$$\mathrm{rot}\ \mathbf{V} = \begin{vmatrix} \mathbf{i} & \mathbf{j} & \mathbf{k} \\ \partial_x & \partial_y & \partial_z \\ V_x & V_y & V_z \end{vmatrix} = \begin{vmatrix} \hat{\mathbf{n}}_1 & \hat{\mathbf{n}}_2 & \hat{\mathbf{n}}_3 \\ \partial_1 & \partial_2 & \partial_3 \\ V_1 & V_2 & V_3 \end{vmatrix} = E^{abc}\,\hat{\mathbf{n}}_a\,\partial_b\,V_c. \tag{6.4}$$

Another useful notation for the rot $\mathbf{V}$ uses its representation as the vector product of ∇ and $\mathbf{V}$

$$\mathrm{rot}\ \mathbf{V} = \nabla \times \mathbf{V}.$$

One can easily prove the following important relations:

$$\textbf{(i)} \qquad \mathrm{rot}\ \mathrm{grad}\ \Psi \equiv 0. \tag{6.5}$$

Proof In the component notations, the *l.h.s.* of this equality looks like

$$E^{abc}\,\hat{\mathbf{n}}_a\,\partial_b\,(\,\mathrm{grad}\,\Psi\,)_c = E^{abc}\,\hat{\mathbf{n}}_a\,\partial_b\,\partial_c\,\Psi.$$

Indeed, this is zero because of the contraction of the symmetric symbol $\partial_b\partial_c = \partial_c\partial_b$ with the antisymmetric one E^{abc} (see Exercise 3.11).

$$\textbf{(ii)} \qquad \mathrm{div}(\,\mathrm{rot}\ \mathbf{V}) \equiv 0. \tag{6.6}$$

Proof In the component notations, the *l.h.s.* of this equality looks like

$$\partial_a\,E^{abc}\,\partial_b\,V_c = E^{abc}\,\partial_a\,\partial_b\,V_c\,,$$

which is zero because of the very same reason: contraction of the symmetric $\partial_b\partial_c$ with the antisymmetric E^{abc} symbols.

Exercises Prove the following identities:
(1) $\mathrm{grad}\,(\varphi\Psi) = \varphi\,\mathrm{grad}\,\Psi + \Psi\,\mathrm{grad}\,\varphi\,;$
(2) $\mathrm{div}\,(\varphi\mathbf{A}) = \varphi\mathrm{div}\mathbf{A} + (\mathbf{A}\cdot\mathrm{grad}\,)\varphi = \varphi\nabla\mathbf{A} + (\mathbf{A}\cdot\nabla)\varphi\,;$
(3) $\mathrm{rot}\,(\varphi\mathbf{A}) = \varphi\,\mathrm{rot}\,\mathbf{A} - [\mathbf{A}, \nabla]\,\varphi\,;$
(4) $\mathrm{div}\,[\mathbf{A}, \mathbf{B}] = \mathbf{B}\cdot\mathrm{rot}\,\mathbf{A} - \mathbf{A}\cdot\mathrm{rot}\,\mathbf{B}\,;$
(5) $\mathrm{rot}\,[\mathbf{A}, \mathbf{B}] = \mathbf{A}\mathrm{div}\mathbf{B} - \mathbf{B}\mathrm{div}\mathbf{A} + (\mathbf{B}, \nabla)\,\mathbf{A} - (\mathbf{A}, \nabla)\,\mathbf{B}\,;$
(6) $\mathrm{grad}\,(\mathbf{A}\cdot\mathbf{B}) = [\mathbf{A},\ \mathrm{rot}\,\mathbf{B}] + [\mathbf{B},\ \mathrm{rot}\,\mathbf{A}] + (\mathbf{B}\cdot\nabla)\,\mathbf{A} + (\mathbf{A}, \nabla)\,\mathbf{B}\,;$
(7) $(\mathbf{C}, \nabla)\,(\mathbf{A}\cdot\mathbf{B}) = (\mathbf{A}, (\mathbf{C}, \nabla)\,\mathbf{B}) + (\mathbf{B}, (\mathbf{C}, \nabla)\,\mathbf{A})\,;$
(8) $(\mathbf{C}\cdot\nabla)\,[\mathbf{A}, \mathbf{B}] = [\mathbf{A}, (\mathbf{C}, \nabla)\,\mathbf{B}] - [\mathbf{B}, (\mathbf{C}, \nabla)\,\mathbf{A}]\,;$
(9) $(\nabla, \mathbf{A})\,\mathbf{B} = (\mathbf{A}, \nabla)\,\mathbf{B} + \mathbf{B}\mathrm{div}\mathbf{A}\,;$
(10) $[\mathbf{A}, \mathbf{B}]\cdot\mathrm{rot}\,\mathbf{C} = (\mathbf{B}, (\mathbf{A}, \nabla)\,\mathbf{C}) - (\mathbf{A}, (\mathbf{B}, \nabla)\,\mathbf{C})\,;$
(11) $[[\mathbf{A}, \nabla], \mathbf{B}] = (\mathbf{A}, \nabla)\,\mathbf{B} + [\mathbf{A},\ \mathrm{rot}\,\mathbf{B}] - \mathbf{A}\cdot\mathrm{div}\mathbf{B}\,;$

(12) $[[\nabla, \mathbf{A}], \mathbf{B}] = -\mathbf{A} \cdot \text{div}\mathbf{B} + (\mathbf{A}, \nabla)\mathbf{B} + [\mathbf{A}, \text{rot}\,\mathbf{B}] + [\mathbf{B}, \text{rot}\,\mathbf{A}]$.

Observation In Exercises (11) and (12), we use the square brackets for the commutators $[\nabla, A] = \nabla A - A \nabla$, while in other cases the same bracket denotes the vector product of two vectors.

6.2   On the Classification of Differentiable Vector Fields

Let us introduce a new notion concerning the classification of differentiable vector fields.

Def. 2 Consider a differentiable vector $\mathbf{C}(\mathbf{r})$. It is called *potential* vector field, if it can be presented as a gradient of some scalar field $\Psi(\mathbf{r})$ (called potential)

$$\mathbf{C}(\mathbf{r}) = \text{grad}\,\Psi(\mathbf{r}). \tag{6.7}$$

Examples of the potential vector field can be easily found in physics. For instance, the potential force $\mathbf{F}$ acting on a particle is defined as $\mathbf{F} = -\text{grad}\,U(\mathbf{r})$, where $U(\mathbf{r})$ is the potential energy of the particle.

Def. 3 A differentiable vector field $\mathbf{B}(\mathbf{r})$ is called *solenoidal*, if it can be presented as a rotor of some vector field $\mathbf{A}(\mathbf{r})$

$$\mathbf{B}(\mathbf{r}) = \text{rot}\,\mathbf{A}(\mathbf{r}). \tag{6.8}$$

The most known physical example of the solenoidal vector field is the magnetic field $\mathbf{B}$, which is derived from the vector potential $\mathbf{A}$ exactly through (6.8).

There is an important theorem that is sometimes called the *Fundamental Theorem of Vector Analysis.*

Theorem *Suppose* $\mathbf{V}(\mathbf{r})$ *is a smooth vector field, defined in the whole* $3D$ *space, which falls sufficiently fast at infinity. Then* $\mathbf{V}(\mathbf{r})$ *has unique (up to a gauge transformation) representation as a sum*

$$\mathbf{V} = \mathbf{C} + \mathbf{B}, \tag{6.9}$$

where $\mathbf{C}$ *and* $\mathbf{B}$ *are potential and solenoidal fields correspondingly.*

Proof We shall accept, without proof (which can be found in many courses of mathematical physics, e.g., [1]), the following statement: The Laplace equation with given boundary conditions at infinity has a unique solution. In particular, the Laplace equation with zero boundary conditions at infinity has unique zero solution.

First, we must prove that the separation of the vector field $\mathbf{V}$ into the solenoidal and potential part is possible. Let us suppose that

$$\mathbf{V} = \text{grad } \Psi + \text{rot } \mathbf{A} \qquad (6.10)$$

and take divergence. Using (6.6), after some calculations, we arrive at the Poisson equation

$$\Delta \Psi = \text{div } \mathbf{V}. \qquad (6.11)$$

Since the solution of this equation is (up to a constant) unique, we now meet a single equation for $\mathbf{A}$. Let us take rot from both parts of Eq. (6.10). Using the identity (6.5), we obtain

$$\text{rot } \mathbf{V} = \text{rot } (\text{rot } \mathbf{A}) = \text{grad } (\text{div} \mathbf{A}) - \Delta \mathbf{A}. \qquad (6.12)$$

Making the transformation

$$\mathbf{A} \rightarrow \mathbf{A}' = \mathbf{A} + \text{grad } f(\mathbf{r})$$

(in physics this is called gauge transformation) one can always provide that $\text{div}\mathbf{A}' \equiv 0$. At the same time, due to (6.5), this transformation does not affect the rot $\mathbf{A}$ and its contribution to $\mathbf{V}(\mathbf{r})$. Therefore, we do not need to distinguish $\mathbf{A}'$ and $\mathbf{A}$ and can simply suppose that $\text{grad } (\text{div} \mathbf{A}) \equiv 0$. Then, Eq. (6.9) has a unique solution with respect to the vectors $\mathbf{B}$ and $\mathbf{C}$, and the proof is complete. Any other choice of the function $f(\mathbf{r})$ leads to a different equation for the remaining part of $\mathbf{A}$, but for any *given choice* of $f(\mathbf{r})$ the solution is also unique. The proof can be easily generalized to the case when the boundary conditions are fixed not at infinity but at some finite closed surface. The reader can find more detailed treatment of this theorem, e.g., in the book [2].

References

1. V.S. Vladimirov, *Equations of Mathematical Physics* (Imported Pubn., 1985)
2. A.I. Borisenko, I.E. Tarapov, *Vector and Tensor Analysis with Applications* (Dover, New York, 1968). Translation from III Russian edn.

Chapter 7
Grad, div, rot, and Δ in Cylindric and Spherical Coordinates

The purpose of this chapter is to calculate the operators grad, rot, div, and Δ in cylindric and spherical coordinates. The method of calculations that will be described below can be applied to more complicated cases. This may include derivation of other differential operators, acting on tensors; also derivation in other, more complicated (in particular, non-orthogonal) coordinates and in the case of dimension $D \neq 3$. In part, our considerations will repeat those we have already performed above, in Chap. 5, but we shall always make calculations in a slightly different manner, expecting that the reader can benefit from the comparison.

Let us start with the *cylindric coordinates*

$$ x = r \cos\varphi , \quad y = r \sin\varphi , \quad z = z. \tag{7.1} $$

In part, we can use here the results for the polar coordinates in $2D$, obtained in Chap. 5. In particular, the metric in cylindric coordinates is

$$ g_{ij} = \text{diag} \left(g_{rr}, g_{\varphi\varphi}, g_{zz} \right) = \text{diag} \left(1, r^2, 1 \right). \tag{7.2} $$

Also, the inverse metric is $g^{ij} = \text{diag} \left(1, r^{-2}, 1 \right)$. Only two components of the connection Γ^i_{jk} are different from zero,

$$ \Gamma^\varphi_{r\varphi} = \frac{1}{r} \quad \text{and} \quad \Gamma^r_{\varphi\varphi} = -r. \tag{7.3} $$

Let us start from the transformation of basic vectors. As always, local transformations of basic vectors and coordinates are performed by the use of the mutually inverse matrices

$$ dx'^i = \frac{\partial x'^i}{\partial x^j} dx^j \quad \Longleftrightarrow \quad \mathbf{e}'_i = \frac{\partial x^j}{\partial x'^i} \mathbf{e}_j. $$

© Springer Nature Switzerland AG 2019
I. L. Shapiro, *A Primer in Tensor Analysis and Relativity*, Undergraduate
Lecture Notes in Physics, https://doi.org/10.1007/978-3-030-26895-4_7

We can present the transformation in the matrix form,

$$
\begin{pmatrix} \hat{\mathbf{i}} \\ \hat{\mathbf{j}} \\ \hat{\mathbf{k}} \end{pmatrix} = \begin{pmatrix} \mathbf{e}_x \\ \mathbf{e}_y \\ \mathbf{e}_z \end{pmatrix} = \frac{D(r, \varphi, z)}{D(x, y, z)} \times \begin{pmatrix} \mathbf{e}_r \\ \mathbf{e}_\varphi \\ \mathbf{e}_z \end{pmatrix} \tag{7.4}
$$

and

$$
\begin{pmatrix} \mathbf{e}_r \\ \mathbf{e}_\varphi \\ \mathbf{e}_z \end{pmatrix} = \frac{D(x, y, z)}{D(r, \varphi, z)} \times \begin{pmatrix} \mathbf{e}_x \\ \mathbf{e}_y \\ \mathbf{e}_z \end{pmatrix}. \tag{7.5}
$$

It is easy to see that in the z-sector the Jacobian is trivial, so one has to concentrate attention to the xy-plane. Direct calculations give us (compare to the Sect. 4.3)

$$
\left(\frac{\partial X^a}{\partial x^i} \right) = \frac{D(x, y, z)}{D(r, \varphi, z)} = \begin{pmatrix} \partial_r x & \partial_r y & \partial_r z \\ \partial_\varphi x & \partial_\varphi y & \partial_\varphi z \\ \partial_z x & \partial_z y & \partial_z z \end{pmatrix} = \begin{pmatrix} \cos\varphi & \sin\varphi & 0 \\ -r\sin\varphi & r\cos\varphi & 0 \\ 0 & 0 & 1 \end{pmatrix}. \tag{7.6}
$$

This is equivalent to the linear transformation

$$
\begin{aligned}
\mathbf{e}_r &= \hat{\mathbf{n}}_x \cos\varphi + \hat{\mathbf{n}}_y \cdot \sin\varphi, \\
\mathbf{e}_\varphi &= -r\hat{\mathbf{n}}_x \sin\varphi + r\hat{\mathbf{n}}_y \cos\varphi \\
\mathbf{e}_z &= \hat{\mathbf{n}}_z.
\end{aligned} \tag{7.7}
$$

As in the $2D$ case, we introduce normalized basic vectors

$$
\hat{\mathbf{n}}_r = \mathbf{e}_r, \qquad \hat{\mathbf{n}}_\varphi = \frac{1}{r}\mathbf{e}_\varphi, \qquad \hat{\mathbf{n}}_z = \mathbf{e}_z.
$$

It is easy to see that the transformation

$$
\begin{pmatrix} \hat{\mathbf{n}}_r \\ \hat{\mathbf{n}}_\varphi \end{pmatrix} = \begin{pmatrix} \cos\varphi & \sin\varphi \\ -\sin\varphi & \cos\varphi \end{pmatrix} \begin{pmatrix} \hat{\mathbf{n}}_x \\ \hat{\mathbf{n}}_y \end{pmatrix}
$$

is nothing else but rotation. Of course, this is due to the fact that the basis $\hat{\mathbf{n}}_r$, $\hat{\mathbf{n}}_\varphi$ is orthonormal.

As before, for the components of vector $\mathbf{A}$, we use the notation

$$
\mathbf{A} = \tilde{A}^r \mathbf{e}_r + \tilde{A}^\varphi \mathbf{e}_\varphi + \tilde{A}^z \mathbf{e}_z = A^r \hat{\mathbf{n}}_r + A^\varphi \hat{\mathbf{n}}_\varphi + A^z \hat{\mathbf{n}}_z,
$$

where $\tilde{A}^r = A^r$, $r\tilde{A}^\varphi = A^\varphi$, $\tilde{A}^z = A^z$.

Now we can calculate the divergence of the vector in cylindric coordinates

$$\text{div } \mathbf{A} = \nabla \mathbf{A} = \nabla_i \tilde{A}^i = \partial_i \tilde{A}^i + \Gamma^i_{ji} \tilde{A}^j = \partial_r \tilde{A}^r + \partial_\varphi \tilde{A}^\varphi + \frac{1}{r} \tilde{A}^r + \partial_z \tilde{A}^z$$

$$= \frac{1}{r} \frac{\partial}{\partial r} \left(r A^r \right) + \frac{1}{r} \frac{\partial}{\partial \varphi} A^\varphi + \frac{\partial}{\partial z} A^z. \tag{7.8}$$

Let us calculate the gradient of a scalar field Ψ in cylindric coordinates. By definition in the Cartesian coordinates

$$\nabla \Psi = \mathbf{e}^i \partial_i \Psi = \left(\hat{\mathbf{i}} \partial_x + \hat{\mathbf{j}} \partial_y + \hat{\mathbf{k}} \partial_z \right) \Psi, \tag{7.9}$$

since for the orthonormal basis, the conjugated basis coincides with the original one. By means of the very same calculations as in the $2D$ case, we arrive at the expression for the gradient in cylindric coordinates

$$\nabla \Psi = \left(\mathbf{e}^r \frac{\partial}{\partial r} + \mathbf{e}^\varphi \frac{\partial}{\partial \varphi} + \mathbf{e}^z \frac{\partial}{\partial z} \right) \Psi.$$

Since

$$\mathbf{e}^r = \hat{\mathbf{n}}_r, \quad \mathbf{e}^\varphi = \frac{1}{r} \hat{\mathbf{n}}_\varphi, \quad \mathbf{e}^z = \hat{\mathbf{n}}_z,$$

we obtain the gradient in the orthonormal basis

$$\nabla \Psi = \left(\hat{\mathbf{n}}_r \frac{\partial}{\partial r} + \hat{\mathbf{n}}_\varphi \frac{1}{r} \frac{\partial}{\partial \varphi} + \hat{\mathbf{n}}_z \frac{\partial}{\partial z} \right) \Psi. \tag{7.10}$$

Exercise 1 Derive the same result by making a rotation of the basis in the expression (7.9).

Consider the derivation of rotation of the vector $\mathbf{V}$ in the cylindric coordinates. First, we note that in Eq. (7.8) for $\nabla_i A^i$ we have considered the divergence of the contravariant vector. The same may be maintained in the calculation of rot $\mathbf{V}$, because only $\left\{ \tilde{V}^r, \tilde{V}^\varphi, \tilde{V}^z \right\}$ components correspond to the basis (and to the change of coordinates) that we are considering. On the other hand, after we expand rot $\mathbf{V}$ in the normalized basis, there is no difference between co- and contravariant components.

In the Cartesian coordinates

$$\text{rot } \mathbf{V} = E^{abc} \hat{\mathbf{n}}_a \partial_b V_c = \begin{vmatrix} \hat{\mathbf{i}} & \hat{\mathbf{j}} & \hat{\mathbf{k}} \\ \partial_x & \partial_y & \partial_z \\ V_x & V_y & V_z \end{vmatrix}. \tag{7.11}$$

We remember that

$$\tilde{V}_r = g_{rr}\tilde{V}^r = \tilde{V}^r \quad , \quad \tilde{V}_\varphi = g_{\varphi\varphi}\tilde{V}^\varphi = r^2\tilde{V}^\varphi \quad , \quad \tilde{V}_z = g_{zz}\tilde{V}^z = \tilde{V}^z \quad (7.12)$$

and therefore

$$V_r = V^r = \tilde{V}_r \quad , \quad V_\varphi = V^\varphi = r\,\tilde{V}^\varphi \quad , \quad V_z = V^z = \tilde{V}^z. \quad (7.13)$$

It is worth noticing that the same relations also hold for the components of the vector rot $\mathbf{V}$.

It is easy to see that the Christoffel symbol here is not relevant, because

$$\text{rot } \mathbf{V} = \varepsilon^{ijk}\mathbf{e}_i \nabla_j \tilde{V}_k = \varepsilon^{ijk}\mathbf{e}_i \left(\partial_j \tilde{V}_k - \Gamma^l_{kj}\tilde{V}_l \right) = \varepsilon^{ijk}\, \mathbf{e}_i\, \partial_j \tilde{V}_k \,, \quad (7.14)$$

where we used an obvious relation $\varepsilon^{ijk}\Gamma^l_{kj} = 0$. Since $g_{ij} = diag\left(1, r^2, 1\right)$, the determinant of the metric is $g = \det\left(g_{ij}\right) = r^2$ and $\sqrt{g} = r$.

Starting calculations in the coordinate basis $\mathbf{e}_r$, $\mathbf{e}_\varphi$, $\mathbf{e}_z$ and then transforming the result into the normalized basis, we obtain

$$(\text{rot } \mathbf{V})^z = \frac{1}{\sqrt{g}}\, E^{zjk}\partial_j\tilde{V}_k = \frac{1}{r}\left(\partial_r\tilde{V}_\varphi - \partial_\varphi\tilde{V}_r \right) \quad (7.15)$$

$$= \frac{1}{r}\,\partial_r\left(\frac{1}{r}\cdot r^2 V^\varphi \right) - \frac{1}{r}\,\partial_\varphi V^r \;=\; \frac{1}{r}\frac{\partial}{\partial r}(r\cdot V^\varphi) - \frac{1}{r}\frac{\partial}{\partial\varphi}V^r.$$

Performing similar calculations for other components of the rotation, we obtain

$$(\text{rot } \mathbf{V})^r = \frac{1}{r}E^{rjk}\partial_j\tilde{V}_k = \frac{1}{r}\left(\partial_\varphi\tilde{V}_z - \partial_z\tilde{V}_\varphi \right)$$

$$= \frac{1}{r}\frac{\partial V^z}{\partial\varphi} - \frac{1}{r}\frac{\partial}{\partial z}(rV^\varphi) = \frac{1}{r}\frac{\partial V^z}{\partial\varphi} - \frac{\partial V^\varphi}{\partial z} \quad (7.16)$$

and

$$(\text{rot } \mathbf{V})^\varphi = r\cdot\frac{1}{r}\,E^{\varphi jk}\partial_j\tilde{V}_k = \partial_z\tilde{V}_r - \partial_r\tilde{V}_z = \frac{\partial V_r}{\partial z} - \frac{\partial V_z}{\partial r}\,, \quad (7.17)$$

where the first factor of r appears because we want the expansion of rot $\mathbf{V}$ in the normalized basis $\hat{\mathbf{n}}_i$ rather than in the coordinate basis $\mathbf{e}_i$.

Finally, in order to complete the list of formulas concerning cylindric coordinates, we write down the result for the Laplace operator (see Chap. 5, where it was calculated for the polar coordinates in $2D$).

$$\Delta\Psi = \frac{1}{r}\frac{\partial}{\partial r}\left(r\frac{\partial\Psi}{\partial r} \right) + \frac{1}{r^2}\frac{\partial^2\Psi}{\partial\varphi^2} + \frac{\partial^2\Psi}{\partial z^2}.$$

Let us now consider the *spherical coordinates.*

$$
\begin{aligned}
x &= r \cos \varphi \sin \theta, & 0 \le \theta \le \pi, \\
y &= r \sin \varphi \sin \theta, & 0 \le \varphi \le 2\pi, \\
z &= r \cos \theta, & 0 \le r \le +\infty.
\end{aligned} \tag{7.18}
$$

The Jacobian matrix has the form

$$
\left(\frac{\partial X^a}{\partial x^i} \right) = \mathcal{U} = \frac{D(x, y, z)}{D(r, \varphi, \theta)}
$$

$$
= \begin{pmatrix}
\cos \varphi \sin \theta & \sin \varphi \sin \theta & \cos \theta \\
-r \sin \varphi \sin \theta & r \cos \varphi \sin \theta & 0 \\
r \cos \varphi \cos \theta & r \sin \varphi \cos \theta & -r \sin \theta
\end{pmatrix} = \begin{pmatrix}
x_{,r} & y_{,r} & z_{,r} \\
x_{,\varphi} & y_{,\varphi} & z_{,\varphi} \\
x_{,\theta} & y_{,\theta} & z_{,\theta}
\end{pmatrix}, \tag{7.19}
$$

where $x_{,r} = \frac{\partial x}{\partial r}$ and so on. The matrix $\mathcal{U}$ is not orthogonal, but one can easily check that

$$
\mathcal{U} \cdot \mathcal{U}^T = \operatorname{diag}\left(1, \ r^2 \sin^2 \theta, \ r^2\right), \tag{7.20}
$$

where the transposed matrix is

$$
\mathcal{U}^T = \begin{pmatrix}
\cos \varphi \sin \theta & -r \sin \varphi \sin \theta & r \cos \varphi \cos \theta \\
\sin \varphi \sin \theta & r \cos \varphi \sin \theta & r \sin \varphi \cos \theta \\
\cos \theta & 0 & -r \sin \theta
\end{pmatrix}. \tag{7.21}
$$

Therefore, we can easily transform this matrix to the unitary form by multiplying it to the corresponding diagonal matrix. Thus, the inverse matrix has the form

$$
\mathcal{U}^{-1} = \frac{D(r, \varphi, \theta)}{D(x, y, z)} = \mathcal{U}^T \cdot \operatorname{diag}\left(1, \frac{1}{r^2 \sin^2 \theta}, \frac{1}{r^2}\right)
$$

$$
= \begin{pmatrix}
\cos \varphi \sin \theta & \frac{-\sin \varphi}{r \sin \theta} & \frac{\cos \varphi \cos \theta}{r} \\
\sin \varphi \sin \theta & \frac{\cos \varphi}{r \sin \theta} & \frac{\sin \varphi \cos \theta}{r} \\
\cos \theta & 0 & -\frac{\sin \theta}{r}
\end{pmatrix}. \tag{7.22}
$$

Exercise 2 Check that $\mathcal{U} \cdot \mathcal{U}^{-1} = \hat{1}$, and compare the above method of deriving the inverse matrix with the standard one used in Chap. 4. Try to describe the situations when we can use the present, more economical, method of inverting the matrix. Is it possible to do so for the transformation to the non-orthogonal basis?

Exercise 3 Check that $g_{rr} = 1$, $g_{\varphi\varphi} = r^2 \sin^2 \theta$, $g_{\theta\theta} = r^2$ and that the other components of the metric are equal to zero.

Hint. Use $g_{ij} = \dfrac{\partial x^a}{\partial x^i} \dfrac{\partial x^b}{\partial x^j} \cdot \delta_{ab}$ and matrix $\mathcal{U} = \dfrac{D(x, y, z)}{D(r, \varphi, \theta)}$. $\qquad$ (7.23)

Exercise 4 Check that $g^{ij} = diag\left(1, \frac{1}{r^2 \sin^2 \theta}, \frac{1}{r^2}\right)$. Try to solve this problem in several distinct ways.

Observation The metric g_{ij} is nondegenerate everywhere except the poles $\theta = 0$ and $\theta = \pi$.

The next step is to calculate the Christoffel symbol in spherical coordinates. The general expression has the form (5.16)

$$\Gamma^i_{jk} = \frac{1}{2} g^{il} \left(\partial_j g_{kl} + \partial_k g_{jl} - \partial_l g_{jk}\right).$$

A useful technical observation is that, since the metric g^{ij} is diagonal, in the sum over index l only the terms with $l = i$ are nonzero. The expressions for the components are

(1) $\Gamma^r_{rr} = \frac{1}{2} g^{rr} \left(\partial_r g_{rr} + \partial_r g_{rr} - \partial_r g_{rr}\right) = 0$;

(2) $\Gamma^r_{\varphi\varphi} = \frac{1}{2} g^{rr} \left(2\partial_\varphi g_{r\varphi} - \partial_r g_{\varphi\varphi}\right) = -\frac{1}{2} g^{rr} \partial_r g_{\varphi\varphi} = -r \sin^2 \theta$,

(3) $\Gamma^r_{\varphi\theta} = \frac{1}{2} g^{rr} \left(\partial_\varphi g_{\theta r} + \partial_\theta g_{\varphi r} - \partial_r g_{\theta\varphi}\right) = 0$,

(4) $\Gamma^r_{\theta\theta} = \frac{1}{2} g^{rr} \left(2\partial_\theta g_{\theta r} - \partial_r g_{\theta\theta}\right) = -\frac{1}{2} g^{rr} \partial_r g_{\theta\theta} = -\frac{1}{2} \frac{\partial}{\partial r} r^2 = -r$,

(5) $\Gamma^\varphi_{\varphi\varphi} = \frac{1}{2} g^{\varphi\varphi} \left(2\partial_\varphi g_{\varphi\varphi} - \partial_\varphi g_{\varphi\varphi}\right) = 0$,

(6) $\Gamma^\varphi_{\varphi r} = \frac{1}{2} g^{\varphi\varphi} \left(\partial_\varphi g_{r\varphi} + \partial_r g_{\varphi\varphi} - \partial_\varphi g_{r\varphi}\right) = \frac{1}{2r^2 \sin^2 \theta} \cdot 2r \sin^2 \theta = \frac{1}{r}$,

(7) $\Gamma^\varphi_{rr} = \frac{1}{2} g^{\varphi\varphi} \left(2\partial_r g_{r\varphi} - \partial_\varphi g_{rr}\right) = 0$,

(8) $\Gamma^\varphi_{\varphi\theta} = \frac{1}{2} g^{\varphi\varphi} \left(\partial_\theta g_{\theta\varphi} + \partial_\theta g_{\varphi\varphi} - \partial_\varphi g_{\varphi\theta}\right) = \frac{2r^2 \sin\theta \cos\theta}{2r^2 \sin^2 \theta} = \cot\theta$,

(9) $\Gamma^\varphi_{r\theta} = \frac{1}{2} g^{\varphi\varphi} \left(\partial_r g_{\theta\varphi} + \partial_\theta g_{r\varphi} - \partial_\varphi g_{r\theta}\right) = 0$,

(10) $\Gamma^\varphi_{\theta\theta} = \frac{1}{2} g^{\varphi\varphi} \left(2\partial_\theta g_{\varphi\theta} - \partial_\varphi g_{\theta\theta}\right) = 0$,

(11) $\Gamma^r_{r\varphi} = \frac{1}{2} g^{rr} \left(\partial_r g_{\varphi r} + \partial_\varphi g_{rr} - \partial_r g_{r\varphi}\right) = 0$,

(12) $\Gamma^r_{r\theta} = \frac{1}{2} g^{rr} \left(\partial_r g_{\theta r} + \partial_\theta g_{rr} - \partial_r g_{\theta r}\right) = 0$,

(13) $\Gamma^\theta_{\theta\theta} = \frac{1}{2} g^{\theta\theta} \left(\partial_\theta g_{\theta\theta}\right) = 0$,

(14) $\Gamma^\theta_{\varphi\varphi} = \frac{1}{2} g^{\theta\theta} \left(2\partial_\varphi g_{\varphi\theta} - \partial_\theta g_{\varphi\varphi}\right) = \frac{1}{2r^2} \left(-r^2 \frac{\partial \sin^2 \theta}{\partial \theta}\right) = -\frac{1}{2} \sin 2\theta$,

(15) $\Gamma^\theta_{rr} = \frac{1}{2} g^{\theta\theta} \left(2\partial_r g_{r\theta} - \partial_\theta g_{rr}\right) = 0$,

(16) $\Gamma^\theta_{\theta\varphi} = \frac{1}{2} g^{\theta\theta} \left(\partial_\theta g_{\varphi\theta} + \partial_\varphi g_{\theta\theta} - \partial_\theta g_{\theta\varphi}\right) = 0$,

(17) $\Gamma^\theta_{\theta r} = \frac{1}{2} g^{\theta\theta} \left(\partial_\theta g_{r\theta} + \partial_r g_{\theta\theta} - \partial_\theta g_{r\theta}\right) = \frac{1}{2} \frac{1}{r^2} \cdot \frac{\partial}{\partial r} r^2 = \frac{1}{r}$,

(18) $\Gamma^\theta_{\varphi r} = \frac{1}{2} g^{\theta\theta} \left(\partial_\varphi g_{r\theta} + \partial_r g_{\varphi\theta} - \partial_\theta g_{\varphi r}\right) = 0$.

We can see that the nonzero coefficients are

$$\Gamma^\varphi_{\varphi\theta} = \cot\theta, \quad \Gamma^\theta_{\theta r} = \Gamma^\varphi_{r\varphi} = \frac{1}{r}, \quad \Gamma^r_{\varphi\varphi} = -r \sin^2 \theta,$$

$$\Gamma^r_{\theta\theta} = -r, \quad \Gamma^\theta_{\varphi\varphi} = -\frac{1}{2} \sin 2\theta. \qquad (7.24)$$

Now we can easily derive   grad Ψ, div $\mathbf{A}$, rot $\mathbf{A}$, and $\Delta \Psi$ in spherical coordinates. According to our practice, we shall express all results in the normalized orthogonal basis $\hat{\mathbf{n}}_r, \hat{\mathbf{n}}_\varphi, \hat{\mathbf{n}}_\theta$. The relations between the components of the normalized basis and the corresponding elements of the coordinate basis are

$$\hat{\mathbf{n}}_r = \mathbf{e}_r, \qquad \hat{\mathbf{n}}_\varphi = \frac{\mathbf{e}_\varphi}{r \sin \theta}, \qquad \hat{\mathbf{n}}_\theta = \frac{\mathbf{e}_\theta}{r}. \tag{7.25}$$

Then

$$\begin{aligned}
\operatorname{grad} \Psi\,(r, \varphi, \theta) &= \mathbf{e}^i \partial_i \Psi = g^{ij} \mathbf{e}_j \partial_i \Psi \tag{7.26}\\
&= g^{rr} \mathbf{e}_r \frac{\partial \Psi}{\partial r} + g^{\varphi\varphi} \mathbf{e}_\varphi \frac{\partial \Psi}{\partial \varphi} + g^{\theta\theta} \mathbf{e}_\theta \frac{\partial \Psi}{\partial \theta}\\
&= \mathbf{e}_r \frac{\partial \Psi}{\partial r} + \frac{\mathbf{e}_\varphi}{r^2 \sin^2 \theta} \frac{\partial \Psi}{\partial \varphi} + \frac{\mathbf{e}_\theta}{r^2} \frac{\partial \Psi}{\partial \theta}\\
&= \hat{\mathbf{n}}_r \frac{\partial \Psi}{\partial r} + \frac{\hat{\mathbf{n}}_\varphi}{r \sin \theta} \frac{\partial \Psi}{\partial \varphi} + \frac{\hat{\mathbf{n}}_\theta}{r} \frac{\partial \Psi}{\partial \theta}.
\end{aligned}$$

And so, we obtained the following components of the gradient of the scalar field Ψ in spherical coordinates:

$$(\operatorname{grad} \Psi)_r = \frac{\partial \Psi}{\partial r}, \qquad (\operatorname{grad} \Psi)_\varphi = \frac{1}{r \sin \theta} \frac{\partial \Psi}{\partial \varphi},$$
$$(\operatorname{grad} \Psi)_\theta = \frac{1}{r} \frac{\partial \Psi}{\partial \theta}. \tag{7.27}$$

Consider the derivation of divergence in spherical coordinates. Using the components of the Christoffel symbol derived above, we obtain

$$\begin{aligned}
\operatorname{div} \mathbf{A} &= \nabla_i \tilde{A}^i = \partial_i \tilde{A}^i + \Gamma^i_{ji} \tilde{A}^j = \partial_i \tilde{A}^i + \cot \theta \cdot \tilde{A}^\theta + \frac{2}{r} \tilde{A}^r \\
&= \frac{\partial \tilde{A}^r}{\partial r} + \frac{\partial \tilde{A}^\theta}{\partial \theta} + \frac{\partial \tilde{A}^\varphi}{\partial \varphi} + \cot \theta \cdot \tilde{A}^\theta + \frac{2 \tilde{A}^r}{r}. \tag{7.28}
\end{aligned}$$

Now we have to rewrite this using the normalized basis $\hat{\mathbf{n}}_i$. The relevant relations are

$$\mathbf{A} = \tilde{A}^i \mathbf{e}_i = A^i \hat{\mathbf{n}}_i,$$

where

$$\tilde{A}^r = A^r, \qquad \tilde{A}^\varphi = \frac{1}{r \sin \theta} A^\varphi, \qquad \tilde{A}^\theta = \frac{1}{r} A^\theta.$$

Then

$$\text{div } \mathbf{A} = \frac{\partial A^r}{\partial r} + \frac{1}{r}\frac{\partial A^\theta}{\partial \theta} + \frac{\cot \theta}{r}A^\theta + \frac{2A^r}{r} + \frac{1}{r \sin \theta}\frac{\partial A^\varphi}{\partial \varphi}$$

$$= \frac{1}{r^2}\frac{\partial}{\partial r}\left(r^2 A^r\right) + \frac{1}{r \sin \theta}\frac{\partial A^\varphi}{\partial \varphi} + \frac{1}{r \sin \theta}\frac{\partial}{\partial \theta}\left(A^\theta \cdot \sin \theta\right). \quad (7.29)$$

The next example is $\Delta \Psi$. There are several different possibilities for this calculations. For instance, one can use the relation $\Delta \Psi = \text{div }(\text{grad } \Psi)$ or derive it directly using covariant derivatives. Let us follow the latter option, which is slightly more economical and also more instructive.[1] We also use the diagonal form of the metric tensor and get

$$\Delta \Psi = g^{ij}\nabla_i \nabla_j \Psi = g^{ij}\left(\partial_i \partial_j - \Gamma_{ij}^k \partial_k\right)\Psi$$

$$= g^{rr}\frac{\partial^2 \Psi}{\partial r^2} + g^{\varphi\varphi}\frac{\partial^2 \Psi}{\partial \varphi^2} + g^{\theta\theta}\frac{\partial^2 \Psi}{\partial \theta^2} - g^{ij}\Gamma_{ij}^k \frac{\partial \Psi}{\partial x^k}. \quad (7.30)$$

Since we already have all components of the Christoffel symbol, it is easy to obtain

$$g^{ij}\Gamma_{ij}^r = g^{\varphi\varphi}\Gamma_{\varphi\varphi}^r + g^{\theta\theta}\Gamma_{\theta\theta}^r = -\frac{2}{r} \quad \text{and} \quad g^{ij}\Gamma_{ij}^\theta = \Gamma_{\varphi\varphi}^\theta g^{\varphi\varphi} = -\frac{1}{r^2}\cot \theta,$$

while $\Gamma_{ij}^\varphi g^{ij} = 0$. After this, we immediately arrive at

$$\Delta \Psi = \frac{\partial^2 \Psi}{\partial r^2} + \frac{1}{r^2 \sin^2 \theta}\frac{\partial^2 \Psi}{\partial \varphi^2} + \frac{1}{r^2}\frac{\partial^2 \Psi}{\partial \theta^2} + \frac{2}{r}\frac{\partial \Psi}{\partial r} + \frac{1}{r^2}\cot an \theta \frac{\partial \Psi}{\partial \theta}$$

$$= \frac{1}{r^2}\frac{\partial}{\partial r}\left(r^2 \frac{\partial \Psi}{\partial r}\right) + \frac{1}{r^2 \sin \theta}\frac{\partial}{\partial \theta}\left(\sin \theta \frac{\partial \Psi}{\partial \theta}\right) + \frac{1}{r^2 \sin^2 \theta}\frac{\partial^2 \Psi}{\partial \varphi^2}. \quad (7.31)$$

Finally, in order to calculate

$$\text{rot } \mathbf{A} = \varepsilon^{ijk}\mathbf{e}_i \nabla_j \tilde{A}_k, \quad (7.32)$$

one has to use the formula $\varepsilon^{ijk} = \frac{1}{\sqrt{g}}E^{ijk}$, where $E^{123} = E^{r\theta\varphi} = 1$. The determinant is of course $g = r^4 \sin^2 \theta$, therefore $\sqrt{g} = r^2 \sin \theta$. Also, due to the identity $\varepsilon^{ijk}\Gamma_{jk}^l = 0$, we can use partial derivatives instead of the covariant ones.

$$\text{rot } \mathbf{A} = \frac{1}{\sqrt{g}}E^{ijk}\mathbf{e}_i \partial_j \tilde{A}_k, \quad (7.33)$$

then substitute $\tilde{A}_k = g_{kl}\tilde{A}^l$ and finally

---

[1] The fastest option is to use $\Delta \Psi = \text{div grad }\Psi$, see Exercise 4.

$$\tilde{A}_r = A^r, \quad \tilde{A}_\varphi = r \sin\theta \cdot A^\varphi, \quad \tilde{A}_\theta = r \cdot A^\theta.$$

Using $E^{r\theta\varphi} = 1$, we obtain (in the normalized basis $\hat{\mathbf{n}}_i$)

$$(\mathrm{rot}\,\mathbf{A})^r = \frac{1}{r^2 \sin\theta}\left(\partial_\theta \tilde{A}_\varphi - \partial_\varphi \tilde{A}_\theta\right) = \frac{1}{r \sin\theta}\left[\frac{\partial}{\partial\theta}(A^\varphi \cdot \sin\theta) - \frac{\partial A^\theta}{\partial\varphi}\right]. \quad (7.34)$$

This component does not need to be renormalized because of the relation $\mathbf{e}_r = \hat{\mathbf{n}}_r$. For other components, the situation is different, and they have to be normalized after the calculations. In particular, the φ-component must be multiplied by $r \sin\theta$ according to (7.25). Thus,

$$(\mathrm{rot}\,\mathbf{A})^\varphi = r\sin\theta \cdot \frac{E^{\varphi ij}}{r^2 \sin\theta}\partial_i \tilde{A}_j$$

$$= \frac{1}{r}\left[\partial_r \tilde{A}_\theta - \partial_\theta \tilde{A}_r\right] = \frac{1}{r}\frac{\partial}{\partial r}(rA^\theta) - \frac{1}{r}\frac{\partial A^r}{\partial\theta}. \quad (7.35)$$

Similarly,

$$(\mathrm{rot}\,\mathbf{A})^\theta = r \cdot \frac{1}{r^2 \sin\theta}E^{\theta ij}\partial_i \tilde{A}_j = \frac{1}{r\sin\theta}\left(\partial_\varphi \tilde{A}_r - \partial_r \tilde{A}_\varphi\right) \quad (7.36)$$

$$= \frac{1}{r\sin\theta}\left[\frac{\partial A^r}{\partial\varphi} - \frac{\partial}{\partial r}(r\sin\theta A^\varphi)\right] = \frac{1}{r\sin\theta}\frac{\partial A^r}{\partial\varphi} - \frac{1}{r}\frac{\partial}{\partial r}(rA^\varphi).$$

And so, we obtained $\mathrm{grad}\,\Psi$, $\mathrm{div}\,\mathbf{A}$, $\mathrm{rot}\,\mathbf{A}$, and $\Delta\Psi$ in both cylindrical and spherical coordinates. Let us present a list of results in the normalized basis.

I. Cylindric coordinates.

$$(\mathrm{grad}\,\Psi)_r = \frac{\partial\Psi}{\partial r}, \quad (\mathrm{grad}\,\Psi)_\varphi = \frac{1}{r}\frac{\partial\Psi}{\partial\varphi}, \quad (\mathrm{grad}\,\Psi)_z = \frac{\partial\Psi}{\partial z},$$

$$\mathrm{div}\,\mathbf{V} = \frac{1}{r}\frac{\partial}{\partial r}(rV^r) + \frac{1}{r}\frac{\partial V^\varphi}{\partial\varphi} + \frac{\partial V^z}{\partial z},$$

$$(\mathrm{rot}\,\mathbf{V})^z = \frac{1}{r}\frac{\partial(rV^\varphi)}{\partial r} - \frac{1}{r}\frac{\partial V^r}{\partial\varphi}, \quad (\mathrm{rot}\,\mathbf{V})^r = \frac{1}{r}\frac{\partial V^z}{\partial\varphi} - \frac{\partial V^\varphi}{\partial z},$$

$$(\mathrm{rot}\,\mathbf{V})^\varphi = \frac{\partial V^r}{\partial z} - \frac{\partial V^z}{\partial r},$$

$$\Delta\Psi = \frac{1}{r}\frac{\partial}{\partial r}\left(r\frac{\partial\Psi}{\partial r}\right) + \frac{1}{r^2}\frac{\partial^2\Psi}{\partial\varphi^2} + \frac{\partial^2\Psi}{\partial z^2}. \quad (7.37)$$

II. Spherical coordinates.

$$\left(\operatorname{grad}\Psi\right)_r = \frac{\partial\Psi}{\partial r}, \qquad \left(\operatorname{grad}\Psi\right)_\varphi = \frac{1}{r\sin\theta}\frac{\partial\Psi}{\partial\varphi}, \qquad \left(\operatorname{grad}\Psi\right)_\theta = \frac{1}{r}\frac{\partial\Psi}{\partial\theta},$$

$$\operatorname{div}\mathbf{V} = \frac{1}{r^2}\frac{\partial}{\partial r}\left(r^2 V^r\right) + \frac{1}{r\sin\theta}\frac{\partial V^\varphi}{\partial\varphi} + \frac{1}{r\sin\theta}\frac{\partial}{\partial\theta}\left(V^\theta\sin\theta\right),$$

$$\Delta\Psi = \frac{1}{r^2}\frac{\partial}{\partial r}\left(r^2\frac{\partial\Psi}{\partial r}\right) + \frac{1}{r^2\sin\theta}\frac{\partial}{\partial\theta}\left(\sin\theta\cdot\frac{\partial\Psi}{\partial\theta}\right) + \frac{1}{r^2\sin^2\theta}\frac{\partial^2\Psi}{\partial\varphi^2},$$

$$\left(\operatorname{rot}\mathbf{V}\right)^r = \frac{1}{r\sin\theta}\left[\frac{\partial}{\partial\theta}\left(V^\varphi\cdot\sin\theta\right) - \frac{\partial V^\theta}{\partial\varphi}\right], \tag{7.38}$$

$$\left(\operatorname{rot}\mathbf{V}\right)^\varphi = \frac{1}{r}\frac{\partial}{\partial r}\left(rV^\theta\right) - \frac{1}{r}\frac{\partial V^r}{\partial\theta}, \qquad \left(\operatorname{rot}\mathbf{V}\right)^\theta = \frac{1}{r\sin\theta}\frac{\partial V^r}{\partial\varphi} - \frac{1}{r}\frac{\partial}{\partial r}\left(rV^\varphi\right).$$

Exercise 5 Use the relation (5.28) from Chap. 5 as a starting point for the derivation of the $\Delta\Psi$ in both cylindric and spherical coordinates. Compare the complexity of this calculation with the one presented above.

Chapter 8
Curvilinear, Surface, and D-Dimensional Integrals

In this chapter, we define and discuss volume integrals in D-dimensional space, curvilinear, and surface integrals of the first and second types in $3D$.

8.1 Brief Reminder Concerning $1D$ Integrals and Area of a Surface in $2D$

Let us start with a brief reminder of the notion of a definite (Riemann) integral in $1D$ and the notion of the area of a surface in $2D$.

Def. 1 Consider a function $f(x)$, which is defined and restricted on $[a, b]$. In order to construct the integral, one has to perform the division of the interval by inserting $n - 1$ intermediate points

$$a = x_0 < x_1 < ... < x_i < x_{i+1} < ... < x_n = b. \tag{8.1}$$

We denote $\Delta x_i = x_{i+1} - x_i$ the length of the interval number i. Let us choose arbitrary points $\xi_i \in [x_i, x_{i+1}]$ inside each of the intervals. The next step is to construct the integral sum

$$\sigma = \sum_{i=0}^{n-1} f(\xi_i) \Delta x_i. \tag{8.2}$$

For a fixed division (8.1) and fixed choice of the intermediate points ξ_i, the integral sum σ is just a number. But, since we have a freedom of changing division (8.1) and ξ_i, this sum can be seen as a function of many variables x_i and ξ_i. Let us add the restriction on a procedure of adding new points x_i and correspondingly ξ_i. These points should be added in such a way that the maximal length $\lambda = \max\{\Delta x_i\}$ should decrease after inserting a new point into the division (8.1). Still we have a freedom

© Springer Nature Switzerland AG 2019
I. L. Shapiro, *A Primer in Tensor Analysis and Relativity*, Undergraduate
Lecture Notes in Physics, https://doi.org/10.1007/978-3-030-26895-4_8

to insert these points in different ways, and hence we have different procedures and sequences of integral sums. Within each procedure, we meet a numerical sequence $\sigma = \sigma_n$. When $\lambda \to 0$, of course $n \to \infty$. If the limit

$$I = \lim_{\lambda \to 0} \sigma \tag{8.3}$$

is finite and universal, in the sense that it does not depend on the choice of the points x_i and ξ_i, this limit is called the definite (or Riemann) integral of the function $f(x)$ over the interval $[a, b]$, denoted as

$$I = \int_a^b f(x)dx. \tag{8.4}$$

The following theorem plays a prominent role in the integral calculus:

Theorem 1   *For continuous function $f(x)$ on $[a, b]$ the integral (8.4) exists. The same is true if $f(x)$ is continuous everywhere except a finite number of points on $[a, b]$.*

Observation 1   The integral possesses many well-known properties, which we will not review here. One can consult the textbooks on real analysis for this purpose.

Def. 2 Consider a restricted region $(S) \subset \mathbb{R}^2$. The word restricted means that (S) is a subset of a finite circle of a finite radius $C > 0$, such that for all points (x, y) of this region

$$x^2 + y^2 \le C.$$

In order to define the area S of the figure (S), we shall cover it by the lattice $x_i = i/n$ and $y_j = j/n$ (see Fig. 8.1) with the size of the cell $\frac{1}{n}$ and its area $\frac{1}{n^2}$. The individual cell may be denoted as

$$\Delta_{ij} = \left\{ x, y \mid x_i < x < x_{i+1} , \; y_j < y < y_{j+1} \right\}.$$

Fig. 8.1 Illustration of the definition of the area in $2D$

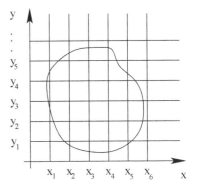

Suppose the number of the cells Δ_{ij}, situated inside (S) is N_{in} and the number of cells that have common points with (S) is N_{out}. One can define an internal area of the figure (S) as $S_{in}(n)$ and an external area of the figure (S) as $S_{out}(n)$, where

$$S_{in}(n) = \frac{N_{in}}{n^2} \quad \text{and} \quad S_{out}(n) = \frac{N_{out}}{n^2}.$$

Taking the limit $n \to \infty$ we meet

$$\lim_{n \to \infty} S_{in}(n) = S_{int} \quad \text{and} \quad \lim_{n \to \infty} S_{out}(n) = S_{ext},$$

where S_{int} and S_{ext} are called internal and external areas of the figure (S). In the case when these two limits coincide $S_{int} = S_{ext} = S$, we shall define the number S to be the area of the figure (S).

Observation 2 Below we consider only such (S) for which S exists. In fact, the existence of the area is guaranteed for any figure (S) with the continuous boundary (∂S).

Observation 3 In the same manner, one can define the volume of the figure (V) in a space of an arbitrary dimension D (in particular for $D = 3$). It is important that (V) is a restricted region in $\mathbb{R}^D$

$$\exists L > 0, \quad (V) \subset \mathcal{O}(0, L) = \{(x_1, x_2, ..., x_D) \,|\, x_1^2 + x_2^2 + ... + x_D^2 \le L^2\} \subset \mathbb{R}^D.$$

8.2 Volume Integrals in Curvilinear Coordinates

The first type of integral we will formulate is the D-dimensional volume integral in curvilinear coordinates $x^i = x^i(X^a)$. For such a formulation, we shall use the metric tensor and Levi-Civita tensor in the curvilinear coordinates.

Let us start from the metric tensor and its geometric meaning. As we already learned in Chap. 2, the square of the distance between the two points X^a and Y^a in Cartesian and arbitrary global coordinates is given by (2.7),

$$s_{xy}^2 = \sum_{a=1}^{D} \left(X^a - Y^a\right)^2 = g_{ij}\left(x^i - y^i\right)\left(x^j - y^j\right)$$

where x^i and y^j are the coordinates of the two points. For the local change of coordinates

$$x^i = x^i\left(X^1, X^2, ..., X^D\right),$$

the similar formula holds for the infinitesimal distances

$$dl^2 = g_{ij}\, dx^i\, dx^j. \tag{8.5}$$

Therefore, if we have a curve in D-dimensional space $x^i = x^i(\tau)$ (here τ is an arbitrary monotonic parameter along the curve), then the length of the curve between the points A with the coordinates $x^i(a)$ and B with the coordinates $x^i(b)$ is

$$l_{AB} = \int_{(AB)} dl = \int_a^b d\tau \sqrt{g_{ij}\, \frac{dx^i}{d\tau}\, \frac{dx^j}{d\tau}}. \tag{8.6}$$

We can make a conclusion that the direct geometric sense of the metric is that it defines a distance between two infinitesimally close points and also the length of the finite curve in the D-dimensional space, in arbitrary curvilinear coordinates.

Exercise 1   Consider the formula (8.6) for the case of Cartesian coordinates, and also for a global non-orthogonal basis. Discuss the difference between these two cases.

Now we are in a position to consider the volume integration in the Euclidean D-dimensional space. The definition of the integral in D-dimensional space

$$I = \int_{(V)} f(X^1, ..., X^D)\, dV \tag{8.7}$$

implies that we can define the volume of the figure (V) bounded by the $(D\text{-}1)$-dimensional continuous surface (∂V). After this, one has to follow the procedure similar to that for the $1D$ Riemann integral: divide the figure $(V) = \bigcup_i (V_i)$ such that each figure (V_i) has a well-defined volume V_i, and the common points of any two sub-figures have zero volume. For each figure (V_i), we have to choose points $M_i \in (V_i)$ and define[1]

$$\lambda = \max\{\operatorname{diam}(V_i)\}. \tag{8.8}$$

The integral is defined as a universal limit of the integral sum

$$I = \lim_{\lambda \to 0} \sigma, \quad \text{where} \quad \sigma = \sum_i f(M_i)\, V_i. \tag{8.9}$$

For a while, one can assume that the volumes in this expression are defined in the Cartesian coordinates in the way that was briefly described in the previous section.

Our purpose is to rewrite the integral (8.7) in terms of curvilinear coordinates x^i. In other words, we have to perform a change of variables $x^i = x^i(X^a)$ in this integral. First of all, the form of the figure (V) may change. In fact, we are performing the

---

[1] Reminder: $\operatorname{diam}(V_i)$ is a maximal (supremum) of the distances between the two points of the figure (V_i).

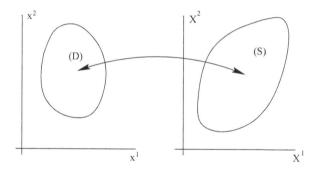

Fig. 8.2 Illustration of the mapping $(G) \to (S)$ in $2D$

mapping of the figure (V) in the space with the coordinates X^a into another figure (D), which lies in the space with the coordinates x^i. It is always assumed that this mapping is invertible everywhere except, at most, some finite number of isolated singular points. After the change of variables, the limits of integrations over the variables x^i correspond to the figure (D), not to the original figure (V) (Fig. 8.2).

The next step is to replace the original function by the composite function of new variables,

$$f(X^a) \to f(X^a(x^i)) = f(X^1(x^i), X^2(x^i), \dots, X^D(x^i)).$$

Looking at the expression for the integral sum in (8.9), it is clear that the transformation of the function $f(M_i)$ may be very important. Of course, we can consider the integral in the case when the function is not necessarily a scalar. It may be, e.g., a component of some tensor. But, in this case, if we consider curvilinear coordinates, the result of the integration may be an object that transforms in a non-regular way. In order to avoid the possible problems, let us suppose that the function f is a scalar. In this case, the transformation of the function f does not pose a problem. Taking the additivity of the volume into account, we arrive at the necessity to transform the infinitesimal volume element

$$dV = dX^1 dX^2 \ \dots \ dX^D \tag{8.10}$$

into the curvilinear coordinates x^i. Thus, the transformation of the volume element is the last step in transforming the integral. When transforming the volume element, we have to provide that it is a scalar, as otherwise the volume V will be different in different coordinate systems. In order to rewrite (8.10) in an invariant form, let us perform transformations

$$dX^1 dX^2 \dots dX^D = \frac{1}{D!} \left| E_{a_1 a_2 \dots a_D} \left| dX^{a_1} dX^{a_2} \dots dX^{a_D} \right. \right.$$

$$= \frac{1}{D!} \left| \varepsilon_{a_1 a_2 \dots a_D} \left| dX^{a_1} dX^{a_2} \dots dX^{a_D} \right. \right. . \tag{8.11}$$

Due to the modulo, all the nonzero terms in (8.11) give identical contributions. Since the number of such terms is $D!$, the two factorials cancel and we have a first identity. In the second equality, we have used the coincidence of the Levi-Civita tensor $\varepsilon_{a_1 a_2 \dots a_D}$ and the coordinate-independent tensor density $E_{a_1 a_2 \dots a_D}$ in the Cartesian coordinates. The last form of the volume element (8.11) is remarkable, because it is explicitly seen that dV is indeed a scalar. Thus, we meet here the contraction of the tensor[2] $\varepsilon_{a_1 \dots a_D}$ with the infinitesimal vectors dX^{a_k}. Since dV is an infinitesimal scalar, it can be easily transformed into other coordinates

$$dV = \frac{1}{D!} \left| \varepsilon_{i_1 i_2 \dots i_D} \left| dx^{i_1} \dots dx^{i_D} \right. \right. = \sqrt{g} \cdot dx^1 dx^2 \dots dx^D . \tag{8.12}$$

In the last step, we have used the same identical transformation as we did in the first equality in (8.11).

The quantity

$$J = \sqrt{g} = \sqrt{\det \|g_{ij}\|} = \left| \det \left(\frac{\partial X^a}{\partial x^i} \right) \right| = \left| \frac{D(X^a)}{D(x^i)} \right| \tag{8.13}$$

is nothing but the well-known Jacobian of the coordinate transformation. The modulo eliminates the possible change of sign that may come from a "wrong" choice of the order of the new coordinates. For example, any permutation of two coordinates, e.g., $x^1 \leftrightarrow x^2$ changes the sign of the determinant $\det \left(\frac{\partial X^a}{\partial x^i} \right)$, but the infinitesimal volume element dV should always be positive, as $J = \sqrt{g}$ always is.

8.3 Curvilinear Integrals

Up to this point, we have considered integrals over flat spaces; we now start to consider integrals over curves and curved surfaces.

Consider first a curvilinear integral of the first type. Suppose we have a curve (L), defined by the vector function of the continuous parameter t :

$$\mathbf{r} = \mathbf{r}(t) \quad \text{where} \quad a \leq t \leq b. \tag{8.14}$$

It is supposed that when the parameter t runs between $t = a$ and $t = b > a$, the point $\mathbf{r}$ moves, monotonically, from the initial point $\mathbf{r}(a)$ to the final one $\mathbf{r}(b)$. In particular, we suppose that in all points of the curve

---

[2]In Exercise 2, we ask to verify that dx^i really transform as contravariant vector components.

$$g_{ij} \, \dot{x}^i \, \dot{x}^j \, > 0,$$

where the over-dot stands for the derivative with respect to the parameter t. As we already learned above, the length of the curve is given by the integral

$$L = \int_{(L)} dl = \int_a^b \sqrt{g_{ij} \, \dot{x}^i \, \dot{x}^j} \, dt.$$

Def. 3 Consider the continuous function $f(\mathbf{r}) = f(x^i)$, where x^i are the coordinates in D-dimensional space, $x^i = \{x^1, x^2, \ldots, x^D\}$ defined along the curve (8.14). One can define the *curvilinear integral of the first type*

$$I_1 = \int_{(L)} f(\mathbf{r}) \, dl \tag{8.15}$$

in a manner quite similar to the definition of the Riemann integral. Namely, one has to divide the interval

$$a = t_0 < t_1 < \ldots < t_k < t_{k+1} < \ldots < t_n = b \tag{8.16}$$

and also choose points $\xi_k \in \left[t_k, \, t_{k+1} \right]$ inside each interval $\Delta t_k = t_{k+1} - t_k$. After that, we have to establish the length of each particular curve

$$\Delta l_k = \int_{t_k}^{t_{k+1}} \sqrt{g_{ij} \, \dot{x}^i \, \dot{x}^j} \, dt \tag{8.17}$$

and construct the integral sum

$$\sigma = \sum_{k=0}^{n-1} f\left(x^i(\xi_k) \right) \Delta l_k. \tag{8.18}$$

It is easy to see that this sum is also an integral sum for the Riemann integral

$$\int_a^b \sqrt{g_{ij} \, \dot{x}^i \, \dot{x}^j} \, f(x^i(t)) \, dt. \tag{8.19}$$

Introducing the maximal length $\lambda = \max \{\Delta l_i\}$, and taking the limit (through the same procedure described in the first section) $\lambda \to 0$, we meet a limit

$$I = \lim_{\lambda \to 0} \sigma. \tag{8.20}$$

If this limit is finite and does not depend on the choice of the points t_i and ξ_i, it is called a curvilinear integral of the first type (8.15).

The existence of the integral is guaranteed for the smooth curve and continuous function in the integrand. Also, the procedure above provides a method of explicit calculation of the curvilinear integral of the first type through the formula (8.19).

The properties of the curvilinear integral of the first type include additivity

$$\int_{AB} + \int_{BC} = \int_{AC} \tag{8.21}$$

and symmetry

$$\int_{AB} = \int_{BA}. \tag{8.22}$$

Both properties can be easily established from (8.19).

Observation In order to provide the coordinate independence of the curvilinear integral of the first type, we have to choose the function $f(x^i)$ to be a scalar field. Of course, the integral may also be defined for the non-scalar quantity. But, in this case, the integral may be coordinate dependent and its geometrical (or physical) sense may be unclear. For example, if the function $f(x^i)$ is a component of a vector, and if we consider only global transformations of the basis, it is easy to see that the integral will also transform as a component of a vector. But, if we allow the local transformations of a basis, the integral may transform in a non-regular way and its geometric interpretation may be lost. The situation is similar to the one for the volume integrals and to the one for other integrals that will be discussed below.

Let us note that the scalar law of transformation for the curvilinear integral may also be achieved in the case when the integrated function is a vector. The corresponding construction is, of course, different from the curvilinear integral of the first type.

Def. 4 The *curvilinear integral of the second type* is a scalar that results from the integration of the vector field along the curve. Consider a vector field $\mathbf{A}(\mathbf{r})$ defined along the curve (L). Let us consider, for the sake of simplicity, the $3D$ case, and keep in mind an obvious possibility to generalize the construction for an arbitrary D. One can construct the following infinitesimal scalar:

$$\mathbf{A} \cdot d\mathbf{r} = A_x d_x + A_y d_y + A_z d_z = A_i dx^i. \tag{8.23}$$

If the curve is parametrized by the continuous monotonic parameter t, the last expression can be presented as

$$A_i dx^i = \mathbf{A} \cdot d\mathbf{r} = \left(A_x \dot{x} + A_y \dot{y} + A_z \dot{z} \right) dt = A_i \dot{x}^i dt. \tag{8.24}$$

This expression can be integrated along the curve (L) to give

$$\int_{(L)} \mathbf{A} \cdot d\mathbf{r} = \int_{(L)} A_x dx + A_y dy + A_z dz = \int_{(L)} A_i dx^i. \qquad (8.25)$$

Each of the three integrals in the *r.h.s.* can be defined as a limit of the corresponding integral sum in the Cartesian coordinates. For instance,

$$\int_{(L)} A_x dx = \lim_{\lambda \to 0} \sigma_x, \qquad (8.26)$$

where

$$\sigma_x = \sum_{i=0}^{n-1} A_x (\mathbf{r}(\xi_i)) \cdot \Delta x_i. \qquad (8.27)$$

Here, we used the previous notations from Eq. (8.16) and below, and also introduced a new one

$$\Delta x_i = x(t_{i+1}) - x(t_i).$$

The limit of the integral sum is taken exactly as in the previous case of the curvilinear integral of the first type. The scalar property of the whole integral is guaranteed by the scalar transformation rule for each term in the integral sum. It is easy to establish the following relation between the curvilinear integral of the second type, the curvilinear integral of the first type, and the Riemann integral:

$$\int_{(L)} A_x dx + A_y dy + A_z dz = \int_{(L)} A_i \frac{dx^i}{dl} dl = \int_a^b A_i \frac{dx^i}{dt} dt, \qquad (8.28)$$

where l is a natural parameter along the curve

$$l = \int_a^t \sqrt{g_{ij} \, \dot{x}^i \dot{x}^j} \, dt, \qquad 0 \leq l \leq L.$$

Let us note that in the case of Cartesian coordinates $X^a = (x, y, z)$, the derivatives in the second integral in (8.28) are nothing but the cosines of the angles between the tangent vector $d\mathbf{r}/dl$ along the curve and the basic vectors $\hat{\mathbf{i}}, \hat{\mathbf{j}}, \hat{\mathbf{k}}$,

$$\cos \alpha = \frac{d\mathbf{r} \cdot \hat{\mathbf{i}}}{dl} = \frac{dx}{dl}, \quad \cos \beta = \frac{d\mathbf{r} \cdot \hat{\mathbf{j}}}{dl} = \frac{dy}{dl}, \quad \cos \gamma = \frac{d\mathbf{r} \cdot \hat{\mathbf{k}}}{dl} = \frac{dz}{dl}. \qquad (8.29)$$

Of course, $\cos^2 \alpha + \cos^2 \beta + \cos^2 \gamma = 1$.

One can rewrite Eq. (8.28) in the following useful form:

$$\int_{(L)} \mathbf{A} \cdot d\mathbf{r} = \int_0^L \left(A_x \cos\alpha + A_y \cos\beta + A_z \cos\gamma \right) dl$$

$$= \int_a^b \left(A_x \cos\alpha + A_y \cos\beta + A_z \cos\gamma \right) \sqrt{\dot{x}^2 + \dot{y}^2 + \dot{z}^2}\, dt, \quad (8.30)$$

where we used $dl^2 = g_{ab} \dot{X}^a \dot{X}^b\, dt^2 = (\dot{x}^2 + \dot{y}^2 + \dot{z}^2)\, dt^2$.

The main properties of the curvilinear integral of the second type are

$$\text{additivity} \quad \int_{AB} + \int_{BC} = \int_{AC} \qquad \text{and antisymmetry} \quad \int_{AB} = -\int_{BA}.$$

Exercise 2   Derive the following curvilinear integrals of the first type:

1. $\oint_{(C)} (x + y)\, dl$,

where (C) is a triangle with the vertices $A(1, 0, 0)$, $B(0, 1, 0)$, $C(0, 0, 1)$.

2. $\int_{(C)} (x^2 + y^2)\, dl$,

$(C) = \{x, y, z \mid x = a\cos t, \ y = a\sin t, \ z = bt\}, \quad \text{where} \ \ 0 \le t \le n\pi$.

3. $\int_{(C)} x\, dl$,

where C is a part of the logarithmic spiral $r = ae^{k\varphi} \quad r \le a$.

Exercise 3   Derive the following curvilinear integral of the second type:

$$\int_{(OA)} x\,dy \pm y\,dx,$$

where $A(1, 2)$ and (OA) are the following curves: (a) Straight line, (b) Part of parabola with Oy-axis of symmetry, (c) Polygonal line OBA where $B(0, 2)$. Explain why for the positive sign, all three cases gave the same integral.

8.4 $2D$ Surface Integrals in a $3D$ Space

Consider the integrals over the $2D$ surface, (S) in the $3D$ space $(S) \subset \mathbb{R}^3$. Suppose the surface is defined by the three smooth functions

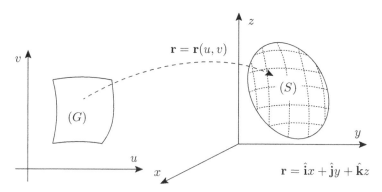

Fig. 8.3 Figure illustrating mapping of (G) into (S)

$$x = x(u, v), \qquad y = y(u, v), \qquad z = z(u, v), \tag{8.31}$$

where u and v are independent variables which are called internal coordinates on the surface. One can regard (8.31) as a mapping of a figure (G) on the plane with coordinates u and v into the $3D$ space with the coordinates x, y, z, as shown in Fig. 8.3.

Example The sphere (S) of the constant radius R and the center at the initial point O of Cartesian coordinate system can be described by internal coordinates (spherical angles) φ, θ. Then the relation (8.31) looks like

$$x = R \cos \varphi \sin \theta, \qquad y = R \sin \varphi \sin \theta, \qquad z = R \cos \theta, \tag{8.32}$$

where φ and θ play the role of internal coordinates u and v. The values of these coordinates are taken from the rectangle

$$(G) = \{(\varphi, \theta) \,|\, 0 \le \varphi < 2\pi, \ 0 \le \theta < \pi\}.$$

The parametric description of the sphere (8.32) is an example of the mapping $(G) \to (S)$. One can express all geometric quantities on the surface of the sphere using the internal coordinates φ and θ, without addressing the coordinates of the external space x, y, z.

In order to establish the geometry of the surface, let us start by considering the infinitesimal line element dl linking two points of the surface. We can suppose that these two infinitesimally close points belong to the same smooth curve $u = u(t)$, $v = v(t)$, situated on the surface. Then the line element is

$$dl^2 = dx^2 + dy^2 + dz^2, \tag{8.33}$$

where

$$
dx = \left(\frac{\partial x}{\partial u} \frac{\partial u}{\partial t} + \frac{\partial x}{\partial v} \frac{\partial v}{\partial t} \right) dt = x'_u du + x'_v dv,
$$

$$
dy = \left(\frac{\partial y}{\partial u} \frac{\partial u}{\partial t} + \frac{\partial y}{\partial v} \frac{\partial v}{\partial t} \right) dt = y'_u du + y'_v dv,
$$

$$
dz = \left(\frac{\partial z}{\partial u} \frac{\partial u}{\partial t} + \frac{\partial z}{\partial v} \frac{\partial v}{\partial t} \right) dt = z'_u du + z'_v dv. \tag{8.34}
$$

Using the last formulas, we obtain

$$
\begin{aligned}
dl^2 &= \left(x'_u du + x'_v dv \right)^2 + \left(y'_u du + y'_v dv \right)^2 + \left(z'_u du + z'_v dv \right)^2 \\
&= \left(x'^2_u + y'^2_u + z'^2_u \right) du^2 + 2 \left(x'_u \cdot x'_v + y'_u \cdot y'_v + z'_u \cdot z'_v \right) du dv \\
&+ \left(x'^2_v + y'^2_v + z'^2_v \right) dv^2 = g_{uu} du^2 + 2 g_{uv} du dv + g_{vv} dv^2, \tag{8.35}
\end{aligned}
$$

where

$$
\begin{aligned}
g_{uu} &= x'^2_u + y'^2_u + z'^2_u, \\
g_{uv} &= x'_u x'_v + y'_u y'_v + z'_u z'_v, \\
g_{vv} &= x'^2_v + y'^2_v + z'^2_v \tag{8.36}
\end{aligned}
$$

are the elements of the matrix which is called *induced metric on the surface*. As we can see, the expression $dl^2 = g_{ij} dx^i dx^j$ is valid for the curved surfaces as well as for the curvilinear coordinates in flat D-dimensional spaces. One can generalize Eq. (8.35) to the case of n-dimensional surface embedded into D-dimensional space. We leave this generalization for later brief discussion and for a while concentrate on the $3D$ space and $2D$ surfaces.

Let us consider the relation (8.35) from another point of view and introduce two basis vectors on the surface in such a way that the scalar products of these vectors are equal to the corresponding metric components. The tangent plane at any point is composed of linear combinations of the two vectors

$$
\mathbf{r}_u = \frac{\partial \mathbf{r}}{\partial u} \quad \text{and} \quad \mathbf{r}_v = \frac{\partial \mathbf{r}}{\partial v}. \tag{8.37}
$$

Below we suppose that at any point of the surface $\mathbf{r}_u \times \mathbf{r}_v \neq 0$. This means that the lines of constant coordinates u and v are not parallel or, equivalently, that the internal coordinates are not degenerate.

For the smooth surface, the infinitesimal displacement may be considered identical to the infinitesimal displacement over the tangent plane. Therefore, for such a displacement, we can write

$$dr = \frac{\partial \mathbf{r}}{\partial u} du + \frac{\partial \mathbf{r}}{\partial v} dv, \tag{8.38}$$

and hence (here we use notations $\mathbf{a}^2$ and $(\mathbf{a})^2$ for the square of the vector $= \mathbf{a} \cdot \mathbf{a}$)

$$dl^2 = (d\mathbf{r})^2 = \left(\frac{\partial \mathbf{r}}{\partial u}\right)^2 du^2 + 2 \frac{\partial \mathbf{r}}{\partial u} \cdot \frac{\partial \mathbf{r}}{\partial v} dudv + \left(\frac{\partial \mathbf{r}}{\partial v}\right)^2 dv^2, \tag{8.39}$$

that is an alternative form of Eq. (8.35). Thus, we see that the tangent vectors $\mathbf{r}_u$ and $\mathbf{r}_v$ may be safely interpreted as basic vectors on the surface, in the sense that the components of the induced metric are nothing but the scalar products of these two vectors,

$$g_{uu} = \mathbf{r}_u \cdot \mathbf{r}_u, \qquad g_{uv} = \mathbf{r}_u \cdot \mathbf{r}_v, \qquad g_{vv} = \mathbf{r}_v \cdot \mathbf{r}_v. \tag{8.40}$$

Exercise 4   (i) Derive the tangent vectors $\mathbf{r}_\varphi$ and $\mathbf{r}_\theta$ for the coordinates φ, θ on the sphere; (ii) Check that these two vectors are orthogonal, $\mathbf{r}_\varphi \cdot \mathbf{r}_\theta = 0$; (iii) Using the tangent vectors, derive the induced metric on the sphere. Check this calculation using the formulas (8.36); (iv) Compare the induced metric on the sphere and the 3D metric in the spherical coordinates. Invent the procedure on how to obtain the induced metric on the sphere from the 3D metric in the spherical coordinates. (v) * Analyze whether the internal metric on any surface may be obtained through a similar procedure.

Exercise 5   Repeat points (i) and (ii) of the previous exercise for the surface of the torus (see Fig. 8.4). The equation of this torus in cylindric coordinates has the form $(r - a)^2 + z^2 = b^2$, where $a > b$. Try to find angular variables (internal coordinates) that are the most useful and verify that in these coordinates, the metric is diagonal. Give a geometric interpretation.

Fig. 8.4 Torus with two radii a and b

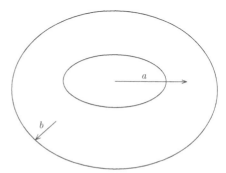

Exercise 6   Derive the basic vectors and metric for the conic surface

$$\frac{z^2}{c^2} = \frac{x^2}{a^2} + \frac{y^2}{b^2}.$$

Try to consider different choices of internal coordinates and verify that the induced metric transforms as a tensor.

Let us now calculate the infinitesimal element of the area of the surface, corresponding to the infinitesimal surface element $dudv$ on the figure (G) in the uv-plane. As far as a curved surface (S) is geometrically different from the figure (G), the elements of the area in two cases are also expected to be different.

After the mapping, on the tangent plane, we have a parallelogram spanned on the vectors $\mathbf{r}_u du$ and $\mathbf{r}_v dv$. According to the analytic geometry, the area of this parallelogram equals to the absolute value of the vector product of the two vectors,

$$dS = |d\mathbf{A}| = |\mathbf{r}_u du \times \mathbf{r}_v dv| = |\mathbf{r}_u \times \mathbf{r}_v| \, dudv. \tag{8.41}$$

In order to evaluate this product, it is useful to take a square of the vector (here α is the angle between $\mathbf{r}_u$ and $\mathbf{r}_v$)

$$
\begin{aligned}
dS = (d\mathbf{A})^2 &= |\mathbf{r}_u \times \mathbf{r}_v|^2 \, du^2 dv^2 = r_u^2 r_v^2 \sin^2 \alpha \cdot du^2 dv^2 \\
&= du^2 dv^2 \left\{ r_u^2 r_v^2 \left(1 - \cos^2 \alpha \right) \right\} = du^2 dv^2 \left\{ \mathbf{r}^2 \cdot \mathbf{r}^2 - (\mathbf{r}_u \cdot \mathbf{r}_v)^2 \right\} \\
&= \left(g_{uu} \cdot g_{vv} - g_{uv}^2 \right) du^2 dv^2 = g \cdot du^2 dv^2, \tag{8.42}
\end{aligned}
$$

where g is the determinant of the induced metric. Once again, we arrived at the conventional formula $dS = \sqrt{g} \, dudv$, where

$$g = \begin{vmatrix} g_{uu} & g_{uv} \\ g_{vu} & g_{vv} \end{vmatrix}$$

is the metric's determinant. Finally, the area S of the whole surface (S) may be defined as an integral

$$S = \iint\limits_{(G)} \sqrt{g} \, du \, dv. \tag{8.43}$$

Geometrically, this definition corresponds to a triangulation of the curved surface with the consequent continuous limit, when the sizes of all triangles simultaneously tend to zero.

Exercise 7   Find g and $\sqrt{g}$ for the sphere, using the induced metric in the (φ, θ) coordinates. Apply this result to calculate the area of the sphere. Try to repeat the same process for the surface of the torus (see Exercise 5).

In a more general situation, we have an n-dimensional surface embedded into the D-dimensional space R^D, with $D > n$. Introducing internal coordinates u^i on the surface, we obtain its parametric equation of the surface in the form

$$x^\mu = x^\mu \left(u^i \right) , \tag{8.44}$$

where

$$\begin{cases} i = 1, ..., n \\ \mu = 1, ..., D \end{cases} .$$

The vector in R^D may be presented in the form $\bar{r} = r^\mu \bar{e}_\mu$, where $\bar{e}_\mu$ are the corresponding basis vectors. The tangent vectors are given by the partial derivatives $\bar{r}_i = \frac{\partial \bar{r}}{\partial u^i}$. Introducing a curve $u^i = u^i(t)$ on the surface

$$dx^\mu = \frac{\partial x^\mu}{\partial u^i} \frac{du^i}{dt} dt = \frac{\partial x^\mu}{\partial u^i} du^i, \tag{8.45}$$

we arrive at the expression for the distance between the two infinitesimally close points,

$$ds^2 = g_{\mu\nu} dx^\mu dx^\nu = \frac{\partial x^\mu}{\partial u^i} \frac{\partial x^\nu}{\partial u^j} g_{\mu\nu} du^i du^j. \tag{8.46}$$

Therefore, in this general case, we meet the same relation as for the usual coordinate transformation

$$g_{ij}(u) = \frac{\partial x^\mu}{\partial u^i} \frac{\partial x^\nu}{\partial u^j} g_{\mu\nu}(x). \tag{8.47}$$

The main difference between this relation and the usual tensor law of transforming metric is that the matrix $\frac{\partial x^\mu}{\partial u^i}$ has dimension $n \times D$ and hence it cannot be inverted.

The condition of the nondegeneracy of the induced metric is

$$\text{rank} \left(g_{ij} \right) = n. \tag{8.48}$$

Sometimes (e.g., for the coordinates on the sphere at the pole points $\theta = 0, \pi$), the metric may be degenerate in some isolated points. The volume element of the area (surface element) in the n-dimensional case also looks as

$$dS = J \, du^1 du^2 ... du^n, \quad \text{where} \quad J = \sqrt{g}. \tag{8.49}$$

Hence, there is no big difference between two-dimensional and n-dimensional surfaces. For the sake of simplicity, we shall concentrate, below, on the two-dimensional case.

Now we are in a position to define the surface integral of the first type. Suppose we have a surface (S) defined as a mapping (8.31) of the closed finite figure (G) in a uv-plane, and a function $f(u, v)$ on this surface. Instead, one can take a function $g(x, y, z)$, which generates a function on the surface through the relation

$$f(u, v) = g(x(u, v), y(u, v), z(u, v)). \tag{8.50}$$

Def. 5 The surface integral of the first type is defined as follows: Let us divide (S) into particular subsurfaces (S_i) such that each of them has a well-defined area

$$S_i = \int_{G_i} \sqrt{g} \, du \, dv. \tag{8.51}$$

Of course, one has to perform the division of (S) into (S_i) such that the intersections of the two sub-figures have zero area. On each of the particular surfaces (S_i), we choose a point $M_i(\xi_i, \eta_i)$ and construct an integral sum

$$\sigma = \sum_{i=1}^{n} f(M_i) \cdot S_i, \tag{8.52}$$

where $f(M_i) = f(\xi_i, \eta_i)$. The next step is to define $\lambda = \max\{\operatorname{diam}(S_i)\}$. If the limit

$$\mathcal{I}_1 = \lim_{\lambda \to 0} \sigma \tag{8.53}$$

is finite and does not depend on the choice of (S_i) and also on the choice of the points M_i, this limit is called *surface integral of the first type*

$$\mathcal{I}_1 = \iint_{(S)} f(u, v) \, dS = \iint_{(S)} g(x, y, z) \, dS. \tag{8.54}$$

From the construction described above, it is clear that this integral can indeed be calculated as a double integral over the figure (G) in the uv-plane

$$\mathcal{I}_1 = \iint_{(S)} f(u, v) \, dS = \iint_{(G)} f(u, v) \sqrt{g} \, du dv. \tag{8.55}$$

Remark The surface integral (8.55) is, by construction, a scalar, if the function $f(u, v)$ is a scalar. Therefore, it can be calculated in any other coordinates, e.g., (u', v'). Of course, when we change the internal coordinates on the surface to (u', v'), the following aspects must be changed: (1) the form of the area $(G) \to (G')$; (2) the surface element $\sqrt{g} \to \sqrt{g'}$; (3) the form of the integrand $f(u, v) =$

$f'(u', v') = f \left(u \left(u', v' \right), v \left(u', v' \right) \right)$. In this case, the surface integral of the first type is a coordinate-independent geometric object.

Consider now a more complicated type of integral, which is called *surface integral of the second type*. The relation between the two types of surface integrals is similar to that between the two types of curvilinear integrals, which we have considered in the previous section. First of all, we need to introduce the notion of an oriented surface. Indeed, for the smooth surface with independent tangent vectors $\mathbf{r}_u = \partial \mathbf{r}/\partial u$ and $\mathbf{r}_v = \partial \mathbf{r}/\partial v$, one has a freedom to choose ordering—that means to decide which of the coordinates is the first and which one is second. This defines the orientation for their vector product and corresponds to the choice of orientation for the surface. In a small vicinity of a point on the surface with the nondegenerate coordinates u, v, there are two opposite directions associated with the normal vectors

$$\pm \hat{\mathbf{n}} = \frac{\mathbf{r}_u \times \mathbf{r}_v}{|\mathbf{r}_u \times \mathbf{r}_v|}. \tag{8.56}$$

Any of these directions could be taken as positive.

Exercise 8 Using the formula (8.56), derive the normal vector for the sphere and torus (see Exercises 4 and 5). Give geometrical interpretation. Discuss to which extent the components of the normal vector depend on the choice of internal coordinates.

The situation changes if we consider the whole surface. One can distinguish two types of surfaces: (1) oriented, or two-sided surface, like the one of the cylinder or sphere and (2) non-oriented (one-sided) surface, like the Möbius band (surface) (see Fig. 8.5).

Below we consider only smooth two-sided oriented surfaces, described by the internal coordinates u, v. If

$$|\mathbf{r}_u \times \mathbf{r}_v| \neq 0 \tag{8.57}$$

in all points on the surface, then the normal vector $\hat{\mathbf{n}}$ never changes its sign. We shall choose, for definiteness, the positive sign in (8.56). According to this formula, the vector $\hat{\mathbf{n}}$ is normal to the surface. It is worth noticing that if we consider the infinitesimal piece of surface that corresponds to the $du\,dv$ area on the corresponding plane (remember that the figure (G) is the inverse image of the "real" surface (S) on the plane with the coordinates u, v), then the area of the corresponding parallelogram

Fig. 8.5 Examples of two-sided and one-sided surfaces

on (S) is $dS = |\mathbf{r}_u \times \mathbf{r}_v|\, dudv$. This parallelogram may be characterized by its area dS and by the normal vector $\hat{\mathbf{n}}$. Using these two quantities, we can construct a new vector $d\mathbf{S} = \hat{\mathbf{n}} \cdot dS = [\mathbf{r}_u, \mathbf{r}_v]\, dudv$. This vector quantity may be regarded as a vector element of the (oriented) surface. This vector contains information about both area of the infinitesimal parallelogram (it is equivalent to the knowledge of the Jacobian $\sqrt{g}$) and its orientation with respect to this surface in the external $3D$ space.

For calculational purposes, it is useful to introduce the components of $\hat{\mathbf{n}}$; using the same cosines, we already considered for the curvilinear second-type integral,

$$\cos\alpha = \hat{\mathbf{n}} \cdot \hat{\mathbf{i}}, \qquad \cos\beta = \hat{\mathbf{n}} \cdot \hat{\mathbf{j}}, \qquad \cos\gamma = \hat{\mathbf{n}} \cdot \hat{\mathbf{k}}. \tag{8.58}$$

Now we are able to formulate the desired definition.

Def. 6 Consider a continuous vector field $\mathbf{A}(\mathbf{r})$ defined on the surface (S). The surface integral of the second type

$$\iint\limits_{(S)} \mathbf{A} \cdot d\mathbf{S} = \iint\limits_{(S)} \mathbf{A} \cdot \mathbf{n}\, dS \tag{8.59}$$

is a universal (if it exists, of course) limit of the integral sum

$$\sigma = \sum_{i=1}^{n} \mathbf{A}_i \cdot \hat{\mathbf{n}}_i\, S_i\,, \tag{8.60}$$

where both vectors must be taken in the same point $M_i \in (S_i)$: $\mathbf{A}_i = \mathbf{A}(M_i)$ and $\hat{\mathbf{n}}_i = \hat{\mathbf{n}}(M_i)$. Other notations here are the same as in the case of the surface integral of the first type, that is

$$(S) = \bigcup_{i=1}^{n} (S_i)\,, \qquad \lambda = \max\{\text{diam}\ (S_i)\}$$

and the integral corresponds to the limit $\lambda \to 0$.

By construction, the surface integral of the second type equals to the following double integral over the figure (G) in the uv-plane:

$$\iint\limits_{(S)} \mathbf{A} \cdot d\mathbf{S} = \iint\limits_{(G)} \left(A_x \cos\alpha + A_y \cos\beta + A_z \cos\gamma\right) \sqrt{g}\, dudv. \tag{8.61}$$

Observation One can prove that the surface integral exists if $\mathbf{A}(u, v)$ and $\mathbf{r}(u, v)$ are smooth functions and the surface (S) has finite area. The sign of the integral changes if we change the orientation of the surface to the opposite one $\hat{\mathbf{n}} \to -\hat{\mathbf{n}}$.

Exercise 9   Calculate the mass of the hemisphere

$$x^2 + y^2 + z^2 = a^2, \qquad z > 0.$$

if its surface mass density at the point $P(x, y, z)$ equals $\frac{\rho z}{a}$.

Exercise 10 *(i)* Calculate the moment of inertia with respect to its axis of symmetry z of the homogeneous spherical shell of the radius a and surface density ρ_0.
(ii) Calculate the moment of inertia of the sphere of radius R, with respect to the axis of symmetry z, if the density of mass of the sphere depends on the distance from the center as $\rho(r) = \rho_0 \left(\frac{r}{R}\right)^n$. Try to solve the problem in two different ways, including using the solution of the part *(i)* and direct integration.
(iii) Try to obtain the result of the part *(i)* using the result of direct integration in the part *(ii)*.

Exercise 11 Consider the spherical shell of radius R that is homogeneously charged such that the total charge of the sphere is Q. Find the electric potential at an arbitrary point outside and inside the sphere.

Hint. Remember that, since the electric potential is a scalar, the result does not depend on the coordinate system you use. Hence, it is recommended to choose the coordinate system that is the most useful, e.g., settle the center of coordinates at the center of the sphere and put the point of interest at the z-axis.

Chapter 9
Theorems of Green, Stokes, and Gauss

In this chapter, we shall obtain important relations between different types of integrals in $3D$. The generalizations to an arbitrary dimension are beyond the scope of this book since the proofs would require too much space. One can find these general versions of Gauss and Stokes theorems in other books, e.g., in [1] or [2].

9.1 Integral Theorems

Let us first remember that any well-defined multiple integral can usually be calculated by reducing to the consequent ordinary definite integrals. Consider, for example, the double integral over the region $(S) \subset \mathbb{R}^2$. We assume that (S) is restricted by the lines

$$\begin{cases} x = a \\ x = b \end{cases} \quad \text{and} \quad \begin{cases} y = \alpha(x) \\ y = \beta(x) \end{cases}, \tag{9.1}$$

where $\alpha(x)$ and $\beta(x)$ are continuous functions. Furthermore, we may suppose that the integrand function $f(x, y)$ is continuous. Then

$$\iint_{(S)} f(x, y) \, dS = \int_a^b dx \int_{\alpha(x)}^{\beta(x)} dy \cdot f(x, y). \tag{9.2}$$

In a more general case, when figure (S) cannot be described by (9.1), this figure should be divided into (S_i) sub-figures of the type (9.1). After that, one can calculate the double integral using the formula (9.2) and additivity of the double integral,

$$\iint_{(S)} = \sum_i \iint_{(S_i)}.$$

© Springer Nature Switzerland AG 2019
I. L. Shapiro, *A Primer in Tensor Analysis and Relativity*, Undergraduate
Lecture Notes in Physics, https://doi.org/10.1007/978-3-030-26895-4_9

Def. 1 We define the positively oriented contour (∂S^+) restricting the figure (S) on the $2D$ plane such that moving along (∂S^+) in the positive direction, one always has the internal part of (S) on the left.

Observation This definition can be easily generalized to the case of the $2D$ surface (S) in the $3D$ space. To this end, one has to fix the orientation $\hat{\mathbf{n}}$ of (S) and then Def. 1 applies directly, by looking from the end of the normal vector $\hat{\mathbf{n}}$.

Now we are in a position to prove the following important statement.

Theorem (Green's formula). *For the surface* (S) *dividable to a finite number of pieces* (S_i) *of the form (9.1), and for the functions* $P(x, y)$, $Q(x, y)$ *that are continuous on* (S) *together with their first derivatives*

$$P'_x, \quad P'_y, \quad Q'_x, \quad Q'_y,$$

the following relation between the double- and second-type curvilinear integrals holds:

$$\iint\limits_{(S)} \left(\frac{\partial Q}{\partial x} - \frac{\partial P}{\partial y} \right) dx dy = \oint\limits_{(\partial S^+)} P\, dx + Q\, dy. \tag{9.3}$$

Proof Consider the region (S) shown in the figure, and take

$$\iint\limits_{(S)} -\frac{\partial P}{\partial y}\, dx dy = -\int_a^b dx \int_{\alpha(x)}^{\beta(x)} \frac{\partial P}{\partial y}\, dy. \tag{9.4}$$

Using the Newton–Leibnitz formula, we transform (9.4) into the form

$$-\int_a^b [P(x, \beta(x)) - P(x, \alpha(x))]\, dx = -\underbrace{\int Pdx}_{(23)} + \underbrace{\int Pdx}_{(14)}$$

$$== \underbrace{\int Pdx}_{(32)} + \underbrace{\int Pdx}_{(14)} = \underbrace{\oint Pdx}_{(14321)} = \underbrace{\oint Pdx}_{(\partial S^+)}. \tag{9.5}$$

In the case when the figure (S) does not have the form presented in Fig. 9.1, it can be cut into a finite number of parts; each of them should have a required geometric form.

Exercise 1 Prove the Q-part of the Green's formula.

   Hint. Consider the region restricted by

$$\begin{cases} y = a \\ y = b \end{cases} \quad \text{and} \quad \begin{cases} x = \alpha(y) \\ x = \beta(y) \end{cases}.$$

Fig. 9.1 Illustration to demonstrating (9.5) of the Green's formula

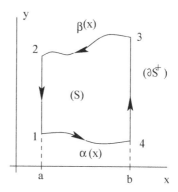

Observation One can use the Green's formula to calculate the area of the surface. To this end, one can take $Q = x$, $P = 0$ or $Q = 0$, $P = -y$, or $Q = x/2$, $P = -y/2$, etc. Then

$$S = \oint_{(\partial S+)} x\,dy = -\oint_{(\partial S+)} y\,dx = \frac{1}{2}\oint_{(\partial S+)} x\,dy - y\,dx. \qquad (9.6)$$

Exercise 2 Calculate, using the above formula, the areas of a square and of a circle.

Let us now consider a more general statement. Consider the oriented region of the smooth surface (S) and the right-oriented contour (∂S^+) that bounds this region. We assume that $x^i = x^i(u, v)$, where $x^i = (x, y, z)$, so that the point $(u, v) \in (G)$ and $(G) \to (S)$ is the mapping of the figures (S) and (G), and also that

$$\left(\partial G^+\right) \longrightarrow \left(\partial S^+\right)$$

is a corresponding mapping for their boundaries.

Stokes's Theorem. For any continuous vector function $\mathbf{F}(\mathbf{r})$ with continuous partial derivatives $\partial_i F_j$, the following relation holds:

$$\iint_{(S)} \mathrm{rot}\,\mathbf{F} \cdot d\mathbf{S} = \oint_{(\partial S+)} \mathbf{F} \cdot d\mathbf{r}. \qquad (9.7)$$

The last integral in (9.7) is called *circulation* of the vector field $\mathbf{F}$ over the closed oriented path (∂S^+).

Proof As before, let us denote

$$\mathbf{F}(u, v) = \mathbf{F}(x(u, v), \ y(u, v), \ z(u, v)).$$

It proves useful to extend the plane with the coordinates u and v into 3D, by introducing the third coordinate w with the unitary basis vector $\hat{\mathbf{n}}_w$ orthogonal to the $\hat{\mathbf{n}}_u$ and $\hat{\mathbf{n}}_v$. Then, even for the region outside the surface (S), one can write

$$\mathbf{F} = \mathbf{F}(x^i) = \mathbf{F}(x^i(u, v, w)).$$

The surface (S) corresponds to the value $w = 0$. The operation of introducing the third coordinate may be regarded as a continuation of the coordinate system $\{u, v\}$ to the region of space outside the initial surface.[1]

At any point of the space

$$\mathbf{F} = \mathbf{F}(u, v, w) = F^u \hat{\mathbf{n}}_u + F^v \hat{\mathbf{n}}_v + F^w \hat{\mathbf{n}}_w. \tag{9.8}$$

The components of the vectors $\mathbf{F}$, rot $\mathbf{F}$, $d\mathbf{S}$, and $d\mathbf{r}$ transform in a standard way, such that the products

$$\text{rot}\,\mathbf{F} \cdot d\mathbf{S} = \varepsilon^{ijk}(dS_i)\,\partial_j F_k \quad \text{and} \quad \mathbf{F} \cdot d\mathbf{r} = F_i dx^i$$

are scalars with respect to the coordinate transformation from (x, y, z) to (u, v, w). Therefore, relation (9.7) can be proved in any particular coordinate system. If it is correct in one coordinate system, it is correct for any other coordinates too. In what follows we shall use the coordinates (u, v, w). Then

$$\oint_{(\partial S^+)} \mathbf{F} \cdot d\mathbf{r} = \oint_{(\partial G^+)} \mathbf{F} \cdot d\mathbf{r} = \oint_{(\partial G^+)} F_u du + F_v dv, \tag{9.9}$$

because the contour (∂G^+) lies in the (u, v) plane.

Further, since the figure (G) belongs to the (u, v) plane, the oriented area vector $d\mathbf{G}$ is parallel to the $\hat{\mathbf{n}}_w$ axis. Therefore, due to Eq. (9.8),

$$(d\mathbf{G})_u = (d\mathbf{G}_v) = 0. \tag{9.10}$$

Then

$$\iint_{(S)} \text{rot}\,\mathbf{F} \cdot d\mathbf{S} = \iint_{(G)} \text{rot}\,\mathbf{F} \cdot d\mathbf{G} = \int_{(G)} (\text{rot}\,\mathbf{F})_w\, dG^w$$

$$= \iint_{(G)} \left(\frac{\partial F_v}{\partial u} - \frac{\partial F_u}{\partial v} \right) du\,dv. \tag{9.11}$$

It is easy to see that the remaining relation

---

[1] As an example we may take internal coordinates φ, θ on the sphere. We can introduce one more coordinate r and the new coordinates φ, θ, r will of course be the spherical ones, which cover all the 3D space. The only one difference is that the new coordinate r, in this example, is not zero on the surface $r = R$. But this can be easily corrected by the redefinition $\rho = r - R$, where $-R < \rho < \infty$.

Fig. 9.2 Two possible
curves of integration
between the points A and B

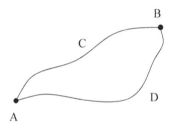

$$\oint_{(\partial G^+)} F_u du + F_v dv = \iint_{(G)} \left(\frac{\partial F_v}{\partial u} - \frac{\partial F_u}{\partial v} \right) du dv \qquad (9.12)$$

is nothing but the Green formula, which has already been proven in the previous theorem. The proof is complete.

There is an important consequence of the Stokes theorem.

Theorem *For the* $\mathbf{F}(\mathbf{r}) = \operatorname{grad} U(\mathbf{r})$, *where* $U(\mathbf{r})$ *is a smooth function of coordinates, the curvilinear integral between two points* A *and* B *doesn't depend on the path* (AB) *and is equal to the difference*

$$\int_{(AB)} \mathbf{F} \cdot d\mathbf{r} = U(B) - U(A). \qquad (9.13)$$

Proof Consider a closed simple path (Γ) that can be always regarded as (∂S^+) for some oriented surface (S). Then, according to the Stokes theorem

$$\oint_{(\Gamma)} \mathbf{F} \cdot d\mathbf{r} = \iint_{(S)} \operatorname{rot} \mathbf{F} \cdot d\mathbf{S}. \qquad (9.14)$$

If $\mathbf{F} = \operatorname{grad} U$, then $\operatorname{rot} \mathbf{F} = 0$ and hence $\oint_{(\Gamma)} \mathbf{F} \cdot d\mathbf{r} = 0$ (Fig. 9.2).

Let us now consider two different paths between the points A and B: (ACB) and (ADB). As we have just seen, $\oint_{(ABCDA)} \mathbf{F} \cdot d\mathbf{r} = 0$. But this also means that

$$\int_{(ACB)} \mathbf{F} \cdot d\mathbf{r} - \int_{(ADB)} \mathbf{F} \cdot d\mathbf{r} = 0, \qquad (9.15)$$

which completes the proof.

Observation 1 For the $2D$ case, this result can be seen just from the Green's formula.

Observation 2 A useful criterion of $\mathbf{F}$ being $\operatorname{grad} U$ is

$$\frac{\partial F_x}{\partial z} = \frac{\partial F_z}{\partial x},$$

and the same for any couple of partial derivatives. The reason is that

$$\frac{\partial F_z}{\partial x} = \frac{\partial^2 U}{\partial x \partial z} = \frac{\partial^2 U}{\partial z \partial x} = \frac{\partial F_x}{\partial z},$$

provided the second derivatives are continuous functions. In total, there are three such relations

$$\frac{\partial F_z}{\partial x} = \frac{\partial F_x}{\partial z}, \qquad \frac{\partial F_y}{\partial x} = \frac{\partial F_x}{\partial y}, \qquad \frac{\partial F_z}{\partial y} = \frac{\partial F_y}{\partial z}. \qquad (9.16)$$

The last of the statements we prove in this section is the

Gauss–Ostrogradsky Theorem.   Consider a $3D$ figure $(V) \subset R^3$ and also define (∂V^+) to be the externally oriented boundary of (V). Consider a vector field $\mathbf{E}(\mathbf{r})$ defined on (V) and on its boundary (∂V^+) and suppose that the components of this vector field are continuous, as are their partial derivatives

$$\frac{\partial E^x}{\partial x}, \qquad \frac{\partial E^y}{\partial y}, \qquad \frac{\partial E^z}{\partial z}.$$

Then these components satisfy the following integral relation:

$$\oiint_{(\partial V^+)} \mathbf{E} \cdot d\mathbf{S} = \iiint_{(V)} \operatorname{div} \mathbf{E} \, dV. \qquad (9.17)$$

Proof Consider z-part of Eq. (9.17), namely,

$$\iiint_{(V)} \partial_z E_z \, dV.$$

Let's suppose, similar to what as we did in the proof of the Green's formula, that (V) is restricted by the surfaces $z = \alpha(x, y)$, $z = \beta(x, y)$ and by some closed cylindric surface with the projection (∂S) (boundary of the figure (S)) on the xy-plane. Then

$$\iiint_{(V)} \frac{\partial E_z}{\partial z} dV = \iint_{(S)} dS \int_{\alpha(x,y)}^{\beta(x,y)} dz \frac{\partial E_z(x, y, z)}{\partial z} \qquad (9.18)$$

$$= \iint_{(S)} dx\, dy \left[-E_z(x, y, \alpha(x, y)) + E_z(x, y, \beta(x, y)) \right] = \oiint_{(\partial V^+)} E_z dS^z.$$

In the same way, one can prove the relations for other two parts of Eq. (9.17), which completes the proof.

9.2 Div, grad, and rot from the New Perspective

Using the Stokes and Gauss–Ostrogradsky theorems, one can give more geometric definitions of divergence and rotation of a vector. Suppose we want to know the projection of $\text{rot}\,\mathbf{F}$ on the direction of the unit vector $\hat{\mathbf{n}}$. Taking an infinitesimal surface vector $d\mathbf{S} = dS\,\hat{\mathbf{n}}$, due to the continuity of all components of the vector $\text{rot}\,\mathbf{F}$, we get

$$\hat{\mathbf{n}} \cdot \text{rot}\,\mathbf{F} = \lim_{dS \to 0} \frac{d}{dS} \oint_{(\partial S^+)} \mathbf{F} \cdot d\mathbf{r}, \tag{9.19}$$

where (∂S^+) is an oriented borderline of (S).

The last relation provides a necessary food for the geometric intuition of what $\text{rot}\,\mathbf{F}$ is. One can say that the presence of nonzero $\text{rot}\,\mathbf{F}$ in a given point means that the vector field forms a vortex with nonzero density at the given point, which can be seen from its circulation.

In the case of $\text{div}\,\mathbf{A}$, one can make similar considerations and arrive at the following relation:

$$\text{div}\,\mathbf{A} = \lim_{V \to 0} \frac{1}{V} \oiint_{(\partial V^+)} \mathbf{A} \cdot d\mathbf{S}, \tag{9.20}$$

where the limit $V \to 0$ must be taken in such a way that also $\lambda \to 0$, where $\lambda = \text{diam}\,(V)$. This formula indicates to the relation between $\text{div}\,\mathbf{A}$ at some point and the presence of the source for this field at this point. If such a source is absent, $\text{div}\,\mathbf{A} = 0$.

The last two formulas make the geometric sense of div and rot operators explicit. In order to complete the story, let us explain the geometric sense of the remaining differential operator grad. For a scalar field $\varphi(\mathbf{r})$, the

$$\text{grad}\,\varphi(\mathbf{r}) = \mathbf{e}^i\,\varphi_{,i} = \hat{\mathbf{i}}\varphi_{,x} + \hat{\mathbf{j}}\varphi_{,y} + \hat{\mathbf{k}}\varphi_{,z}. \tag{9.21}$$

Let us multiply this vector by an arbitrary unitary vector $\hat{\mathbf{n}}$. Such a vector can always be seen as a tangent vector of a smooth curve parameterized by a natural parameter l, then

$$\frac{d\varphi}{dl} = \frac{\partial\varphi}{\partial x}\frac{dx}{dl} + \frac{\partial\varphi}{\partial y}\frac{dy}{dl} + \frac{\partial\varphi}{\partial y}\frac{dy}{dl} = \hat{\mathbf{n}} \cdot \text{grad}\,\varphi. \tag{9.22}$$

Using the last formula, by setting different directions for the unitary vector $\hat{\mathbf{n}}$, we can explore the change of the field φ in different directions. From Eq. (9.22), it follows that grad φ is a vector field that satisfies the following conditions:

(1) It points to the direction of the maximal growth of φ.

(2) It has the absolute value equal to the maximal derivative of $\varphi(\mathbf{r})$ with respect to the natural parameter l along the corresponding curve.

References

1. A.T. Fomenko, S.P. Novikov, B.A. Dubrovin, *Modern Geometry-Methods and Applications, Part I: The Geometry of Surfaces, Transformation Groups, and Fields* (Springer, Berlin, 1992)
2. W. Rudin, *Principles of Mathematical Analysis*, 3 edn. (McGraw-Hill, New York, 1976)

Chapter 10
Solutions to the Exercises from Part 1

In this chapter, we collect brief solutions of selected exercises of the first part of the book. The few exercises that remain without solutions are either too simple, or (in a very few cases) a little bit more difficult—those are left as small challenges to the reader.

10.1 Solutions to Exercises from Chap. 1

Exercise 1 *(a)* Number 2 has no positive opposite number.
(b) Consider $z = x + iy$, $x < 0$. The number $-z = -x - iy$ has positive real part, and hence on the subset with $x < 0$ the number z has no opposite.
(c) $ix_1 + ix_2 = ix_2 + ix_1$, etc.

Exercise 2 *(a)* $\mathbf{0} \cdot c + 0 \cdot \mathbf{a}_1 + \cdots + 0 \cdot \mathbf{a}_n = 0$ even it $c \neq 0$.
(b) Suppose $\mathbf{a}_1 = \mathbf{i} + \mathbf{k}$, $\mathbf{a}_2 = \mathbf{i} - \mathbf{k}$, $\mathbf{a}_3 = \mathbf{i} - 3\mathbf{k}$. Consider

$$\alpha_1\mathbf{a}_1 + \alpha_2\mathbf{a}_2 + \alpha_3\mathbf{a}_3 = \mathbf{i}(\alpha_1 + \alpha_2 + \alpha_3) + \mathbf{k}(\alpha_1 - \alpha_2 - 3\alpha_3) = 0,$$
$$\text{Then} \qquad \alpha_1 + \alpha_2 + \alpha_3 = 0, \qquad \alpha_1 - \alpha_2 - 3\alpha_3 = 0.$$

Taking $\alpha_1 = \alpha_3$, $\alpha_2 = -2\alpha_3$ with $\forall$ $\alpha_3 \neq 0$ we get zero linear combination with nonzero coefficients $\alpha_{1,2,3}$.

Exercise 3 Assume that $\mathbf{a}_1 = \mathbf{i} + \mathbf{k}$, $\mathbf{a}_2 = \mathbf{i} - \mathbf{k}$. Consider

$$\alpha_1\mathbf{a}_1 + \alpha_2\mathbf{a}_2 = \mathbf{i}(\alpha_1 + \alpha_2) + \mathbf{k}(\alpha_1 - \alpha_2) = 0,$$
$$\text{therefore} \qquad \alpha_1 + \alpha_2 = 0, \quad \alpha_1 - \alpha_2 = 0,$$

with the unique solution $\alpha_{1,2} = 0$.

© Springer Nature Switzerland AG 2019
I. L. Shapiro, *A Primer in Tensor Analysis and Relativity*, Undergraduate
Lecture Notes in Physics, https://doi.org/10.1007/978-3-030-26895-4_10

Exercise 4 Consider $\chi = (\mathbf{r}, \mathbf{v})$. Then, for $\chi_1 = (\mathbf{r}_1, \mathbf{v}_1)$, $\chi_2 = (\mathbf{r}_2, \mathbf{v}_2)$, and $\chi_3 = (\mathbf{r}_3, \mathbf{v}_3)$, we have

$$\chi_1 + \chi_2 = \chi_2 + \chi_1 = (\mathbf{r}_1 + \mathbf{r}_2, \mathbf{v}_1 + \mathbf{v}_2),$$
$$\text{also}\quad (\chi_1 + \chi_2) + \chi_3 = \chi_1 + (\chi_2 + \chi_3).$$

Furthermore, $\chi + 0 = \chi$, where $0 = (\mathbf{0}, \mathbf{0})$; $\chi + (-\chi) = 0$, where $-\chi = (-\mathbf{r}, -\mathbf{v})$; $\chi \cdot 1 = (\mathbf{r} \cdot 1, \mathbf{v} \cdot 1) = \chi$; $e(k\chi) = (e \cdot k) \cdot \chi$, since this is true for $\mathbf{r}$ and $\mathbf{v}$. And finally, $k(\chi_1 + \chi_2) = k\chi_1 + k\chi_2$.

Exercise 5 $\mathbf{e}_{i'} = \wedge_{i'}^j \mathbf{e}_j$, $\mathbf{e}_i = (\wedge^{-1})_i^k \mathbf{e}_{k'}$, $a^{i'} = (\wedge^{-1})_k^{i'} a^k$, $a^j = \wedge_{i'}^j a_{i'}$. Consider

$$\mathbf{e}_{i'} \cdot a^{i'} = \mathbf{e}_j \wedge_{i'}^j (\wedge^{-1})_k^{i'} a_k = \mathbf{e}_i \delta_k^j a^k = \mathbf{e}_j a^j,$$

the same for the second product $\mathbf{e}_i \cdot a^i = \mathbf{e}_{k'} a^{k'}$.

Exercise 6

(a) $x' = 5x$ and $y = x^2 = \left(\dfrac{1}{5} x'\right)^2 = y'(x')$.

(b) $x' = ax$, that is, a linear homogeneous transformation, hence

$$y(x) = y\left(\frac{x'}{a}\right) = \frac{(x')^2}{a^2} = y'(x').$$

(c) $y(x) = y\left(\dfrac{x'}{a}\right) = \dfrac{(x')^2}{a^2} = y'(x')$.

Example, if $y(x) = \cot g\, x$, then $y'(x') = \cot g\left(\dfrac{x'}{a}\right)$.

Exercise 7 The numbers T, p, ρ do not form a contravariant vector because they transform as scalars. For instance, $T'(\mathbf{r}') = T(\mathbf{r})$, which is not the transformation rule of a vector component.

Exercise 8a According to Eq. (1.17), we have $\varphi(x') = \varphi(x) + \xi^i \partial_i \varphi$, while Eq. (1.18) gives $\varphi'(x) = \varphi(x) - \xi^i \partial_i \varphi = \varphi + \delta\varphi$. Then we can claim that $\varphi'(x') = \varphi(x') + \delta\varphi$ because the next order terms should be ignored in the linear approximation. Thus, $\varphi'(x') = \varphi(x) + \xi^i \partial_i \varphi - \xi^i \partial_i \varphi = \varphi(x)$.

Exercise 8b* In the case of $\varphi(x')$, we need just the Taylor expansion,

$$\varphi(x') = \varphi(x) + \xi^i \partial_i \varphi + \frac{1}{2} \xi^i \xi^j \partial_i \partial_j \varphi + \cdots$$

In the case of $\varphi'(x)$ we use

$$\varphi'(x) = \varphi'(x'^i - x^i) = \varphi'(x') - \xi^i \frac{\partial \varphi'(x')}{\partial x'^i} + \frac{1}{2} \xi^i \xi^j \frac{\partial^2 \varphi'(x')}{\partial x'^i \partial x'^j} + \cdots$$
$$= \varphi'(x') - \xi^i \frac{\partial \varphi(x)}{\partial x^j} \frac{\partial x^j}{\partial x'^i} + \frac{1}{2} \xi^i \xi^j \frac{\partial^2 \varphi(x)}{\partial x^i \partial x^j},$$

because the next order terms are beyond the $\mathcal{O}(\xi^2)$-approximation. In the same way, we get

$$\frac{\partial x^j}{\partial x'^i} = \frac{\partial}{\partial x'^i}(x'^j - \xi^j) = \delta_i^j - \partial_i\xi^j$$

and hence

$$\varphi'(x) = \varphi(x) - \xi^i\partial_j\varphi \cdot (\delta_i^j - \partial_i\xi^j) + \frac{1}{2}\xi^i\xi^j\frac{\partial^2\varphi(x)}{\partial x^i\partial x^j}$$

$$= \varphi(x) - \xi^i\partial_i\varphi + \xi^i(\partial_i\xi^j)(\partial_j\varphi) + \frac{1}{2}\xi^i\xi^j\partial_i\partial_j\varphi.$$

In the case of a constant ξ^i, the terms with its derivatives vanish, of course. Let us now check that the definition of scalar is satisfied. We have

$$\varphi'(x') = \varphi'(x + \xi) = \varphi'(x) + \xi^i\partial_i\varphi'(x) + \frac{1}{2}\xi^i\xi^j\partial_i\partial_j\varphi$$

$$= \varphi(x) - \xi^i\partial_i\varphi + \xi^i(\partial_i\xi^j)\partial_j\varphi + \frac{1}{2}\xi^i\xi^j\partial_i\partial_j\varphi$$

$$+ \xi^i\left[\partial_i\varphi(x) - \xi^j\partial_i\partial_j\varphi - (\partial_i\xi^j)\partial_j\varphi\right] + \frac{1}{2}\xi^i\xi^j\partial_i\partial_j\varphi = \varphi(x),$$

in the $\mathcal{O}(\xi^2)$-approximation.

Exercise 9

$$a^i(x')b^j(x') = \frac{\partial x^{i'}}{\partial x^l}a^l(x)\frac{\partial x^{j'}}{\partial x^k}b^k(x)$$

is the tensor transformation rule.

Exercise 10

$$x' = \frac{x+y}{\sqrt{2}} + 3, \quad y' = \frac{x-y}{\sqrt{2}} - 5.$$

The change of initial point $(0, 0) \to (3, -5)$ does not have relation to the change of basis. Then

$$x'\hat{i}' + y'\hat{j}' = \left(\frac{x+y}{\sqrt{2}}\right)\hat{i}' + \left(\frac{x-y}{\sqrt{2}}\right)\hat{j}'$$

$$= \frac{x}{\sqrt{2}}(\hat{i}' + \hat{j}') + \frac{y}{\sqrt{2}}(\hat{i}' - \hat{j}') = x\hat{i} + y\hat{j}.$$

Thus, $\hat{i} = \frac{1}{\sqrt{2}}(\hat{i}' + \hat{j}')$, $\hat{j} = \frac{1}{\sqrt{2}}(\hat{i}' - \hat{j}')$. Obviously, $\hat{i}^2 = \hat{j}^2 = 1$ and $\hat{i} \cdot \hat{j} = 0$.

Exercise 11 The possible transformation includes rotation, inversion of axes (parity transformation), and change of the central point, like in Exercise 9. In $2D$ parity transformation can be reduced to rotation.

Exercise 12

$$\mathbf{r} = x^i \mathbf{e}_i = \frac{\partial x^i}{\partial X^a} \cdot X^a \mathbf{e}_i = \hat{n}_a X^a,$$

hence $\hat{n}_a = \frac{\partial x^i}{\partial X^a} \mathbf{e}_i$. Similarly, $\mathbf{e}_i = \frac{\partial X^a}{\partial x^i} \hat{n}_a$.

Exercise 13

$$\mathbf{e}_i = \frac{\partial X^a}{\partial x^i} \hat{n}_a \quad \text{and} \quad \mathbf{e}_j = \frac{\partial x^j}{\partial X^b} \hat{n}_b.$$

Then the conditions are

$$\mathbf{e}_i = \frac{\partial X^a}{\partial x^i} \hat{n}_a = \frac{\partial x^i}{\partial X^a} \hat{n}^a \Rightarrow \left\| \frac{\partial X^a}{\partial x^i} \right\| = \left\| \frac{\partial x^i}{\partial X^a} \right\|^T.$$

This means the transformation is orthogonal, and hence A reduces to rotations and parity transformations. Thus, x^i are Cartesian coordinates.

Exercise 14 $\mathbf{e}_1 = \hat{i} + \hat{j}, \quad \mathbf{e}_2 = \hat{i}$. Thus,

$$\mathbf{e}^1 \cdot \mathbf{e}_2 = 0 \Rightarrow \mathbf{e}^1 = \alpha \hat{j} \quad \text{and} \quad \alpha \hat{j} \cdot (\hat{i} + \hat{j}) = 1,$$

which gives $\alpha = 1$. Furthermore, $\mathbf{e}^2 \cdot \mathbf{e}_1 = 0 \Rightarrow \mathbf{e}^2 = \beta(\hat{j} - \hat{i})$ and $\beta(\hat{j} - \hat{i}) \cdot \hat{i} = 1$ gives $\beta = 1 \Rightarrow \mathbf{e}^1 = \hat{j}, \mathbf{e}^2 = \hat{i} - \hat{j}$.

Exercise 15 The first part has been already done in Exercise 9. The answer to the second question is negative, because an arbitrary tensor $c_{ij}(x)$ has nine components and $A_i(x)$ and $B_j(x)$ together only have six.

Exercise 16 It directly follows from Def. 15.

Exercise 17

$$\delta^{i'}_{j'} = \frac{\partial x^{i'}}{\partial x^k} \frac{\partial x^l}{\partial x^{j'}} \delta^k_l = \frac{\partial x^{i'}}{\partial x^{j'}}.$$

Exercise 18

$$A^{i'}(x') A_{j'}(x') = \frac{\partial x^{i'}}{\partial x^k} A^k(x) \frac{\partial x^l}{\partial x^{j'}} B_l(x) = \frac{\partial x^{i'}}{\partial x^k} \frac{\partial x^l}{\partial x^{j'}} A^k(x) B_l(x).$$

Exercise 19 For an arbitrary $\mathbf{a}(x) = a^i(x) \mathbf{e}_i$ we have

$$a^{j'}(x') = \frac{\partial x^{j'}}{\partial x^i} a^i(x), \quad \mathbf{e}_{j'}(x') = \mathbf{e}_k(x)\frac{\partial x^k}{\partial x^{j'}}.$$

Then

$$a^{j'}(x')\mathbf{e}_{j'}(x') = \frac{\partial x^{j'}}{\partial x^i} a^i(x)\mathbf{e}_k(x)\frac{\partial x^k}{\partial x^{j'}} = \delta_l^k a^i(x)\mathbf{e}_k(x) = a^i(x)\mathbf{e}_i(x).$$

Exercise 21

$$\hat{\wedge}_x(\beta) = \begin{pmatrix} 1 & 0 & 0 \\ 0 & \cos\beta & -\sin\beta \\ 0 & \sin\beta & \cos\beta \end{pmatrix}, \qquad \hat{\wedge}_y(\gamma) = \begin{pmatrix} \cos\gamma & 0 & -\sin\gamma \\ 0 & 1 & 0 \\ \sin\gamma & 0 & \cos\gamma \end{pmatrix}.$$

Exercise 24

$$(i) \quad \hat{\pi}_y = \begin{pmatrix} 1 & 0 & 0 \\ 0 & -1 & 0 \\ 0 & 0 & 1 \end{pmatrix}, \quad \hat{\pi}_x = \begin{pmatrix} -1 & 0 & 0 \\ 0 & 1 & 0 \\ 0 & 0 & 1 \end{pmatrix}.$$

(ii) $\hat{\wedge}_y\left(\frac{\pi}{2}\right)$; (iii) Inversion of the corresponding axes.

Exercise 25 $\mathbf{e}_k = \wedge_k^i \hat{\mathbf{n}}_i = \wedge_k^1 \hat{\mathbf{n}}_1 + \wedge_k^2 \hat{\mathbf{n}}_2 + \wedge_k^3 \hat{\mathbf{n}}_3.$

Exercise 26 $\hat{\mathbf{n}}_k' = R_k^i \hat{\mathbf{n}}_i$ and also $\hat{\mathbf{n}}'' = (R^{-1})_j^i \hat{\mathbf{n}}^j$. Since $\hat{\mathbf{n}}_k' = \hat{\mathbf{n}}^k$ and $\hat{\mathbf{n}}_i = \hat{\mathbf{n}}^i$, we also have $(R^{-1})_j^i = (R^T)_j^i$.

Exercise 27

$$(i) \quad \begin{pmatrix} x \\ y \\ z \end{pmatrix} = \hat{\wedge}_z(\alpha)\begin{pmatrix} x' \\ y' \\ z' \end{pmatrix}, \quad \begin{pmatrix} x' \\ y' \\ z' \end{pmatrix} = \hat{\wedge}_z(-\alpha)\begin{pmatrix} x \\ y \\ z \end{pmatrix}.$$

$$(iii) \quad a_{x'} = a_x\frac{\partial x}{\partial x'} + a_y\frac{\partial y}{\partial x'} + a_z\frac{\partial z}{\partial x'} = a_x\cos\alpha + a_y\sin\alpha = \cos\alpha + 2\sin\alpha,$$

$$a_{y'} = a_x\frac{\partial x}{\partial y'} + a_y\frac{\partial y}{\partial y'} + a_z\frac{\partial z}{\partial y'} = -a_x\sin\alpha + a_y\cos\alpha = -\sin\alpha + 2\cos\alpha,$$

$$a_{z'} = a_z = 3.$$

$$(iv) \quad b_{x'y'} = b_{xx}\frac{\partial x}{\partial x'}\frac{\partial x}{\partial y'} + b_{xy}\frac{\partial x}{\partial x'}\frac{\partial y}{\partial y'} + b_{yx}\frac{\partial y}{\partial x'}\frac{\partial x}{\partial y'} + b_{yy}\frac{\partial y}{\partial x'}\frac{\partial y}{\partial y'}$$

$$= -b_{xx}\cos\alpha\sin\alpha + b_{xy}\cos^2\alpha - b_{yx}\sin^2\alpha + b_{yy}\sin\alpha\cos\alpha$$

$$= \cos^2\alpha + 3\sin\alpha\cos\alpha.$$

Others components can be obtained in a similar way.

Exercise 28

(i) $t = t' \cosh \psi + x' \sinh \psi$, $x = -t' \sinh \psi + x' \cosh \psi$,

(ii) Change $\psi \to -\psi$, $t' = t \cosh \psi - x \sinh \psi$, $x = -t \sinh \psi + x \cosh \psi$,

(iv) $a_{x'} = \dfrac{\partial x}{\partial x'} a_x + \dfrac{\partial t}{\partial x'} a_t = a_x \cosh \psi + a_t \sinh \psi$

$a_{t'} = \dfrac{\partial x}{\partial t'} a_x + \dfrac{\partial t}{\partial t'} a_t = a_x \sinh \psi + a_t \cosh \psi.$

10.2 Solutions to Exercises from Chap. 2

Exercise 4

$$a^{i'}(x')b^{k}{}_{j'}(x') = \frac{\partial x^{i'}}{\partial x^m} a^m(x) \frac{\partial x^{k'}}{\partial x^l} \frac{\partial x^n}{\partial x^{j'}} b^l{}_{.n}(x).$$

Exercise 6

$$b^{i'}(x') = \frac{\partial x^{i'}}{\partial x^l} b^l(x), \qquad a_{i'}(x') = \frac{\partial x^k}{\partial x^{i'}} a_k(x).$$

Then

$$a_{i'}(x')b^{i'}(x') = \frac{\partial x^{i'}}{\partial x^l} \frac{\partial x^k}{\partial x^{i'}} b^l(x)a_k(x) = \delta^k_l b^l(x)a_k(x) = b^k(x)a_k(x).$$

Exercise 7

$$g_{i'j'} z^{i'} z^{j'} = g_{kl} \frac{\partial x^k}{\partial x^{i'}} \frac{\partial x^l}{\partial x^{j'}} z^m \frac{\partial x^{i'}}{\partial x^m} z^n \frac{\partial x^{j'}}{\partial x^n} = g_{kl} z^m z^n \frac{\partial x^k}{\partial x^m} \frac{\partial x^l}{\partial x^n} = g_{kl} z^k z^l.$$

Exercise 8

$$S^2_{12} = g_{ab}(X^a - Y^a)(X^b - Y^b) = (\mathbf{e}_a, \mathbf{e}_b)(X^a - Y^a)(X^b - Y^b)$$

$$= \frac{\partial x^i}{\partial X^a} \frac{\partial x^j}{\partial X^b} (\mathbf{e}_i, \mathbf{e}_j) \cdot \frac{\partial X^a}{\partial x^k} \frac{\partial X^b}{\partial x^l} (x^k - y^k)(x^l - y^l)$$

$$= \delta^i_k \delta^j_l g_{ij} (x^k - y^k)(x^l - y^l) = g_{ij}(x^i - y^i)(x^j - y^j).$$

Exercise 9 The proof is completely analogous to the one of property **3**.

Exercise 10 Can be easily done starting from the obvious relation $\hat{\mathbf{n}}_a g^{ab} = \hat{\mathbf{n}}^b$, and making coordinate transformation.

Exercise 11 For instance, consider $\mathbf{r} = x_j \mathbf{e}_i (\mathbf{e}^i, \mathbf{e}^j)$ and prove it is $\mathbf{r} = x^i \mathbf{e}_i$.

Exercise 12   Consider $\mathbf{a} = a_x \hat{i} + a_y \hat{j}$, where $\mathbf{e}_1 = 3\hat{i} + \hat{j}$ and $\mathbf{e}_2 = \hat{i} - \hat{j}$.

$$(i)\quad \mathbf{e}_1 + \mathbf{e}_2 = 4\hat{i} \;\Rightarrow\; \hat{i} = \frac{1}{4}(\mathbf{e}_1 + \mathbf{e}_2)$$

$$\mathbf{e}_1 - 3\mathbf{e}_2 = 4\hat{j} \;\Rightarrow\; \hat{j} = \frac{1}{4}(\mathbf{e}_1 - 3\mathbf{e}_2).$$

Then

$$\mathbf{a} = \frac{1}{4}(\mathbf{e}_1 + \mathbf{e}_2)a_x + (\mathbf{e}_1 - 3\mathbf{e}_2)a_y = \mathbf{e}_1 \frac{a_x + a_y}{4} + \mathbf{e}_2 \frac{a_x - 3a_y}{4}. \tag{10.1}$$

$$(ii)\quad \mathbf{e}_1 = \frac{\partial X}{\partial x^1}\hat{i} + \frac{\partial Y}{\partial x^1}\hat{j} \;\Rightarrow\; \frac{\partial X}{\partial x^1} = 3, \quad \frac{\partial Y}{\partial x^1} = 1,$$

$$\mathbf{e}_2 = \hat{i} - \hat{j} = \frac{\partial X}{\partial x^2}\hat{i} + \frac{\partial Y}{\partial x^2}\hat{j} \;\Rightarrow\; \frac{\partial X}{\partial x^2} = 1, \quad \frac{\partial Y}{\partial x^2} = -1.$$

After that, one has to use

$$\begin{pmatrix} \frac{\partial X}{\partial x^1} & \frac{\partial X}{\partial x^2} \\ \frac{\partial Y}{\partial x^1} & \frac{\partial Y}{\partial x^2} \end{pmatrix} \begin{pmatrix} \frac{\partial x^1}{\partial X} & \frac{\partial x^1}{\partial Y} \\ \frac{\partial x^2}{\partial X} & \frac{\partial x^2}{\partial Y} \end{pmatrix} = \begin{pmatrix} 1 & 0 \\ 0 & 1 \end{pmatrix}.$$

Alternatively, from (10.1) we can write $\mathbf{r} = x\hat{i} + y\hat{j} = \frac{x+y}{4}\mathbf{e}_1 + \frac{x-3y}{4}\mathbf{e}_2$, hence $x^1 = \frac{x+y}{4}, x^2 = \frac{x-3y}{4}$, and $\frac{\partial x^1}{\partial x} = \frac{1}{4}, \frac{\partial x^2}{\partial x} = \frac{1}{4}, \frac{\partial x^1}{\partial y} = \frac{1}{4}, \frac{\partial x^2}{\partial y} = -\frac{3}{4}$. It is easy to verify that

$$\begin{pmatrix} 3 & 1 \\ 1 & -1 \end{pmatrix} \begin{pmatrix} \frac{1}{4} & \frac{1}{4} \\ \frac{1}{4} & -\frac{3}{4} \end{pmatrix} = \begin{pmatrix} 1 & 0 \\ 0 & 1 \end{pmatrix}.$$

After that the calculation is quite simple.

10.3   Solutions to Exercises from Chap. 3

Exercise 1 $g^{ij} = (\mathbf{e}^i, \mathbf{e}^j) = (\mathbf{e}^j, \mathbf{e}^i) = g^{ji}$.

Exercise 2   See Exercise 4 of Chap. 2. Also, $a^i a^j = a^j a^i$ because real numbers commute.

Exercise 3 $a^{(i}b^{j)} = \frac{1}{2}(a^i b^j + a^j b^i)$, which is called a symmetrized product.

Exercise 4 $c^{(ij)} = \frac{1}{2}(c^{ij} + c^{ji})$, which is the symmetric part of the tensor c^{ij}.

Exercise 5 $\frac{D(D+1)}{2}$.

Exercise 6* $N_D = D + D(D-1) + \frac{D(D-1)(D-2)}{3!} = \frac{D(D+1)(D+2)}{6}$.

Exercise 7 $c^{[ij]} = \frac{1}{2}(c^{ij} - c^{ji})$, which is the antisymmetric part of the tensor c^{ij}.

Exercise 8 $c^{ij} = c^{(ij)} + c^{[ij]}$. For symmetric $c^{ij} = c^{ji}$ we have $c^{[ij]} = 0$ and for antisymmetric $c^{ij} = -c^{ji}$ we have $c^{(ij)} = 0$. It works for $D \geq 2$.

Exercise 10 For $A^{ij} = -A^{ji}$ we have

$$n_D = \frac{D(D-1)}{2}.$$

For $D = 2$, this gives $n_2 = 1$ and for $D = 3$, $n_3 = 3$. For $A^{ijk} = A^{[ijk]}$ (this means absolute antisymmetry)

$$n_D = \frac{D(D-1)(D-2)}{3!}.$$

Exercise 12 $xa_{ij} + ya_{ji} = 0$ can be recast as

$$x(a_{(ij)} + a_{[ij]}) + y(a_{(ij)} - a_{[ij]}) = (x+y)a_{(ij)} + (x-y)a_{[ij]}.$$

After that the solution is obvious.

Exercise 13* c^{ijk} is symmetric in the first two and antisymmetric in the last two indices. Then

$$c^{ijk} = c^{jik} = -c^{jki} = -c^{kji} = c^{kij} = c^{ikj} = -c^{ijk},$$

hence it is zero.

Exercise 14 See Exercise 6 of Chap. 2. *(iii)* $b^{ij} = b^{[ij]}$.

Exercise 17 *(b)* The result follows from combinatorial arguments. In $E^{a_1 a_2 \cdots a_D}$ with nonzero components all indices are different. Hence, a_1 takes D values, a_2 takes $D-1$ values, etc. Then $E^{a a_2 \cdots a_D} E_{b a_2 \cdots a_D} = x\delta^a_b$, where $Dx = D!$, thus $x = (D-1)!$. repeating the same procedure, and taking antisymmetry into account, finally leads us to Eq. (3.16).

Exercise 19

(i) $(\mathbf{U}, \mathbf{V}, \mathbf{W}) = E_{abc}U^a V^b W^c = E_{cab}U^c V^a W^b = -E_{acb}U^c V^a W^b$
        $= E_{abc}U^c V^a W^b = (\mathbf{V}, \mathbf{W}, \mathbf{U})$.

(ii) $(\mathbf{U}, \mathbf{V}, \mathbf{W}) = E_{abc}U^a V^b W^c$
        $= E_{bca}U^b V^a W^c = -E_{abc}U^b V^a W^c = -(\mathbf{V}, \mathbf{U}, \mathbf{W})$.

Exercise 20

(i) $[[\mathbf{U}, \mathbf{V}], [\mathbf{W}, \mathbf{Y}]] = \hat{\mathbf{n}}^a \varepsilon_{abc} \, \varepsilon^{bed} U_e V_d \cdot \varepsilon^{cfh} W_f Y_h$

$= \hat{\mathbf{n}}^a (\delta_a^d \delta_c^e - \delta_c^d \delta_a^e) U_e V_d W_f Y_h \, \varepsilon^{cfh} = \mathbf{V}(\mathbf{U}, \mathbf{W}, \mathbf{Y}) - \mathbf{U}(\mathbf{V}, \mathbf{W}, \mathbf{Y}).$

(ii) $[\mathbf{U} \times \mathbf{V}] \cdot [\mathbf{W} \times \mathbf{Y}] = \varepsilon_{abc} U^b V^c \cdot \varepsilon^{ade} W_d Y_e$

$= (\delta_b^d \delta_c^e - \delta_b^e \delta_c^d) V^c U^b W_d Y_e = (\mathbf{V}, \mathbf{Y})(\mathbf{U}, \mathbf{W}) - (\mathbf{V}, \mathbf{W})(\mathbf{U}, \mathbf{Y}).$

(iv) $[\mathbf{A}, [\mathbf{B}, \mathbf{C}]] = \hat{\mathbf{n}}^a \varepsilon_{abc} A^b \varepsilon^{ced} B_e C_d$

$= \hat{\mathbf{n}}^a (\delta_a^e \delta_b^d - \delta_a^d \delta_b^e) A^b B_e C_d = \mathbf{B}(\mathbf{A}, \mathbf{C}) - \mathbf{C}(\mathbf{A}, \mathbf{B}).$

Exercise 21

$[\mathbf{U}, [\mathbf{V}, \mathbf{W}]] + [\mathbf{W}, [\mathbf{U}, \mathbf{V}]] + [\mathbf{V}, [\mathbf{W}, \mathbf{U}]]$

$= \hat{\mathbf{n}}^a \varepsilon_{abc} (U^b \varepsilon_{ced} V^e W^d + W^b \varepsilon_{ced} U^e V^d) + V^b \varepsilon_{ced} W^e U^d)$

$= \hat{\mathbf{n}}^a (\delta_a^e \delta_b^d - \delta_a^d \delta_b^e)(U^b V_e W_d + W^b U_e V_d + V^b W_e U_d)$

$= \mathbf{V}(\mathbf{U}, \mathbf{W}) + \mathbf{U}(\mathbf{V}, \mathbf{W}) + \mathbf{W}(\mathbf{U}, \mathbf{V}) - \mathbf{W}(\mathbf{U}, \mathbf{V}) - \mathbf{V}(\mathbf{W}, \mathbf{U}) - \mathbf{U}(\mathbf{V}, \mathbf{W}) = 0.$

Exercise 22

(i) $\varepsilon_{i'j'k'} = \dfrac{\partial x^a}{\partial x^{i'}} \dfrac{\partial x^b}{\partial x^{j'}} \dfrac{\partial x^c}{\partial x^{k'}} E_{abc} = \dfrac{\partial x^a}{\partial x^l} \dfrac{\partial x^l}{\partial x^{i'}} \dfrac{\partial x^b}{\partial x^m} \dfrac{\partial x^m}{\partial x^{j'}} \dfrac{\partial x^c}{\partial x^n} \dfrac{\partial x^n}{\partial x^{k'}} E_{abc}$

$= \dfrac{\partial x^l}{\partial x^{i'}} \dfrac{\partial x^m}{\partial x^{j'}} \dfrac{\partial x^n}{\partial x^{k'}} E_{lmn};$

(ii) $\varepsilon_{ikj} = \dfrac{\partial x^a}{\partial x^i} \dfrac{\partial x^b}{\partial x^k} \dfrac{\partial x^c}{\partial x^j} E_{abc} = \dfrac{\partial x^a}{\partial x^i} \dfrac{\partial x^c}{\partial x^k} \dfrac{\partial x^b}{\partial x^j} E_{abc}$

$= -\dfrac{\partial x^a}{\partial x^i} \dfrac{\partial x^b}{\partial x^j} \dfrac{\partial x^c}{\partial x^k} E_{abc} = -\varepsilon_{ijk}.$

Exercise 23 The first two points are almost identical to the solution of Exercise 22.

(iii) $\varepsilon^{123} = \dfrac{\partial x^1}{\partial x^a} \dfrac{\partial x^2}{\partial x^b} \dfrac{\partial x^3}{\partial x^c} \varepsilon^{abc} = \det \left| \dfrac{\partial x^i}{\partial x^a} \right|.$

Assuming $\det \left(\dfrac{\partial x^i}{\partial x^a} \right) > 0$, we have $\det \left\| \dfrac{\partial x^i}{\partial x^a} \right\| = \dfrac{1}{\sqrt{g}}.$

Exercise 25 $A^{ijklm} = g^{ij} \varepsilon^{klm}.$

Exercise 26

$$\varepsilon_{i_1 i_2 \cdots i_D} = \dfrac{\partial x^{a_1}}{\partial x^{i_1}} \dfrac{\partial x^{a_2}}{\partial x^{i_2}} \cdots \dfrac{\partial x^{a_D}}{\partial x^{i_D}} E_{a_1 a_2 \cdots a_D}.$$

Since

$$g_{ij} = \frac{\partial x^a}{\partial x^i} \frac{\partial x^b}{\partial x^j} \delta^a_b,$$

assuming that the orientation of axes is such that $\det \left(\frac{\partial x^a}{\partial x^i} \right) > 0$, we get

$$\det \left\| \frac{\partial x^a}{\partial x^i} \right\| = \sqrt{g}.$$

Then

$$\varepsilon_{123\cdots D} = \frac{\partial x^{a_1}}{\partial x^1} \frac{\partial x^{a_2}}{\partial x^2} \cdots \frac{\partial x^{a_D}}{\partial x^D} E_{a_1 a_2 \cdots a_D} = \det \left(\frac{\partial x^a}{\partial x^i} \right) = \sqrt{g}.$$

Similarly, $\varepsilon^{12 \cdots D} = \frac{1}{\sqrt{g}}$.

10.4 Solutions to Exercises from Chap. 4

Exercise 1 Consider $\mathbf{a} = a^i(x)\mathbf{e}_i(x) = a^{j'}(x')\mathbf{e}_{j'}(x') = a^i(x)\frac{\partial x^{j'}}{\partial x^i} \cdot \mathbf{e}_{j'}(x')$. Since the expansion in the given basis is unique, we get the solution of this equation in form (4.8).

Exercise 2

$$g_{ij} = \mathbf{e}_i(x) \cdot \mathbf{e}_j(x) = \left(\hat{\mathbf{n}}_a \frac{\partial x^a}{\partial x^i} \right) \cdot \left(\hat{\mathbf{n}}_b \frac{\partial x^b}{\partial x^j} \right) = g_{ab} \frac{\partial x^a}{\partial x^i} \frac{\partial x^b}{\partial x^j}.$$

Exercise 3

$$(ii) \quad \begin{pmatrix} 1 & 0 \\ 0 & r \end{pmatrix} \begin{pmatrix} \cos \varphi & \sin \varphi \\ -\sin \varphi & \cos \varphi \end{pmatrix} = \begin{pmatrix} \cos \varphi & \sin \varphi \\ -r \sin \varphi & r \cos \varphi \end{pmatrix}.$$

Exercise 5 In the formula for $\mathbf{V}$, the first term is radial acceleration and the second is angular, orthogonal to $\mathbf{r}$. In the formula for $\mathbf{a}$, the term $-r\hat{\mathbf{n}}_r\ddot{\varphi} = -r\ddot{\varphi}$ is the centripetal acceleration. The term $2\dot{r}\dot{\varphi}\hat{\mathbf{n}}_\varphi$ appears only in the case when particle is moving with a nonzero radial speed. Radial motion changes the angular component of the velocity and gives corresponding acceleration.

Exercise 6 *(a)* $\mathbf{e}_r = \hat{i} \cos \varphi + \hat{j} \sin \varphi + \hat{k}z$, $\mathbf{e}_z = \hat{k}$, $\mathbf{e}_\varphi = r(-\hat{i} \sin \varphi + \hat{j} \cos \varphi)$.
(b) $g_{ij} = \mathrm{diag}\,(1, r^2, 1)$.
(c) $\hat{\mathbf{n}}_r = \mathbf{e}_r$, $\hat{\mathbf{n}}_\varphi = \frac{1}{r}\mathbf{e}_\varphi$, $\hat{\mathbf{n}}_z = \mathbf{e}_z$.

Exercise 10 The correct order in case of cylindric coordinates is (r, φ, z), for the spherical case it is (r, θ, φ). With these choices the Jacobian of the transformation from Cartesian to curved coordinates is positive. Then for cylindric and spherical coordinates we have $\sqrt{g} = r$ and $\sqrt{g} = r^2 \sin\theta$, correspondingly, defining the Levi-Civita tensors.

10.5 Solutions to Exercises from Chap. 5

Exercise 1

$$\partial_{k'}a^{i'} = \frac{\partial}{\partial x^{k'}}\left(\frac{\partial x^{i'}}{\partial x^j}a^j\right) = \frac{\partial x^{i'}}{\partial x^j}\frac{\partial x^l}{\partial x^{k'}}\partial_l a^j + \frac{\partial^2 x^{i'}}{\partial x^j \partial x^l}\frac{\partial x^l}{\partial x^{k'}}a^j,$$

$$\partial_{k'}T_{j'}^{i'} = \frac{\partial}{\partial x^{k'}}\left(\frac{\partial x^{i'}}{\partial x^l}\frac{\partial x^m}{\partial x^{j'}}T_m^l\right)$$

$$= \frac{\partial x^n}{\partial x^{k'}}\frac{\partial x^{i'}}{\partial x^l}\frac{\partial x^m}{\partial x^{j'}}\partial_n T_m^l + \frac{\partial^2 x^{i'}}{\partial x^n \partial x^l}\frac{\partial x^n}{\partial x^{k'}}\frac{\partial x^m}{\partial x^{j'}}T_m^l + \frac{\partial x^{i'}}{\partial x^l}\frac{\partial^2 x^m}{\partial x^{k'}\partial x^{j'}}T_m^l.$$

Exercise 2

$$\nabla_i S^i = \frac{\partial X^a}{\partial x^i}\frac{\partial x^j}{\partial X^b}\partial_a S^b = \frac{\partial X^a}{\partial x^i}\frac{\partial x^j}{\partial X^b}\frac{\partial}{\partial X^a}\left(\frac{\partial X^b}{\partial x^k}S^k\right)$$

$$= \frac{\partial X^a}{\partial x^i}\frac{\partial x^l}{\partial X^a}\partial_l S^j + S^k \frac{\partial x^j}{\partial X^b}\frac{\partial^2 X^b}{\partial x^i \partial x^k} = \partial_i S^j + \frac{\partial x^j}{\partial X^b}\frac{\partial^2 X^b}{\partial x^i \partial x^k}S^k.$$

Thus,

$$\Gamma_{ik}^j = \frac{\partial x^j}{\partial X^b}\frac{\partial^2 X^b}{\partial x^i \partial x^k},$$

exactly as in Eq. (5.9).

Exercise 4 Let us make it only for one of the representations,

$$\Gamma_{m'n'}^{l'} = \frac{\partial x^{l'}}{\partial X^a}\frac{\partial^2 X^a}{\partial x^{m'}\partial x^{n'}} = \frac{\partial x^{l'}}{\partial x^i}\frac{\partial x^i}{\partial X^a}\frac{\partial}{\partial x^{m'}}\left[\frac{\partial X^a}{\partial x^j}\frac{\partial x^j}{\partial x^{n'}}\right]$$

$$= \Gamma_{kj}^i \frac{\partial x^{l'}}{\partial x^i}\frac{\partial x^k}{\partial x^{m'}}\frac{\partial x^j}{\partial x^{n'}} + \frac{\partial x^{l'}}{\partial x^i}\frac{\partial^2 x^i}{\partial x^{m'}\partial x^{n'}}.$$

Exercise 7

$$\nabla_i A^i = \partial_i \tilde{A}^i + \Gamma_{ki}^i \tilde{A}^k = \partial_r \tilde{A}^r + \partial_\varphi \tilde{A}^\varphi + \tilde{A}^r \partial_r \ln\sqrt{g},$$

where we used the result of the subsequent Exercise 10. Indeed, one can achieve the same result directly, but this requires extra effort. Since $g = r$, we have $\partial_r \ln\sqrt{g} = \frac{1}{2}\partial_r \ln g = \frac{1}{2r}$ and

$$\nabla_i A^i = \partial_r A^r + \frac{1}{2r} A^r + \frac{1}{r} \partial_\varphi A^\varphi.$$

Exercise 8 For the sake of brevity, consider a second-rank tensor

$$[\nabla_i, \nabla_j] T_l^k = \frac{\partial X^a}{\partial x^i} \frac{\partial X^b}{\partial x^j} \frac{\partial x^k}{\partial X^c} \frac{\partial X^d}{\partial x^l} [\nabla_a, \nabla_b] T_d^c.$$

Since in the Cartesian coordinates $\nabla_a = \partial_a$, $[\nabla_a, \nabla_b] T_d^c \equiv 0$ and therefore we also have $[\nabla_i, \nabla_j] T_l^k \equiv 0$. Let us stress that this proof is completely based on the existence of global Cartesian coordinates. If the space is such that there are no global Cartesian coordinates, the result will be different, as we shall see in the second part of the book.

Exercise 9

$$\text{div grad } \Psi = \nabla_i (\text{grad } \Psi)^i = \nabla_i (g^{ij} \nabla_j \Psi) = \Delta \Psi.$$

Exercise 10 Consider

$$\text{grad } (\text{div } \mathbf{A}) = \mathbf{e}^r \partial_r (\text{div } \mathbf{A}) + \mathbf{e}^\varphi \partial_\varphi (\text{div } \mathbf{A}).$$

Then

$$\text{div } \mathbf{A} = \nabla_i A^i = \partial_i A^i + \Gamma^i_{ji} A^i = \frac{\partial \tilde{A}^r}{\partial r} + \frac{\partial \tilde{A}^\varphi}{\partial \varphi} + \frac{1}{2} \partial_j \ln g \cdot A^j.$$

In polar coordinates $g = r^2$, therefore

$$\text{div } \mathbf{A} = \frac{\partial \tilde{A}^r}{\partial r} + \frac{\partial \tilde{A}^\varphi}{\partial \varphi} + \frac{1}{r} \tilde{A}^r.$$

Then

$$\text{grad } (\text{div } \mathbf{A}) = \mathbf{e}^r \frac{\partial}{\partial r} \left[\frac{\partial \tilde{A}^r}{\partial r} + \frac{\partial \tilde{A}^\varphi}{\partial \varphi} + \frac{1}{r} \tilde{A}^r \right] + \mathbf{e}^\varphi \frac{\partial}{\partial \varphi} \left[\frac{\partial \tilde{A}^r}{\partial r} + \frac{\partial \tilde{A}^\varphi}{\partial \varphi} + \frac{1}{r} \tilde{A}^r \right]$$

$$= \mathbf{e}^r \left[\frac{\partial^2 \tilde{A}^r}{\partial r^2} + \frac{\partial^2 \tilde{A}^\varphi}{\partial r \partial \varphi} + \frac{1}{r} \frac{\partial \tilde{A}^r}{\partial r} - \frac{1}{r^2} \tilde{A}^r \right]$$

$$+ \mathbf{e}^\varphi \left[\frac{\partial^2 \tilde{A}^r}{\partial \varphi \partial r} + \frac{\partial^2 \tilde{A}^\varphi}{\partial \varphi^2} + \frac{1}{r} \frac{\partial \tilde{A}^r}{\partial \varphi} - \frac{1}{r^2} \tilde{A}^r \right].$$

In order to use normalized basis, one has to remember that $\mathbf{e}^\varphi = \frac{1}{r} \hat{n}_\varphi$, $\mathbf{e}^r = \hat{n}_r$ and also $\tilde{A}^r = A^r$ and $\tilde{A}^\varphi = \frac{1}{r} A^\varphi$.

Exercise 11

$$\Gamma^i_{ij} = \frac{1}{2}\delta^j_k\Gamma^k_{ij} = \frac{1}{2}\delta^j_k\left(\partial_i g_{jl} + \partial_j g_{il} - \partial_l g_{ij}\right) = \frac{1}{2}g^{jl}\partial_i g_{jl} = \frac{1}{2g}\partial_i g,$$

where $g = \det(g_{ij})$. Then,

$$\Gamma^j_{ij} = \frac{1}{2}\partial_i \ln g = \partial_i \ln \sqrt{g}.$$

Exercise 12

$$\mathrm{div}\,\mathbf{A} = \partial_i A^i + \Gamma^i_{ji} A^j = \partial_i A^i + \frac{1}{2}A^j\partial_j \ln g.$$

For elliptic coordinates we have

$$\mathbf{e}_r = \frac{\partial x}{\partial r}\hat{\mathbf{i}} + \frac{\partial y}{\partial r}\hat{\mathbf{j}} = a\hat{\mathbf{i}}\cos\varphi + b\hat{\mathbf{j}}\sin\varphi \qquad (10.2)$$

$$\mathbf{e}_\varphi = \frac{\partial x}{\partial\varphi}\hat{\mathbf{i}} + \frac{\partial y}{\partial\varphi}\hat{\mathbf{j}} = r(-a\hat{\mathbf{i}}\sin\varphi + b\hat{\mathbf{j}}\cos\varphi). \qquad (10.3)$$

Thus,

$$g_{rr} = \mathbf{e}_r \cdot \mathbf{e}_r = a^2\cos^2\varphi + b^2\sin^2\varphi,$$
$$g_{\varphi\varphi} = \mathbf{e}_\varphi \cdot \mathbf{e}_\varphi = r^2(a^2\sin^2\varphi + b^2\cos^2\varphi),$$
$$g_{\varphi r} = \mathbf{e}_\varphi \cdot \mathbf{e}_r = r(-a^2\cos\varphi\sin\varphi + b^2\cos\varphi\sin\varphi) = \frac{b^2 - a^2}{2}r\sin 2\varphi,$$
$$\begin{aligned}g = g_{rr} \cdot g_{\varphi\varphi} - g^2_{\varphi r}\\ = r^2[(a^2\cos^2\varphi + b^2\sin^2\varphi)(a^2\sin^2\varphi + b^2\cos^2\varphi) - (a^2 - b^2)^2\sin^2\varphi\cos^2\varphi]\\ = r^2 a^2 b^2.\end{aligned}$$

Observation: this is a natural result, as one can understand after reading Chap. 8. Then

$$\frac{1}{2}\partial_j \ln g = \delta^r_j\frac{\partial}{\partial r}\ln(abr) = \frac{1}{r}\delta^r_j$$

and

$$\mathrm{div}\,\mathbf{A} = \partial_r\tilde{A}^r + \partial_\varphi\tilde{A}^\varphi + \frac{1}{r}\tilde{A}^r,$$

exactly like in polar coordinates. For hyperbolic coordinates

$$\mathbf{e}_\rho = \frac{\partial x}{\partial \rho}\hat{\mathbf{i}} + \frac{\partial y}{\partial \rho}\hat{\mathbf{j}} = \hat{\mathbf{i}}\cosh\chi + \hat{\mathbf{j}}\sinh\chi,$$

$$\mathbf{e}_\chi = \frac{\partial x}{\partial \chi}\hat{\mathbf{i}} + \frac{\partial y}{\partial \chi}\hat{\mathbf{j}} = \rho(\hat{\mathbf{i}}\sinh\chi + \hat{\mathbf{j}}\cosh\chi),$$

and hence

$$g_{\rho\rho} = \cosh^2\chi + \sinh^2\chi,$$
$$g_{\chi\chi} = \rho^2(\cosh^2\chi + \sinh^2\chi),$$
$$g_{\rho\chi} = \rho(\cosh\chi\sinh\chi + \sinh\chi\cosh\chi) = 2\rho\cosh\chi\sinh\chi,$$
$$g = g_{\rho\rho}g_{\chi\chi} - g_{\chi\rho}^2 = \rho^2.$$

Then we get

$$\mathrm{div}\,\mathbf{A} = \partial_\rho\tilde{A}^\rho + \partial_\chi\tilde{A}^\chi + \frac{1}{\rho}\tilde{A}^\rho,$$

in the coordinate basis.

Exercise 13

$$g^{ij}\Gamma_{ij}^k = \frac{1}{2}g^{ij}g^{kl}(\partial_i g_{jl} + \partial_j g_{il} - \partial_l g_{ij}) = g^{ij}g^{kl}\partial_i g_{jl} - \frac{1}{2}(g^{ij}\partial_l g_{ij})g^{kl}$$

$$= -\partial_i g^{ik} - g^{kl}\partial_l \ln\sqrt{g} = -\frac{1}{\sqrt{g}}\partial_l(\sqrt{g}\,g^{kl}).$$

Obviously,

$$\Delta\Psi = g^{ij}\nabla_i\nabla_j\Psi = g^{ij}(\partial_i\partial_j\Psi - \Gamma_{ij}^k\partial_k\Psi) = g^{ij}\partial_i\partial_j\Psi + \frac{1}{\sqrt{g}}\partial_i(\sqrt{g}\,g^{ij})\partial_j\Psi,$$

making required calculations a trivial (albeit not very short for elliptic and hyperbolic cases) exercise. Let us say that finding an inverse metric from the conjugate vector basis may be a relatively economic approach.

10.6   Solutions to Exercises from Chap. 6

Let us give the solutions just for a few examples.

Exercise 3

$$\mathrm{rot}\,(\varphi\mathbf{A}) = \hat{\mathbf{n}}^a\varepsilon_{abc}\partial^b(\varphi A^c) = \varphi\,\mathrm{rot}\,\mathbf{A} + \nabla\varphi\times\mathbf{A} = \varphi\,\mathrm{rot}\,\mathbf{A} - [\mathbf{A},\nabla]\varphi.$$

Exercise 6

$$\text{grad}\,(\mathbf{A} \cdot \mathbf{B}) = \hat{\mathbf{n}}^a \partial_a (A_b B^b) = \hat{\mathbf{n}}^a (A_b \partial_a B^b + B^b \partial_a A_b)$$
$$= \hat{\mathbf{n}}^a A^b \partial_a B_b + (\mathbf{A} \leftrightarrow \mathbf{B}).$$

On the other hand,

$$[\mathbf{A},\, \text{rot}\,\mathbf{B}] + (A, \nabla)\mathbf{B} + (\mathbf{A} \leftrightarrow \mathbf{B})$$
$$= \hat{\mathbf{n}}^a \varepsilon^{abc} A_b \varepsilon_{cde} \partial^d B^e + A^a \partial_a B^b \cdot \hat{\mathbf{n}}^b + (\mathbf{A} \leftrightarrow \mathbf{B})$$
$$= \hat{\mathbf{n}}^a A_b (\delta_d^a \delta_e^b - \delta_e^a \delta_d^b) \delta^d B^e + A^a \partial_a B_b \hat{\mathbf{n}}^b + (\mathbf{A} \leftrightarrow \mathbf{B})$$
$$= A^b \hat{\mathbf{n}}^a \partial_a B^b + (\mathbf{A} \leftrightarrow \mathbf{B}),$$

which is the same.

Exercise 10

$$[\mathbf{A}, \mathbf{B}] \cdot \text{rot}\,\mathbf{C} = \varepsilon^{abc} A_b B_c \, \varepsilon_{ade} \partial^d C^e = (\delta_d^b \delta_e^c - \delta_e^b \delta_d^c) A_b B_c \partial^d C^e$$
$$= \mathbf{B} \cdot (\mathbf{A}, \nabla)\mathbf{C} - \mathbf{A} \cdot (\mathbf{B}, \nabla)\mathbf{C}.$$

Exercise 11

$$[[\mathbf{A}, \nabla], \mathbf{B}] = \hat{\mathbf{n}}^a \varepsilon_{abc} \varepsilon^{bde} A_d \partial_e B^c$$
$$= \hat{\mathbf{n}}^a (\delta_a^e \delta_c^d - \delta_a^d \delta_c^e) A_d \partial_e B^c = A^c \nabla B_c - \mathbf{A} \cdot \text{div}\,\mathbf{B}.$$

On the other hand,

$$(A, \nabla)\mathbf{B} + [\mathbf{A},\, \text{rot}\,\mathbf{B}] - \mathbf{A} \cdot \text{div}\,\mathbf{B} = (A, \nabla)\mathbf{B} + \hat{\mathbf{n}}^a \varepsilon_{abc} A^b \varepsilon^{cde} \partial_d B_e - \mathbf{A} \cdot \text{div}\,\mathbf{B}$$
$$= (A, \nabla)\mathbf{B} + \hat{\mathbf{n}}^a (\delta_a^d \delta_b^e - \delta_a^e \delta_b^d) A^b \partial_d B_e - \mathbf{A} \cdot \text{div}\,\mathbf{B} = A^c \nabla B_c - \mathbf{A}\,\text{div}\,\mathbf{B}.$$

Exercise 12

$$[[\nabla, \mathbf{A}], \mathbf{B}] = \hat{\mathbf{n}}^a \varepsilon_{abc} \varepsilon^{cde} \partial_d A_e B^c = \hat{\mathbf{n}}^a (\delta_a^d \delta_c^e - \delta_a^e \delta_c^d)(B^c \partial_d A_e + A_e \partial_d B^c)$$
$$= \nabla(\mathbf{A} \cdot \mathbf{B}) - (\mathbf{B}, \nabla)\mathbf{A} - \mathbf{A}\,\text{div}\,\mathbf{B}.$$

On the other hand,

$$\mathbf{A}\,\text{div}\,\mathbf{B} - (A, \nabla)\mathbf{B} - [\mathbf{A},\, \text{rot}\,\mathbf{B}] - [\mathbf{B},\, \text{rot}\,\mathbf{A}]$$
$$= \mathbf{A}\,\text{div}\,\mathbf{B} - (A, \nabla)\mathbf{B} - \hat{\mathbf{n}}^a \varepsilon_{abc} A^b \varepsilon^{cde} \partial_d B_e - \hat{\mathbf{n}}^a \varepsilon_{abc} B^b \varepsilon^{cde} \partial_d A_e$$
$$= \mathbf{A}\,\text{div}\,\mathbf{B} - (A, \nabla)\mathbf{B} - \hat{\mathbf{n}}^a (\delta_a^d \delta_b^e - \delta_a^e \delta_b^d)(A^b \partial_d B_e + B_e \partial_d A_e)$$
$$= \mathbf{A}\,\text{div}\,\mathbf{B} - \nabla(\mathbf{A} \cdot \mathbf{B}) + (\mathbf{B}, \nabla)\mathbf{A},$$

which is the same.

10.7   Solutions to Exercises from Chap. 7

Exercise 1 Consider the transformation of partial derivatives,

$$\frac{\partial}{\partial r} = \frac{\partial x}{\partial r}\frac{\partial}{\partial x} + \frac{\partial y}{\partial r}\frac{\partial}{\partial y} = \cos\varphi\,\frac{\partial}{\partial x} + \sin\varphi\,\frac{\partial}{\partial y},$$

$$\frac{\partial}{\partial\varphi} = \frac{\partial x}{\partial\varphi}\frac{\partial}{\partial x} + \frac{\partial y}{\partial\varphi}\frac{\partial}{\partial y} = -r\sin\varphi\,\frac{\partial}{\partial x} + r\cos\varphi\,\frac{\partial}{\partial y}.$$

Similarly, one can easily derive

$$\frac{\partial}{\partial x} = \cos\varphi\,\frac{\partial}{\partial r} - \frac{1}{r}\sin\varphi\,\frac{\partial}{\partial\varphi},$$

$$\frac{\partial}{\partial y} = \sin\varphi\,\frac{\partial}{\partial r} + \frac{1}{r}\cos\varphi\,\frac{\partial}{\partial\varphi}.$$

On the other hand,

$$\begin{cases} \hat{\mathbf{n}}_r = \hat{\mathbf{i}}\cos\varphi + \hat{\mathbf{j}}\sin\varphi \\ \hat{\mathbf{n}}_\varphi = -\hat{\mathbf{i}}\sin\varphi + \hat{\mathbf{j}}\cos\varphi \end{cases} \implies \begin{cases} \hat{\mathbf{i}} = \hat{\mathbf{n}}_r\cos\varphi - \hat{\mathbf{n}}_\varphi\sin\varphi \\ \hat{\mathbf{j}} = \hat{\mathbf{n}}_r\sin\varphi + \hat{\mathbf{n}}_\varphi\cos\varphi. \end{cases}$$

Replacing these formulas, and also $\hat{\mathbf{n}}_z = \hat{\mathbf{k}}$, into (7.9), we get

$$\hat{\mathbf{i}}\frac{\partial}{\partial x} + \hat{\mathbf{j}}\frac{\partial}{\partial y} + \hat{\mathbf{k}}\frac{\partial}{\partial z} = (\hat{\mathbf{n}}_r\cos\varphi - \hat{\mathbf{n}}_\varphi\sin\varphi)\left(\cos\varphi\,\frac{\partial}{\partial r} - \frac{1}{r}\sin\varphi\,\frac{\partial}{\partial\varphi}\right)$$

$$+ (\hat{\mathbf{n}}_r\sin\varphi + \hat{\mathbf{n}}_\varphi\cos\varphi)\left(\sin\varphi\,\frac{\partial}{\partial r} + \frac{1}{r}\cos\varphi\,\frac{\partial}{\partial\varphi}\right) + \hat{\mathbf{k}}\frac{\partial}{\partial z}$$

$$= \hat{\mathbf{n}}_r\frac{\partial}{\partial r} + \frac{1}{r}\hat{\mathbf{n}}_\varphi\frac{\partial}{\partial\varphi} + \hat{\mathbf{k}}\frac{\partial}{\partial z}.$$

Exercise 2 Can be easily done using $\mathbf{e}_r,\ \mathbf{e}_\theta,\ \mathbf{e}_\varphi$ from Chap. 4 and the definition of the metric tensor, $g_{ij} = \mathbf{e}_i \cdot \mathbf{e}_j$.

Exercise 3 For instance, it is possible to use Eqs. (7.25) and derive $\hat{\mathbf{n}}_r = \mathbf{e}^r,\ \hat{\mathbf{n}}_\varphi = r\sin\theta\,\mathbf{e}^\varphi,\ \hat{\mathbf{n}}_\theta = r\mathbf{e}^\theta$. Alternatively, one can invert g_{ij}, which may be even simple.

10.8   Solutions to Exercises from Chap. 8

Exercise 1 For Cartesian coordinates

$$g_{ij}\frac{dx^i}{d\tau}\frac{dx^j}{d\tau} = \delta_{ab}\frac{dX^a}{d\tau}\frac{dX^b}{d\tau},$$

while in a nonconstant basis the expression can be more complicated.

Exercise 2 Let us just give the answers.

$$1. \quad \oint_c (x+y)\,dl = 2\sqrt{2};$$

$$2. \quad \int_{(c)} (x^2+y^2)\,dl = a^2\sqrt{a^2+b^2}\,\pi n;$$

$$3. \quad \int_{(c)} x\,dl = \frac{a^2\sqrt{k^2+1}}{4k^2+1}\cdot 2k.$$

Exercise 3 First of all, for the positive sign we have $xdy + ydx = d(xy)$. Then if the curve is parameterized by the continuous parameter $\tau \in [\tau_1, \tau_2]$, we have

$$\int_{(OA)} d(xy) = \int_{\tau_1}^{\tau_2} \frac{d(xy)}{d\tau}d\tau = xy\Big|_{\tau_2} - xy\Big|_{\tau_1}, \qquad (10.4)$$

independent of the path. For the point 0, we have $xy = 0$ and in another limit, $xy\big|_A = 2$. Hence, for positive sign the integral equals 2. Consider now the negative sign and the straight line,

$$x = \frac{l}{\sqrt{5}}, \qquad y = \frac{2l}{\sqrt{5}}, \qquad 0 \le l \le 1.$$

Then

$$\int_{(OA)} xdy - ydx = 0.$$

For parabola $y = kx^2$ and therefore $k = 2$. Then, $dy = 2kxdx$ and we have

$$\int_{(OA)} xdy - ydx = \frac{k}{3}.$$

Finally, for $OB + BA$ one can easily arrive at $\int_{(OA)} = -2$.

Exercise 4 We have $\mathbf{r} = a\hat{\mathbf{n}}_r$, where a is the radius of the sphere. According to Eq. (4.30), $\mathbf{n}_r = \sin\theta(\hat{\mathbf{i}}\cos\varphi + \hat{\mathbf{j}}\sin\varphi) + \hat{\mathbf{k}}\cos\theta$. Then

$$\mathbf{r}_\varphi = \frac{d\mathbf{r}}{d\varphi} = a\sin\theta(-\hat{\mathbf{i}}\sin\varphi + \hat{\mathbf{j}}\cos\varphi), \qquad (10.5)$$

$$\mathbf{r}_\theta = \frac{d\mathbf{r}}{d\theta} = a\cos\theta(\hat{\mathbf{i}}\cos\varphi + \hat{\mathbf{j}}\sin\varphi) - a\hat{\mathbf{k}}\sin\theta.$$

Obviously, $g_{\varphi\varphi} = a^2 \sin^2 \theta$, $g_{\theta\theta} = a^2$, $g_{\varphi\theta} = 0$. This metric is an angular part of the $3D$ metric of space in spherical coordinates on the surface $r = a$.

Exercise 5 A useful parameterization is

$$x = \cos \varphi (a + b \cos \theta), \qquad y = \sin \varphi (a + b \cos \theta), \qquad \text{and} \qquad z = b \sin \theta,$$

where $0 \le \theta \le 2\pi$ and $0 \le \varphi \le 2\pi$. Then

$$\begin{aligned}
\mathbf{r}_\varphi &= (-\hat{\mathbf{i}} \sin \varphi + \hat{\mathbf{j}} \cos \varphi)(a + b \cos \theta), \\
\mathbf{r}_\theta &= -b \sin \theta (\hat{\mathbf{i}} \cos \varphi + \hat{\mathbf{j}} \sin \varphi) + \hat{\mathbf{k}} \cos \theta.
\end{aligned} \tag{10.6}$$

Obviously, $g_{\varphi\varphi} = (a + b \cos \theta)^2$, $g_{\theta\theta} = b^2$ and $g_{\varphi\theta} = 0$.

Exercise 7 Obviously, $g = a^4 \sin^2 \theta$, and hence the area is

$$S = \int_0^{2\pi} d\varphi \int_0^\pi a^2 \sin^2 \theta d\theta = 4\pi a^2.$$

Exercise 8 For the sphere $\hat{\mathbf{n}}_\theta \times \hat{\mathbf{n}}_\varphi = \hat{\mathbf{n}}_r$. For the torus, we have the vector product $\mathbf{r}_\varphi \times \mathbf{r}_\theta = a^2 \left[\hat{\mathbf{k}} \cos^2 - \sin^2 \theta (\hat{\mathbf{i}} \cos \varphi + \hat{\mathbf{j}} \sin \varphi) \right]$, and the normalized version can be derived by using $|\mathbf{r}_\varphi \times \mathbf{r}_\theta| = a^2$.

Exercise 10 The parameterization is $x^2 + y^2 + z^2 = a^2$ with $z > 0$ and $\rho(x, y, z) = \frac{\rho z}{a}$. In the spherical coordinates

$$M = \int_0^{2\pi} d\varphi \int_0^{\frac{\pi}{2}} a \frac{\rho}{a} \cos \theta \, a^2 \sin \theta d\theta = \pi \rho a^2. \tag{10.7}$$

Exercise 11

$(i) \quad I = \int_0^{2\pi} d\varphi \int_0^\pi d\theta \sin \theta a^2 \rho a^2 \sin^2 \theta = \frac{8\pi}{3} \rho_0 a^4 = \frac{2}{3} M a^2,$

where $M = 4\varphi a^2 \rho_0$.

$(ii) \quad \rho(r) = \rho_0 \left(\frac{r}{a} \right)^n.$

Consider a thin shell with the radius $R \in (0, a]$. Then

$$dm = 4\pi R^2 \rho(R) dR.$$

According to (i), we have

$$dI = \frac{8\pi}{3} R^4 \rho_0 \left(\frac{R}{a} \right)^n dR.$$

The corresponding integral is

$$I = \frac{8\pi}{3}\rho_0 a^{-n} \int_0^a R^{4+n} dR = \frac{8\pi\rho_0}{3(5+n)} a^5.$$

By direct calculation,

$$dI = \rho \cdot dV \cdot r^2 \sin^2\theta = \rho(r) \cdot r^2 \sin\theta \cdot r^2 \sin^2\theta \, dr d\theta d\varphi = \rho_0 \frac{r^{n+4}}{a^n} \sin^3\theta \, dr d\theta d\varphi.$$

Then,

$$I = \frac{8\pi\rho_0}{3} \frac{a^5}{5+n}.$$

Furthermore, taking $n = 0$ we get the expression for the particular case,

$$I = \frac{2Ma^2}{5} = \frac{8\pi}{15}\rho_0 a^5.$$

Finally, the contribution of the thin shell at $r = a$ can be easily obtained from the last expression,

$$(iii) \quad dI_a = \frac{8\pi}{15}\rho_0 \cdot 5a^4 da = \frac{2}{3}m_a a^2,$$

where $m_a = 4\pi a^2 \rho_0$. This is exactly the result of the part (i).

Part II
Elements of Electrodynamics and Special Relativity

In this part of the book, we consider one of the most important applications of tensor methods and describe the backgrounds of special relativity and electrodynamics. Due to the specific interest in the fundamentals of the theory, we consider only vacuum electrodynamics. For this and other reasons, this part cannot replace the standard textbooks on electrodynamics. In this respect, we can especially recommend [1, 2, 3, 4], which are selected through the personal preferences of the author. The last item in this list is the book of problems. Many good exercises can be also found in the book of problems on general relativity [5].

References

1. J.D. Jackson, Classical Electrodynamics, 3rd edn. (Wiley, 1998).
2. L.D. Landau, and E.M. Lifshits, The Classical Theory of Fields-Course of Theoretical Physics, Volume 2, (Butterworth-Heinemann, 1987).
3. D.J. Griffiths, Introduction to Electrodynamics, 4th edn. (Pearson, 2012).
4. I.N. Toptygin, V.V. Batygin, Problems in Electrodynamics, (Academic Press, 1965-Translation from Russian. Last Russian Edition in 2010).
5. A.P. Lightman, W.H. Press, R.H. Price, and S.A. Teukolsky, Problem book in Relativity and Gravitation, (Princeton University Press, 1975).

Chapter 11
Maxwell Equations and Lorentz Transformations

11.1 Maxwell Equations and Lorentz Force

Maxwell's equations were obtained by generalization of several empirical laws of electromagnetic phenomena, especially the Gauss law (an alternative form of Coulomb's law), Ampère's law, and Faraday's law of induction. The complete, unified form of these equations credited to the British physicist James Clerk Maxwell (1831–1879), who published the theory of electromagnetic phenomena between 1855 and 1873. One of the main points in this development was that light has an electromagnetic origin.

The classical electromagnetic theory describes charged particles interacting with the electromagnetic field, which has electric $\mathbf{E}$ and magnetic $\mathbf{H}$ parts. The particle with the charge q and velocity $\mathbf{v}$ is the subject of the Lorentz force

$$\mathbf{F} = q\mathbf{E} + \frac{q}{c}\,\mathbf{v} \times \mathbf{H}\,. \tag{11.1}$$

This equation (with all that follow) is written in the useful Gaussian system of units, where electric and magnetic fields have the same dimension. The quantity c in Eq. (11.1) is a universal constant called the speed of light.

The electric $\mathbf{E}$ and magnetic $\mathbf{H}$ fields satisfy Maxwell equations,

$$\operatorname{div}\mathbf{E} = 4\pi\rho\,, \quad \operatorname{rot}\mathbf{H} = \frac{1}{c}\frac{\partial\mathbf{E}}{\partial t} + \frac{4\pi}{c}\,\mathbf{j}\,, \tag{11.2}$$

$$\operatorname{div}\mathbf{H} = 0\,, \quad \operatorname{rot}\mathbf{E} = -\frac{1}{c}\frac{\partial\mathbf{H}}{\partial t}\,. \tag{11.3}$$

In these equations, $\rho = \rho(t, \mathbf{r})$ and $\mathbf{j} = \mathbf{j}(t, \mathbf{r})$ are the densities of change and electric current. In order to understand how these quantities are related, let us consider a fluid that has density of charge $\rho = \rho(t, \mathbf{r})$. This means that in the infinitesimal volume ΔV at the point $\mathbf{r}$ and at the instant of time t, there is a charge $\Delta Q = \rho(t, \mathbf{r}) \cdot \Delta V$. Without loss of generality we can assume that the volume ΔV is so small that all

© Springer Nature Switzerland AG 2019
I. L. Shapiro, *A Primer in Tensor Analysis and Relativity*, Undergraduate
Lecture Notes in Physics, https://doi.org/10.1007/978-3-030-26895-4_11

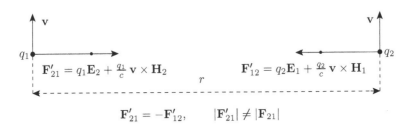

Fig. 11.1 Diagrams of forces in a static frame (upper part) and in another inertial frame moving homogeneously with the velocity $-\mathbf{v}$ (lower part). These diagrams were constructed using standard logic of Newtonian mechanics and the expression for the Lorentz force (11.1). The forces between the static charges and between the ones moving with constant speed should be different, which contradicts the rule that the forces should be the same in all inertial reference frames

particles in this volume have the same velocity $\mathbf{v}$. Then the current density is defined as

$$\mathbf{j} = \mathbf{v} \cdot \rho(t, \mathbf{r}). \tag{11.4}$$

We will not be interested in the physical justification of the Maxwell laws. The interested reader can find this part in the books such as [1] or [2]. Since our purpose is to use tensor methods to clarify the relation between electrodynamics and relativity, we will simply regard the Lorentz force and Maxwell equations as generalizations of well-established experimental results—that means we take these equations as granted.

One of the most remarkable features of Eqs. (11.1) (11.2) and (11.3) is that they contradict the laws of Newtonian mechanics. In order to see this, consider a simple thought experiment, which requires two static, electric, point-like charges q_1 and q_2 that interact by means of the Coulomb force. Let us follow the standard Newtonian logic and consider the same two charges in another inertial frame that moves with the constant speed $-\mathbf{v}$ in the direction orthogonal to the line between the two charges. The situation is illustrated in Fig. 11.1.

According to the standard logic of classical mechanics, the electric Coulomb force should not change because of the choice of inertial frame. At the same time, in the new frame, each of the charges will move with a constant speed $+\mathbf{v}$. According to (11.4), the two charges become currents and hence each of them creates a magnetic field, which can be evaluated from (11.2). This magnetic field will act on another charge according to (11.1); therefore, there will be an increase in the force acting on each of the charges. On the other hand, according to the Galilean principle, the force cannot change when we switch from one inertial frame to another such frame—hence this example shows that there is a contradiction.

The particular example considered above illustrates the general property of Eqs. (11.1)–(11.3). This system of equations is not invariant under the Galilean transformations between inertial reference frames,

$$\mathbf{r}' = \mathbf{r} + \mathbf{v}t, \quad \mathbf{t}' = \mathbf{t}. \tag{11.5}$$

The most natural resolution of the contradiction between classical mechanics and Maxwell's electromagnetism is to assume that Eqs. (11.1)–(11.3) must be supplemented by some extra terms that would correct them. In the epoch between the original works of Maxwell and the advent of special relativity, there were quite a lot of works devoted to this idea, trying to introduce a new fluid called "ether". Some of these works were very interesting and absolutely cannot be seen as something useless. For instance, the Lorentz transformations between different inertial reference frames, which represent the mathematical core of Einstein's special relativity, were found by Joseph Larmor in the work entitled "Aether and Matter" in 1896. Later on, more detailed and mathematically complete derivations of these transformations as a symmetry of Maxwell's equations were given by Hendrik Antoon Lorentz in 1904 and Henri Poincare in 1905.

The final successful solution of the conflict between mechanics and electromagnetism does not require any kind of ether. Instead, one has to generalize the Galilean transformations (11.5) in such a way that the new transformation laws are compatible with the system of equations (11.1)–(11.3). Being compatible here means that changing from one inertial frame to another one does not change the form of these equations, while the quantities such as $\mathbf{H}$, $\mathbf{E}$, ρ, $\mathbf{j}$, and $\mathbf{v}$ do transform in a controllable tensorial way. For the small velocities, the Galilean transformation law should remain a good approximation. The Lorentz transformations that we will obtain soon, passed great number of tests and verifications, and they are certainly correct within their area of applicability. At the same time, these transformations (as with everything what humans could create, until now) are not perfect, and should be viewed as an approximation of a more general theories that take into account, e.g., the effects of gravity.

An interesting question is of course whether the solution we still have to find cannot be traded for another solution. Let us refer the interested reader to the excellent exposition of the uniqueness of special relativity for electromagnetism in [1]. Here we will try to solve a more simple problem, namely, to find an economical shortcut that will show that special relativity resolves the conflict between electromagnetism and mechanics.

(i) Using Maxwell equations, we derive the equation for the electromagnetic wave and show that it propagates with the same speed c in any inertial frame.

(ii) Formulate the combination of time and space coordinates that should remain invariant under the change from one inertial frame to another.

(iii) Using this invariant, find an explicit form of the Lorentz transformations that replace (11.5) and, at the same time, reduce to it for the small speeds, with $|\mathbf{v}| \ll c$.

(iv) Prove that the whole system of equations (11.1)–(11.3) remains invariant under Lorentz transformations.

Through the above steps, we shall achieve a solid background in special relativity and show its compatibility with electrodynamics.

11.2 Universal Speed of Electromagnetic Wave

Consider Maxwell equations without sources, meaning both ρ and $\mathbf{j}$ are equal to zero. The result looks as follows:

$$\operatorname{div} \mathbf{E} = 0, \quad \operatorname{rot} \mathbf{H} = \frac{1}{c} \frac{\partial \mathbf{E}}{\partial t}, \tag{11.6}$$

$$\operatorname{div} \mathbf{H} = 0, \quad \operatorname{rot} \mathbf{E} = -\frac{1}{c} \frac{\partial \mathbf{H}}{\partial t}. \tag{11.7}$$

In both cases, Eqs. (11.2) and (11.3), or Eqs. (11.6) and (11.7) are linear differential equations in partial derivatives, with one independent time variable and three independent space variables. The solution of this type of equation, without sources, is called *wave*.

In order to obtain the wave equation, we remember the formula from the vector analysis (see Chap. 1) $\operatorname{rot} (\operatorname{rot} \mathbf{E}) = \operatorname{grad} (\operatorname{div} \mathbf{E}) - \Delta \mathbf{E}$ and apply the rotation operator to both sides of the second equation in (11.7). Taking into account that $\operatorname{div} \mathbf{E} = 0$ due to the first of (11.6), one arrives at the equation

$$-\Delta \mathbf{E} = -\frac{1}{c} \operatorname{rot} \frac{\partial \mathbf{H}}{\partial t} = -\frac{1}{c} \frac{\partial \operatorname{rot} \mathbf{H}}{\partial t} = -\frac{1}{c^2} \frac{\partial^2 \mathbf{E}}{\partial t^2}, \tag{11.8}$$

where we used that the time and space partial derivatives commute on the functions of physical interest (which are supposed to have continuous second partial derivatives).

Equation (11.8) can be written in a compact form using a new differential operator $\Box$, called the d'Alembert operator,

$$\Box \mathbf{E} = 0, \quad \Box = \frac{1}{c} \frac{\partial^2}{\partial t^2} - \Delta. \tag{11.9}$$

Equation (11.9) describes the wave that propagates with the velocity c. The numerical value of this quantity, called speed of light, is huge: $c \approx 3 \times 10^{10}$ cm/s = 300,000 km/s. For the sake of comparison, the escape velocity from the Earth is

11.2 km/s, from the Sun 617.5 km/s, and from the average galaxy about 1000 km/s. Obviously, the speed of light is much greater than all these quantities.

Exercise 1 Starting from the same Eqs. (11.6) and (11.7), show that the magnetic field **H** satisfies similar equation $\Box\mathbf{H} = 0$.

Exercise 2 Consider rotation in the space of free electromagnetic fields

$$\begin{pmatrix} \mathbf{E} \\ \mathbf{H} \end{pmatrix} = \begin{pmatrix} \cos\alpha & -\sin\alpha \\ \sin\alpha & \cos\alpha \end{pmatrix} \begin{pmatrix} \mathbf{E}' \\ \mathbf{H}' \end{pmatrix}. \tag{11.10}$$

(i) Show that the form of Eqs. (11.6) and (11.7) does not change and that the new fields with primes satisfy the same d'Alembert equations $\Box\mathbf{E}' = \Box\mathbf{H}' = 0$.
(ii) Consider the same rotation for Eqs. (11.2) and (11.3) and find how they should be modified for being invariant under the rotation (11.10).

Let us make the last two very important observations. Remember that our main purpose is to find the modified form of the transformation (11.5) that will provide the invariance of Eqs. (11.3)–(11.1). As far as (11.9) is a direct consequence of these equations, we conclude that the desired transformation should preserve the form of the wave equation (11.9) and, in particular, provide that the speed c of the wave solution would remain the same in any inertial frame. This fact shows that the transformations we are looking for will be very different from the ones one meets in the classical mechanics. At the same time, since the speed of light is so great, the difference for the change between frames with the speed $v \ll c$ is supposed to be very small, and hence we are still within the framework of our plan for constructing the new theory.

Another observation is that analysis of Maxwell equations shows that the lines of the magnetic field **H** are always closed and do not have their ends on some sources. According to what we learned in the first part of the book, this feature can be provided by assuming that there is a *vector potential* field **A** such that

$$\mathbf{H} = \operatorname{rot}\mathbf{A}. \tag{11.11}$$

Contrary to this, in the presence of sources the lines of electric field **E** always have their ends on the point-like sources or at the infinity. This feature corresponds to the definition

$$\mathbf{E} = -\frac{1}{c}\frac{\partial\mathbf{A}}{\partial t} - \operatorname{grad}\varphi, \tag{11.12}$$

where **A** is the same potential as in (11.11) and φ is an additional scalar potential.

Exercise 3 Analyze the number of apparently independent equations in (11.2) and (11.3) and compare it to the six degrees of freedom of the two vectors **H** and **E** from one side and with the four degrees of freedom in φ and **A** from another side.

Observation The explanation of the apparently strange result of this exercise will be given later on, when we discuss gauge invariance.

Exercise 4 Explore whether the scalar and vector potentials satisfy the same wave equations, namely, $\Box \mathbf{A} = 0$ and $\Box \varphi = 0$.

Hint. The solution of this exercise is not completely trivial; the details can be found in Sect. 13.4.

Exercise 5 Obtain the most general plane wave solution of the wave equation $\Box \varphi = 0$ for the wave moving in a direction $\hat{\mathbf{n}}$.

Solution. Consider first the wave that is propagating along the axis X, then the equation $\Box \varphi = 0$ becomes

$$\left(\frac{1}{c^2} \frac{\partial^2}{\partial t^2} - \frac{\partial^2}{\partial x^2} \right) \varphi(x, t) = 0. \tag{11.13}$$

This equation can be rewritten in the new variables $\xi = x + ct$ and $\eta = x - ct$ as

$$\frac{\partial^2 \varphi}{\partial \xi \partial \eta} = 0, \tag{11.14}$$

which has an obvious general solution of the form

$$\varphi(\xi, \eta) = \varphi_1(\xi) + \varphi_2(\eta) = \varphi_1(x + ct) + \varphi_2(x - ct) = \varphi(x, t). \tag{11.15}$$

The last two functions are called advanced and retarded potentials, respectively. Those can be any twice differentiable functions. For the wave propagating along the direction defined by a unitary vector $\hat{\mathbf{n}}$, the solution is

$$\varphi(\mathbf{r}, t) = \varphi_1(\mathbf{r} \cdot \hat{\mathbf{n}} + ct) + \varphi_2(\mathbf{r} \cdot \hat{\mathbf{n}} - ct). \tag{11.16}$$

This formula follows directly from the possibility to chose $\hat{\mathbf{n}} = \hat{\mathbf{i}}$ and isotropy of $3D$ space.

In conclusion, we have found that the covariant form of Maxwell equations with respect to some, yet unknown to us, generalized transformation requires that the speed of light (electromagnetic wave) has the same value c in inertial reference frames. From the point of view of classical mechanics, this is a very unusual property that certainly contradicts our experience and intuition. But since the speed of light is so huge, we can assume that our experience and intuition may actually fail because they were formed to describe much smaller speeds. From this perspective, the most important requirement to the new transformation law is that it gives (11.5) as a good approximation for the velocities $v \ll c$. In what follows we shall see that this is exactly the case with the Lorentz transformations.

11.3 Invariant Interval and Minkowski Space

Let us start by defining the Minkowski space, which has the coordinates

$$x^0 = ct, \quad x^1 = x, \quad x^2 = y, \quad x^3 = z. \tag{11.17}$$

Here x^1, x^2, x^3 are Cartesian space coordinates and x^0 is the time coordinate measured in the same units as space coordinates x, y, z. The coefficient that provides the change of the units is the universal speed of light c. The standard abbreviations for Minkowski space are $M_{1,3}$ or M_4.

The point in Minkowski space is frequently called the *event*. The event is defined by time and space coordinates. For instance, there may be events 1 and 2, which have coordinates

$$x_1^\mu = \left(x_1^0, x_1^1, x_1^2, x_1^3\right) \quad \text{and} \quad x_2^\mu = \left(x_2^0, x_2^1, x_2^2, x_2^3\right), \tag{11.18}$$

meaning the event 1 occurs at the instant $t_1 = x_1^0/c$ at the space point with coordinates x_1^1, x_1^2, x_1^3, and similarly for the second event, which occurs at the instant $t_2 = x_2^0/c$ at the space point with coordinates x_2^1, x_2^2, x_2^3. In what follows we shall use Greek indices for the four-dimensional notations in Minkowski space, e.g., $\alpha, \beta, \ldots, \mu \cdots = 0, 1, 2, 3$ and Latin indices for the space variables, e.g., $a, b, \ldots, i\, j, \cdots = 1, 2, 3$. One can write, for instance, $x^\mu = \left(x^0, x^i\right)$ or $x^\mu = \left(x^0, \mathbf{r}\right)$.

It is not difficult to see that the frame invariance of the speed of light under the change between two inertial frames provides the invariance of the object, called *interval*. The interval between the two points 1 and 2 is defined as

$$
\begin{aligned}
s_{12}^2 &= \left(x_2^0 - x_1^0\right)^2 - \left(x_2^1 - x_1^1\right)^2 - \left(x_2^2 - x_1^2\right)^2 - \left(x_2^3 - x_1^3\right)^2 \\
&= c^2 \left(t_2 - t_1\right)^2 - \left(x_2 - x_1\right)^2 - \left(y_2 - y_1\right)^2 - \left(z_2 - z_1\right)^2.
\end{aligned}
\tag{11.19}
$$

Correspondingly, the interval between two infinitesimally separated events has the form

$$ds^2 = \left(dx^0\right)^2 - \left(dx^1\right)^2 - \left(dx^2\right)^2 - \left(dx^3\right)^2 = c^2 dt^2 - dl^2, \tag{11.20}$$

where $dl^2 = dx^2 + dy^2 + dz^2$ is the space interval between the two close events. The last formula can be also written in the form

$$ds^2 = \eta_{\mu\nu}\, dx^\mu\, dx^\nu, \quad \text{where} \quad \eta_{\mu\nu} = \text{diag}\,(1, -1, -1, -1) \tag{11.21}$$

is the metric of the *pseudo-Euclidean* Minkowski space, which is also called *space–time*.

In the conventional Euclidean space, one can always make a change of variable, i.e., choose curvilinear coordinates. From the very beginning, there are some restrictions on the choice of such coordinates in Minkowski space. Namely, the nonlinear changes of coordinates that involve both space and time variables can be consistently considered in general relativity, but better avoided in special relativity, which we are dealing with now. The reason is that this type of the change in the space–time coordinates means that we go to the non-inertial reference frame. We shall discuss how to consistently deal with this case in the last part of the book since doing it within special relativity may be difficult.

At the same time, a linear mixing of space and time coordinates is allowed in special relativity and is in fact one of our main objects of interest. It is important to stress that such mixing does not make the time and space coordinates the same. Indeed, after any such change of coordinates, they remain different mathematically, as well as physically.

In this chapter, we will be especially interested in the transformation of coordinates in $M_{1,3}$, which preserves the form of the metric (11.21), because in physics these changes of coordinates turn out to correspond to the transition between inertial frames. This kind of transformation always leaves one time-like and three space-like coordinates, showing that time and space mix, but never become the same thing.

There is a clear classification of the intervals between the two events into three distinct groups. The illustration of these three cases can be found in Fig. 11.2. The events of the first type have a positive square of the interval, $s_{12}^2 > 0$, and are called time-like. The motion from one event to another one in this case can be done with the constant velocity $v < c$. In particular, the two events that occur at the same space point are separated by a time-like interval. The second type of separation is the light-like interval, $s_{12}^2 = 0$. The motion with constant velocity between the two intervals can be performed only with the speed of light, $v = c$. The third type of separation between the two events is by the space-like interval, $s_{12}^2 < 0$; in this case, the connection between the events requires constant velocity $v > c$. In what follows we will see that the massive particles of normal origin[1] cannot move with the speed $v \geq c$, and that the massless particles (e.g., the photon, the quantum of electromagnetic wave, or of light) can move only with the speed c. Therefore, the events separated by space-like intervals cannot have causal connection.

Let us follow the book [3] and show that the interval between the two events does not change its value when we move from the inertial frame K to another inertial frame K'. It is supposed that K' moves with the constant velocity $\mathbf{v}_1$ with respect to K. The statement we intend to prove is that $s_{12}'^2 = s_{12}^2$.

Since the expression for the interval (11.19) is bilinear in coordinates, it is sufficient to prove this statement for the infinitesimal intervals (11.20). In the case of infinitesimal changes of variables, all dependencies between the quantities can be considered linear, and therefore we can assume that

$$ds'^2 = a_1 ds^2. \tag{11.22}$$

---

[1]From the theoretical side, besides normal particles there may be other particles called tachyons, which move faster than light. We shall comment on some of their features below, after introducing the backgrounds of relativistic dynamics.

Fig. 11.2 Two-dimensional
diagram represents a
$y = z = 0$ cut of the
space–time. Dashed lines
show the light cone formed
by the lines $x = \pm ct$. One
can distinguish time-like $0B$,
light-like $0A$, and space-like
$0C$ intervals on this plot

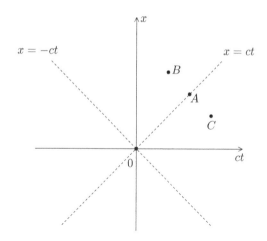

On the other hand, consider another inertial reference frame K'', which moves with
the constant velocity $\mathbf{v}_2$ with respect to K' and with the constant velocity $\mathbf{v}_3$ with
respect to K. Then, along with (11.22), there are two other similar relations

$$ds''^2 = a_2 \, ds'^2 \quad \text{and} \quad ds''^2 = a_3 \, ds^2 . \tag{11.23}$$

As a consequence, there is the third relation

$$ds'^2 = \frac{a_3}{a_2} \, ds^2 , \quad \text{and hence} \quad a_1 = \frac{a_3}{a_2} . \tag{11.24}$$

Due to the isotropy of space, the coefficients $a_{1,2,3}$ cannot depend on the directions
of the velocities, but only on their absolute values, therefore

$$a_1 = a_2(v_1), \quad a_2 = a_2(v_2), \quad a_3 = a_2(v_3) . \tag{11.25}$$

Since we do not know the transformation law that is going to replace (11.5), it is not
possible to state that $\mathbf{v}_1 = \mathbf{v}_3 - \mathbf{v}_2$. However, we can certainly state that the absolute
value v_1 depends on the angle between the directions of the velocities v_2 and v_3.
There is a situation when the *l.h.s.* of the last formula in (11.24) does not depend on
this angle, while the *r.h.s.* of the same formula does. The only possible resolution
of the paradox is to assume that the coefficient a_1 is a universal constant, which
doesn't depend on the speed v_1. Then a_2 and a_3 should also be constants, and the
same relation (11.24) tells us that $a_{1,2,3} \equiv 1$. Thus, we arrive at the invariance of the
interval under the change of inertial frame,

$$ds'^2 = ds^2 . \tag{11.26}$$

The last relation is remarkable, because it shows that the change from one inertial frame to another should involve the time variable. It is clear that this formula is not compatible with the Galilean transformation (11.5), because the last requires an absolute invariance of time, which is obviously incompatible with Eq. (11.26). At the same time, the defect is proportional to the square $(v/c)^2$, which becomes negligible for $v \ll c$.

Exercise 6   Verify the last statement, first for the case when the velocity has direction along the axis X and then for an arbitrary direction of $\mathbf{v}$.

11.4   Lorentz Transformations

One can easily derive the form of the coordinate transformations that is compatible with the invariance of the interval (11.26). Consider the two inertial reference frames, the static one K and another K', which are moving with the constant velocity $\mathbf{v}$ with respect to K. Since the space is isotropic, without loss of generality it is possible to assume that the direction of the velocity is along the axis X, that is, $\mathbf{v} = \hat{\mathbf{i}}v$.

A very simple thought experiment shows that in this case the transformation of coordinates in the transverse direction is trivial, $y' = y$ and $z' = z$. Imagine a trolley that moves along X with the speed v and makes pencil marks on the wall at height y with respect to the trolley and with height y' with respect to the wall, while another pencil on the wall makes marks on the trolley at y. Assuming that $y' \neq y$ we arrive at the contradiction.

Now we have to resolve Eq. (11.26) in the two-dimensional case, where it becomes

$$c^2 dt'^2 - dx'^2 = c^2 dt^2 - dx^2 . \tag{11.27}$$

The solution can be easily found by analogy with the Euclidean rotations. We know that the continuous transformation that preserves the length, $x'^2 + y'^2 = x^2 + y^2$, is rotation,

$$\begin{pmatrix} x' \\ y' \end{pmatrix} = \begin{pmatrix} \cos\alpha & -\sin\alpha \\ \sin\alpha & \cos\alpha \end{pmatrix} \begin{pmatrix} x \\ y \end{pmatrix} .$$

One can make an analytic continuation of time $ct = ix^4$ and remember that $\cos(iz) = \cosh z$ and $\sin(iz) = \sinh z$. In this way, it is easy to show that the general solution that preserves the interval (11.27) is the hyperbolic rotation,[2]

$$\begin{pmatrix} ct' \\ x' \end{pmatrix} = \begin{pmatrix} \cosh\psi & \sinh\psi \\ \sinh\psi & \cosh\psi \end{pmatrix} \begin{pmatrix} ct \\ x \end{pmatrix} . \tag{11.28}$$

---

[2]Due to the linearity of differentials, there is no need to indicate them here.

It is good to note that (different from the Euclidean case) in Eq. (11.28), both signs of sinh's are positive or negative.

Exercise 7 Explore relation (11.28) in more detail. (a) Verify that it satisfies the condition (11.27). (b) Find the inverse transformation, which can be obtained by changing the sign of ψ. (c) Consider the two consequent transformations along the same axis X with the angles ψ_1 and ψ_2, and verify that this is equivalent to a single transformation with the angle $\psi_1 + \psi_2$. The last two properties mean that the transformations from one inertial frame to another, along the axis X form a group. Indeed, the transformations between the frames that move in arbitrary dimensions also form a group (called a Lorentz group), but this one is much more complicated and will not be considered here.

The next step is to understand the relation between the angle ψ and the speed v of the frame K' with respect to the frame K. To this end, let us take a particle that is static in K. Then in the differential version of (11.28)

$$\begin{pmatrix} cdt' \\ dx' \end{pmatrix} = \begin{pmatrix} \cosh \psi & \sinh \psi \\ \sinh \psi & \cosh \psi \end{pmatrix} \begin{pmatrix} cdt \\ dx \end{pmatrix}, \tag{11.29}$$

there is $dx = 0$ and therefore

$$v = c\beta = \frac{dx'}{dt'} = -c \tanh \psi. \tag{11.30}$$

The last relation can be easily transformed by using

$$\sinh^2 \psi = \frac{\sinh^2 \psi}{\cosh^2 \psi - \sinh^2 \psi} = \frac{\tanh^2 \psi}{1 - \tanh^2 \psi} = \frac{\beta^2}{1 - \beta^2},$$

and

$$\cosh^2 \psi = \frac{\cosh^2 \psi}{\cosh^2 \psi - \sinh^2 \psi} = \frac{1}{1 - \tanh^2 \psi} = \frac{1}{1 - \beta^2},$$

such that

$$\cosh \psi = \frac{1}{\sqrt{1 - \beta^2}}, \quad \sinh \psi = -\frac{\beta}{\sqrt{1 - \beta^2}}, \quad \text{where } \beta = \frac{v}{c}. \tag{11.31}$$

Now we are in a position to write the Lorentz transformations in terms of v or β,

$$ct' = \frac{ct - \beta x}{\sqrt{1 - \beta^2}}, \quad x' = \frac{x - \beta ct}{\sqrt{1 - \beta^2}}, \quad y' = y, \quad z' = z. \tag{11.32}$$

We still have to check that transformations (11.32) reduce to the Galilean version (11.5) in the limit $\beta \to 0$. As far as $\beta \ll 1$, one can approximately set

$$\frac{1}{\sqrt{1-\beta^2}} = 1 + \frac{\beta^2}{2} + \mathcal{O}(\beta^4). \tag{11.33}$$

Then we can immediately see that rule (11.5) is satisfied with the precision of $\mathcal{O}(\beta^2)$. This fact explains why our intuition and life experience always confirm the correctness of Galilean transformations and are "protesting" against relativistic Lorentz transformations, regardless of their correctness.

Exercise 8 (i) Verify that the transformation inverse to (11.31) can be obtained by changing the sign of v,

$$ct = \frac{ct' + \beta x'}{\sqrt{1-\beta^2}}, \quad x = \frac{x' + \beta ct'}{\sqrt{1-\beta^2}}, \quad y = y', \quad z = z'. \tag{11.34}$$

Is it necessary to verify the group property of Lorentz transformations in these notations?

(ii) Verify that the Lorentz transformations do not modify the metric $\eta_{\alpha\beta}$, and that the same is true for the space rotations and inversions (parity transformations).

Exercise 9 (i) Using Lorentz transformations (11.34), show that the apparent length of the object is contracting in the direction of motion according to the formula

$$l = l_0\sqrt{1-\beta^2}, \tag{11.35}$$

where l_0 is the length of the object in rest. Show that the transverse sizes (in the direction orthogonal to the velocity) is not changing. (ii) Explain why the correct formula (11.35) for relativistic contraction cannot be obtained directly from the inverse transformation (11.32). (iii) Using the result (11.35) show that the space volume contracts for the moving observer according to

$$V = V_0\sqrt{1-\beta^2}, \tag{11.36}$$

where V_0 is the volume at rest. (iv) Consider the gas of N particles in the volume V. Calculate the relation between concentration of particles $n = N/V$ in the rest frame and in the moving frame.

References

1. J.D. Jackson, *Classical Electrodynamics*, 3rd edn. (Wiley, Hoboken, 1998)
2. D.J. Griffiths, *Introduction to Electrodynamics*, 4th edn. (Pearson, London, 2012)
3. L.D. Landau, E.M. Lifshits, *The Classical Theory of Fields - Course of Theoretical Physics*, vol. 2 (Butterworth-Heinemann, New York, 1987)

Chapter 12
Laws of Relativistic Mechanics

12.1 Relativistic Kinematics

Let us start with the basic elements of relativistic kinematics.

The Lorentz transformations (also called boosts) and orthogonal space transformations (rotations and inversion of coordinates) do not change the form of the metric. At the same time, they are relevant and have physical sense, as we have seen in the previous section. It is certainly interesting to keep the relativistic kinematics under control when making such changes of variables. To this end, we have to require that the physical quantities are scalars, vectors, and tensors in Minkowski space, because in this case the transformation rules are under control. The definitions of all types of tensors change trivially as compared to the $3D$ Euclidean space; therefore, we leave it to the reader to write down these definitions. For example, the metric $\eta_{\mu\nu}$ and its inverse $\eta^{\mu\nu}$ are tensors of the second rank, which satisfy $\eta_{\mu\nu}\eta^{\nu\alpha} = \delta^\alpha_\mu$. The interval is a scalar, which is clear from its main property (11.26). An important relation concerning ds for a massive particle that moves with the speed $v = \beta c \le c$ is as follows:

$$ds = \sqrt{c^2 dt^2 - dl^2} = \sqrt{c^2 dt^2 - v^2 dt^2} = c\sqrt{1 - \beta^2} dt \,. \tag{12.1}$$

Along with the interval, one can introduce a *proper time* of the particle, which is defined as $d\tau = ds/c = \sqrt{1 - \beta^2} dt$.

The evolution of a particle in the Minkowski space M_4 is described by the *world line* $x^\mu(\lambda)$ where λ is a parameter along the curve. The tangent vector along the curve is defined as $k^\mu = dx^\mu/d\lambda$. One can choose λ in different ways, for example, take $d\lambda = dx^0$. However, in many cases, it is most useful to choose $d\lambda = ds$. Indeed, this choice works well only for the time-like curves, where the tangent vector k^μ to the *world line* of a particle is time-like, $k^\mu k_\mu \ge 0$. Certainly, the interval is not the best choice in the case of a light-like tangent vector, because in this case $ds^2 = 0$. One of the possibilities for this case is to use the time variable, t or x^0.

© Springer Nature Switzerland AG 2019
I. L. Shapiro, *A Primer in Tensor Analysis and Relativity*, Undergraduate
Lecture Notes in Physics, https://doi.org/10.1007/978-3-030-26895-4_12

Consider the world line of a massive particle, which (as we shall see in what follows) has a time-like tangent vector. Taking $d\lambda = ds$, we arrive at the four-velocity of the particle,

$$u^\mu = \frac{dx^\mu}{ds}. \tag{12.2}$$

The square of the four-velocity is always equals to one,

$$u_\mu u^\mu = \eta_{\mu\nu} u^\mu u^\nu = \frac{ds^2}{ds^2} \equiv 1. \tag{12.3}$$

It is interesting to find the relation between four-velocity and usual $3D$ velocity of the particle, $\mathbf{v} = d\mathbf{r}/dt$. Indeed,

$$u^i = \frac{dx^i}{ds} = \frac{dx^i}{cd\tau} = \frac{1}{c\sqrt{1-\beta^2}} \frac{dx^i}{dt} = \frac{v^i}{c\sqrt{1-\beta^2}}, \tag{12.4}$$

where v^i are Cartesian components of the $3D$ vector $\mathbf{v}$. In a similar way, one can verify the relation

$$u^0 = \frac{dx^0}{ds} = \frac{1}{\sqrt{1-\beta^2}}. \tag{12.5}$$

The next relevant quantity is four-acceleration,

$$\omega^\mu = \frac{du^\mu}{ds} = \frac{d^2 x^\mu}{ds^2}. \tag{12.6}$$

It is easy to use relation (12.3) to prove that $u_\mu \omega^\mu \equiv 0$.

Exercise 1   Verify the following relations for the components of four-acceleration:

$$\omega^0 = \frac{cv\dot{v}}{(c^2-v^2)^2}, \qquad \omega^i = \frac{a^i}{c^2-v^2} + \frac{v^i v\dot{v}}{(c^2-v^2)^2}. \tag{12.7}$$

where a^i are the components of the $3D$ vector of acceleration, $\mathbf{a} = d\mathbf{v}/dt$.

The last thing about kinematics is to obtain the formula for summing velocities. Let us suppose that the inertial reference frame K' moves with velocity $\mathbf{V}$ with respect to the static inertial frame K. Consider the particle that moves with the constant velocity $\mathbf{v}'$ with respect to K'. What is its velocity with respect to K? In the framework of classical mechanics, the answer follows from the Galilean form (11.5) of the transformation between the two frames, $\mathbf{v} = \mathbf{V} + \mathbf{v}'$. Let us see how this relation modifies in special relativity.

In the frame K', we have the relations

$$dx' = v'_x dt', \quad dy' = v'_y dt', \quad dz' = v'_z dt'. \tag{12.8}$$

Without loss of generality one can assume that the velocity $\mathbf{V}$ has only X component, $\mathbf{V} = V\hat{\mathbf{i}}$. Then, according to the differential version of (11.34),

$$cdt = \frac{cdt' + \beta dx'}{\sqrt{1 - \beta^2}}, \quad dx = \frac{dx' + \beta cdt'}{\sqrt{1 - \beta^2}}, \quad dy = dy', \quad dz = dz', \tag{12.9}$$

where $\beta = V/c$. Combining (12.8) and (12.9), we arrive at the result

$$v_x = \frac{v'_x + V}{1 + Vv'_x/c^2}, \quad v_y = \frac{\sqrt{1 - \beta^2}\,v'_y}{1 + Vv'_x/c^2}, \quad v_z = \frac{\sqrt{1 - \beta^2}\,v'_z}{1 + Vv'_x/c^2}. \tag{12.10}$$

Exercise 2 Verify and analyze expressions (12.10). (a) Consider the lowest order nonrelativistic approximation to (12.10) and show that it exactly coincides with the Galilean rule $\mathbf{v} = \mathbf{V} + \mathbf{v}'$. (b) In the first order in β, verify the relation

$$\mathbf{v} = \mathbf{v}' + \mathbf{V} - \frac{(\mathbf{v}' \cdot \mathbf{V})}{c^2}\mathbf{v}'. \tag{12.11}$$

(c) Estimate numerically the difference between relativistic and nonrelativistic results for $|\mathbf{v}'| = |\mathbf{V}| = 10^4$ km/s, taking into account the angle between these two vectors.

Exercise 3 Derive the transformation rule for an arbitrary covariant vector A_μ and for an arbitrary contravariant vector B^μ under the Lorentz transformation (11.32) and under the inverse transformation (11.34). Use these results to verify the rule (12.10) and to derive the inverse transformation. Explain why only the contravariant formula can be used directly.

Exercise 4 Repeat the calculation of the first part of the previous exercise for the case of a completely antisymmetric tensor of second rank, in the covariant and contravariant versions, $F_{\mu\nu}$ and $F^{\mu\nu}$.

Exercise 5 Consider the pure space transformations (e.g., rotations and parity transformation, that is, the simultaneous inversion of all space coordinates) of $F_{\mu\nu} = -F_{\nu\mu}$. Show that with respect to these particular transformations, the components of $F_{\mu\nu}$ transform as the components of usual V_i and axial A_i vectors, such that

$$F_{\mu\nu} = \begin{pmatrix} 0 & V_1 & V_2 & V_3 \\ -V_1 & 0 & A_3 & -A_2 \\ -V_2 & -A_3 & 0 & A_1 \\ -V_3 & A_2 & -A_1 & 0 \end{pmatrix}, \tag{12.12}$$

or $F_{0i} = V_i$ and $F_{ij} = \epsilon_{ijk} A_k$.

Solution. Consider the transformation of the desired type, with $x^{0'} = x^0$. Then

$$F'_{0i}(x') = \frac{\partial x^\alpha}{\partial x^{0'}} \frac{\partial x^\beta}{\partial x^{i'}} F_{\alpha\beta}(x).$$

The first factor is nonzero only when $\alpha = 0$, and the second only when $\beta = j$, therefore

$$F'_{0i} = \frac{\partial x^j}{\partial x^{i'}} F_{0j}.$$

The last formula means that the components F_{0i} form a vector under continuous transformations, such as rotations. Under the parity transformation, the matrix of derivatives is

$$P = \frac{\partial x^j}{\partial x^{i'}} = \text{diag}\,(-1, -1, -1), \tag{12.13}$$

and it is easy to see that F_{0i} behaves like a normal vector $V'_i = -V_i$. The second part of the exercise can be most easily solved by noting that $A_k = \frac{1}{2}\,\epsilon_{ijk}F^{ij}$. As we know from the first part of the book, in this case, the components of A_k do not change under parity, and hence they form an axial vector. Sometimes the notation $F_{\mu\nu} = (\mathbf{V}, \mathbf{A})$ is used for (12.12).

12.2 Relativistic Dynamics of a Free Particle

In order to establish the laws of relativistic dynamics, let us first define the action of a free particle with the mass m. Indeed, this problem can be solved in many different ways, but we shall follow the simplest approach.

Consider the particle moving between the instants of time t_1 and t_2. In order to have Lorentz-covariant equations of motion, the action of the particle should be a constant scalar with respect to the Lorentz rotations (boosts), and also with respect to orthogonal transformations of space coordinates. The simplest scalar expression that satisfies this condition has the form

$$S = -\alpha \int_{t_1}^{t_2} ds, \tag{12.14}$$

where the integral is taken via the world line of the particle. The negative sign is taken just for convenience. α is an unknown constant and one can find it by requiring that the nonrelativistic limit of the expression (12.14) coincides with the action of a free particle in classical mechanics,

$$S_{cl} = \int_{t_1}^{t_2} dt\, L_{cl}, \qquad L_{cl} = T_{cl} = \frac{mv^2}{2}, \tag{12.15}$$

which is just a classical expression for kinetic energy.

In order to compare the expressions (12.14) and (12.15) in the nonrelativistic limit, let us assume that $\beta \ll 1$ and make a series expansion,

$$ds = c\sqrt{1 - \beta^2}dt = c\left(1 - \frac{\beta^2}{2} - \frac{\beta^4}{8} + \dots\right)dt. \tag{12.16}$$

Replacing this expression into (12.14) and using $\beta = v/c$, we arrive at the nonrelativistic limit,

$$S_{nr} = \int_{t_1}^{t_2} \left(-\alpha c + \frac{\alpha v^2}{2c} + \frac{\alpha v^4}{8c^3} + \dots\right)dt. \tag{12.17}$$

The first term in the parentheses is an irrelevant constant. The second term becomes exactly the desired expression (12.15) if we assume that $\alpha = mc$. Then the third and next terms become small $\mathcal{O}(\beta^4)$ corrections to (12.15), which we certainly should expect to meet here. Finally, since the value of α has been fixed, we can write down the final form of the action (12.14),

$$S = -mc \int_{t_1}^{t_2} ds = \int_{t_1}^{t_2} dt \, L, \tag{12.18}$$

where

$$L = -mc^2\sqrt{1 - \beta^2}. \tag{12.19}$$

The next steps will be to derive the expressions for canonical momenta and energy, and construct the Lagrange equations. It is instructive to find the dynamical equations in both four-dimensional and three-dimensional formalisms and compare the results.

Let us start from the $4D$ formalism. Remember that $ds = \sqrt{\eta_{\mu\nu}dx^\mu dx^\nu}$ and consider an arbitrary variation of four-coordinate $\delta x^\mu(t)$, such that

$$\delta x^\mu(t_1) = \delta x^\mu(t_2) = 0. \tag{12.20}$$

The variation of the action is

$$\delta S = -mc \int_{t_1}^{t_2} \frac{1}{2\,ds} \cdot \left(2\,\eta_{\mu\nu}\,dx^\mu \delta dx^\nu\right) = -mc \int_{t_1}^{t_2} ds\,\eta_{\mu\nu}\frac{dx^\mu}{ds}\frac{\delta dx^\nu}{ds}$$

$$= -mc\,\eta_{\mu\nu}\,u^\mu \delta x^\nu \Big|_{t_1}^{t_2} + mc \int_{t_1}^{t_2} ds\,\omega_\mu \delta x^\mu. \tag{12.21}$$

In the last expression, we integrated by parts, used the standard relation $\delta dx^\nu = d\delta x^\nu$, and also the definitions of four-velocity and four-acceleration from (12.2) and (12.6). It is easy to see that the first term in the last expression vanishes due to (12.20). As far as δS should be zero for an arbitrary δx^μ, the equations of motion for a free relativistic particle have the form

$$\omega^\mu = \frac{du^\mu}{ds} = \frac{d^2 x^\mu}{ds^2} = 0.$$ (12.22)

Now we can compare this result with the usual $3D$ treatment. Starting from the Lagrange function (12.19), one can take the derivative with respect to velocity,

$$p_i = \frac{\partial L}{\partial v^i} = \frac{mv^i}{\sqrt{1 - \beta^2}} \implies \mathbf{p} = \frac{m\mathbf{v}}{\sqrt{1 - \beta^2}},$$ (12.23)

which is the relativistic expression for the momentum of the particle. According to classical mechanics, the energy of the particle is

$$\varepsilon = \mathbf{p} \cdot \mathbf{v} - L = \frac{mc^2}{\sqrt{1 - \beta^2}}.$$ (12.24)

The Lagrange equation for the particle and energy conservation provide us with the relations

$$\frac{d\mathbf{p}}{dt} = 0, \quad \frac{d\varepsilon}{dt} = 0.$$ (12.25)

Both expressions (12.23) and (12.24) are very interesting and tell us a lot about relativity. First of all, the nonrelativistic ($\beta \ll 1$) limits for momentum and energy are compatible with classical mechanics, plus small corrections,

$$\mathbf{p} = m\mathbf{v} + \frac{1}{2} m\beta^2 \mathbf{v} + \mathcal{O}(\beta^4), \quad \varepsilon = mc^2 + \frac{mv^2}{2} + \mathcal{O}(\beta^4).$$ (12.26)

Let us note the second formula we already saw in the expansion of Lagrange function. The expression for the energy has the remarkable first term $\varepsilon_0 = mc^2$, corresponding to the *rest energy* of the particle. This is perhaps the most symbolic formula of relativity, which shows that even the static particle has this energy. From the viewpoint of classical mechanics, this term is completely irrelevant since it does not depend on the coordinates or velocity of the particle. However, this formula is indeed very important, because in relativistic quantum theory one can consistently describe how to transform the rest energy of the composite or even elementary particle into other forms of energy. It is interesting that the formula $\varepsilon_0 = mc^2$ can be seen as a definition of mass. For instance, the baryonic matter (which is everything we can see with eyes) is mainly made of neutrons and protons (electrons are much lighter). Both neutrons and protons are composed of quarks, which strongly interact by exchanging massless particles called gluons. It is good to know that the mass of the quarks inside the neutron or proton is only about 5% of the masses of these composite particles. In terms of classical mechanics, we can say that what we observe as the mass of neutron or proton is mainly the potential energy of the interaction of quarks divided by the square of the speed of light.

In the next sections and especially in the last part of the book, we will discuss important applications of the rest energy, e.g., in cosmology.

The second important feature of the expressions (12.23) and (12.24) is that they grow rapidly when $\beta \to 1$, meaning in the ultrarelativistic motions with $v \to c$. This means that one needs an infinite energy and infinite force to make a massive particle move with the speed of light. Also, the superluminal velocities $v > c$ lead to the imaginary energy and momentum of the particle. The particles with such velocities can be consistently interpreted only if their masses are also imaginary. There is an extensive literature on the features of such particles, which are called tachyons (see, e.g., [1, 2]), but the detailed discussion of this interesting subject is out of the scope of the present book.

Finally, let us compare the results of the $4D$ and $3D$ treatments. The expressions (12.4) and (12.5) can be compared with (12.23) and (12.24) to yield the following identification for the four-momentum:

$$p^\mu = mcu^\mu = \left(\frac{\varepsilon}{c}, \mathbf{p}\right). \tag{12.27}$$

Then we can readily check that Eqs. (12.22) and (12.25) are equivalent. An important additional note is that the energy and momentum of the free particle form the four-vector according to (12.27). Also, the identity $u^\mu u_\mu = 1$ for the four-velocity can be cast into the formula called *dispersion relation*,

$$p^2 = p^\mu p^\nu \eta_{\mu\nu} = \frac{\varepsilon^2}{c^2} - \mathbf{p}^2 = m^2 c^2. \tag{12.28}$$

An interesting feature of this formula is that it applies not only to normal massive particles but also to massless particles and even to tachyons (assuming their masses are complex). Of course, formula (12.28) alone does not let us to make a conclusion that it gives the correct universal relation between energy and momentum since both expressions for momentum and energy, (12.23) and (12.24), do not admit the limit $\beta \to 1$. However, later on, we shall meet a solid evidence that formula (12.28) really works for the $m = 0$ case.

Exercise 6 Consider the $4D$ vector of relativistic force

$$g^\alpha = \frac{dp^\alpha}{ds} = mc\frac{du^\alpha}{ds} \tag{12.29}$$

and show that it is related to the usual $3D$ vector of force

$$\mathbf{f} = \frac{d\mathbf{p}}{dt} \tag{12.30}$$

$$\text{by the relations} \quad g^0 = \frac{\mathbf{f} \cdot \mathbf{v}}{c^2\sqrt{1 - \beta^2}} \quad \text{and} \quad g^i = \frac{f^i}{c\sqrt{1 - \beta^2}}. \tag{12.31}$$

Try to establish the relation with the kinematic formulas (12.7).

12.3   Charged Particle in an External Electromagnetic Field

Consider how one can include interaction with an external electromagnetic field into the action (12.18). The first question is how such a field can be parameterized. We know that the electromagnetic field in vacuum has electric and magnetic components **E** and **H**. However, it proves more useful to introduce a four-vector of electromagnetic potential $A^\mu(t, \mathbf{r}) = (\varphi, \mathbf{A})$ and define electric and magnetic fields according to Eqs. (11.12) and (11.11). Some observations are in order here. The two vectors **E** and **H** have six degrees of freedom (that means the number of independent functions), while $A^\mu(t, \mathbf{r})$ has only four of them. This reduction is one manifestation of the fact that the electric and magnetic fields are not independent, but form the unique electromagnetic field. This mixed dynamic can already be seen from Maxwell equations (11.2) and (11.3), and in the solutions of these equations in electrodynamics one can meet many other manifestations of the same feature. Furthermore, we shall see that Maxwell equations possess gauge symmetry, which reduce the number of degrees of freedom to just two.

The fact that scalar φ and vector **A** potentials form the four-vector A^μ is important, as it enables one to construct a covariant form of the interacting term for the charged particle in the form

$$S_{\text{int}} = -\frac{e}{c} \int_{t_1}^{t_2} A_\mu \, dx^\mu. \tag{12.32}$$

Let us note that $A_\mu = (\varphi, -\mathbf{A})$ and hence $A_\mu \, dx^\mu = -\mathbf{A} \cdot d\mathbf{r} + \varphi c \, dt$. Therefore, in the $3D$ notations the total action of the particle together with the interaction term can be written as follows:

$$S = -\int_{s_1}^{s_2} \left\{ mc + \frac{e}{c} A_\mu \frac{dx^\mu}{ds} \right\} ds = \int_{t_1}^{t_2} dt \, L,$$

$$L = -mc^2\sqrt{1 - \beta^2} + \frac{e}{c} \mathbf{A} \cdot \mathbf{v} - e\varphi. \tag{12.33}$$

Here in the first integral the values of the interval $s_{1,2}$ correspond to the instants of time $t_{1,2}$ and $\beta^2 = (\mathbf{v} \cdot \mathbf{v})/c^2$, as usual.

Before deriving the corresponding equations of motion, let us calculate the canonically conjugated momentum and energy of the particle. The first one is

$$P = \frac{\partial L}{\partial v} = \frac{m v}{\sqrt{1 - \beta^2}} + \frac{e}{c} A = p + \frac{e}{c} A, \qquad (12.34)$$

where p is the momentum of the free particle. It is worth remembering that $\frac{\partial f}{\partial v}$ means the gradient of the function f in the space of velocities, while $\frac{\partial f}{\partial r}$ denotes the gradient of the function f in the space of coordinates. These notations are especially useful in the case when f depends on both r and v, such that using the notation grad f may lead to confusion.

The energy of the particle is

$$\varepsilon = P \cdot v - L = \frac{m c^2}{\sqrt{1 - \beta^2}} + e\varphi. \qquad (12.35)$$

Making the Legendre transformation, the last formula can be expressed in terms of momentum (12.34) to give

$$H = \sqrt{m^2 c^4 + c^2 \left(P - \frac{e}{c} A \right)^2} + e\varphi. \qquad (12.36)$$

Using Eq. (12.36), one can trade

$$P \to -\frac{\partial S}{\partial r}$$

and arrive at the Hamilton–Jacobi equation in the form

$$\left(\nabla S - \frac{e}{c} A \right)^2 - \frac{1}{c^2} \left(\frac{\partial S}{\partial t} + e\varphi \right)^2 + m^2 c^2 = 0. \qquad (12.37)$$

Let us now use the variational principle to obtain the equations of motion. To this end consider an arbitrary variation of the coordinates of the particle $\delta x^\mu(t)$, which is frozen at the ends of the interval of integration, $\delta x^\mu(t_1) = \delta x^\mu(t_2) = 0$. Then, as before,

$$\delta s = u_\mu \frac{d \delta x^\mu}{ds}, \qquad (12.38)$$

hence the first variation of the action (12.33) is

$$\delta S = - \int_{s_1}^{s_2} ds \left\{ m c u_\mu \frac{d \delta x^\mu}{ds} + \frac{e}{c} u^\mu (\partial_\alpha A_\mu) \delta x^\alpha + \frac{e}{c} A_\mu \frac{d \delta x^\mu}{ds} ds \right\}. \qquad (12.39)$$

Integrating by parts and changing the names of indices, we arrive at

$$\delta S = \int_{s_1}^{s_2} ds \left\{ m c \frac{d u_\mu}{ds} + \frac{e}{c} u^\alpha (\partial_\alpha A_\mu) - \frac{e}{c} (\partial_\mu A_\alpha) u^\alpha \right\} \delta x^\mu, \qquad (12.40)$$

which provides the following equation of motion:

$$mc\frac{du_\mu}{ds} - \frac{e}{c}u^\alpha F_{\mu\alpha} = 0, \tag{12.41}$$

In the last formula, we introduced a very important object

$$F_{\mu\nu} = \partial_\mu A_\nu - \partial_\nu A_\mu, \tag{12.42}$$

which is called the *tensor of electromagnetic field*. The origin of this name is that the explicit calculation (we leave it to the reader as an exercise) shows that

$$F_{\mu\nu} = \begin{pmatrix} 0 & E_x & E_y & E_z \\ -E_x & 0 & H_z & -H_y \\ -E_y & -H_z & 0 & H_x \\ -E_z & H_y & -H_x & 0 \end{pmatrix} = (\mathbf{E}, \mathbf{H}), \tag{12.43}$$

where E_i and H_i are components of the electric and magnetic fields defined in (11.12) and (11.11).

In Eq. (12.43), we have started to use the following useful notations: $E_{x,y,z}$ and $H_{x,y,z}$ stand for the components of Euclidean $3D$ vectors, while $A_{1,2,3} = -A^{1,2,3}$, $H_{1,2,3} = -H^{1,2,3}$, and $E_{1,2,3} = -E^{1,2,3}$ are regarded as components of $4D$ vector in Minkowski space. The first index μ in the *l.h.s.* of (12.43) indicates the number of lines, while ν numbers the columns. It is clear from (12.43) that the tensor under discussion is antisymmetric $F_{\mu\nu} = -F_{\nu\mu}$. For the space components F_{ik}, there is the following useful relation:

$$F_{ik} = \sum_{l=1,2,3} \varepsilon_{ikl} H_l, \quad \text{where} \quad H_l = H_x, H_y, H_z. \tag{12.44}$$

Let's check it for one particular example. Omitting the vanishing terms, we get

$$F_{12} = \varepsilon_{123} H_3 = \sum_k \varepsilon_{12k} H_k,$$

which is exactly Eq. (12.44). Another two cases can be verified in the same way. Thus, we can use (12.44), and furthermore, the result of the exercise (12.12) shows that $\mathbf{H}$ is axial vector while $\mathbf{E}$ is the usual vector, which changes the sign of its components under parity transformation.

Equation (12.41) is nothing but the $4D$ version of the Lorentz force acting on the particle with the charge e. Later on, in one of the exercises, we shall see that the familiar form of the Lorentz force is recovered in the $3D$ notations.

Exercise 7 Verify the dispersion relation between energy and momentum of a particle interacting with electromagnetic potential,

$$\left(\frac{\varepsilon - e\varphi}{c}\right)^2 = m^2c^2 + \left(\mathbf{P} - \frac{e}{c}\mathbf{A}\right)^2. \tag{12.45}$$

Exercise 8 Verify that in the nonrelativistic limit, expressions (12.35) and (12.36) boil down to

$$\varepsilon_{nr} = mc^2 + \frac{mv^2}{2} + e\varphi + \dots . \tag{12.46}$$

$$H_{nr} = \frac{1}{2m}\left(\mathbf{P} - \frac{e}{c}\mathbf{A}\right)^2 + e\varphi + \dots . \tag{12.47}$$

Exercise 9 Verify the formula (12.43), paying special attention to the distinction between indices in $4D$ (which are raised and lowered with Minkowski metric) and indices in $3D$, which are raised and lowered with δ^{ij} and δ_{ij}. The indices in the matrix in the r.h.s. of (12.43) are $3D$ Euclidean ones. After this calculation, check that $F_{ik} = \epsilon_{ikl}H^l$, once again in Euclidean notations with $H_l = H^l$.

Exercise 10 Remember that the four-momentum is defined as

$$p^\mu = mcu^\mu = \left(\frac{\varepsilon_{kin}}{c}, \mathbf{p}\right), \quad \text{where} \quad \varepsilon_{kin} = \frac{mc^2}{\sqrt{1-\beta^2}} \quad \text{and} \quad \mathbf{p} = \frac{m\mathbf{v}}{\sqrt{1-\beta^2}}.$$

Show that Eq. (12.41) can be cast into the form

$$\frac{1}{c\sqrt{1-\beta^2}}\frac{dp_\alpha}{dt} = \frac{e}{mc^2}p^\mu F_{\alpha\mu} \tag{12.48}$$

and furthermore as $\quad \mu = 0 : \quad \frac{d\varepsilon_{kin}}{dt} = e\mathbf{v}\cdot\mathbf{E}, \tag{12.49}$

$$\mu = 1, 2, 3 : \quad \frac{d\mathbf{p}}{dt} = e\mathbf{E} + \frac{e}{c}\mathbf{v}\times\mathbf{H}. \tag{12.50}$$

The last equation is a familiar $3D$ form of the Lorentz force (11.1).

Observation Deriving (12.50) as Lagrange equation from (12.33) requires the identity

$$\text{grad}\,(\mathbf{A}\cdot\mathbf{B}) = [\mathbf{A},\,\text{rot}\,\mathbf{B}] + [\mathbf{B},\,\text{rot}\,\mathbf{A}] + (\mathbf{A},\nabla)\mathbf{B} + (\mathbf{B}\cdot\nabla)\mathbf{A},$$

which was suggested as Exercise 6 in Sect. 1.6.

Exercise 11 By raising indices, verify that $F^{\mu\nu} = (\mathbf{E}, -\mathbf{H})$.

$$F^{\mu\nu} = \begin{pmatrix} 0 & -E_x & -E_y & -E_z \\ E_x & 0 & H_z & -H_y \\ E_y & -H_z & 0 & H_x \\ E_z & H_y & -H_x & 0 \end{pmatrix} = (-\mathbf{E}, \mathbf{H}). \tag{12.51}$$

Calculate also the matrix forms of $F^\alpha_{\cdot\,\nu}$ and $F_{\mu\cdot}^{\,\,\beta}$ and discuss the difference between them.

Exercise 12 It proves useful to introduce an absolutely antisymmetric Levi-Civita symbol $\varepsilon_{\alpha\beta\mu\nu}$ in Minkowski space. It is customary to define $\varepsilon^{0123} = 1$. Assuming that it is a tensor under Lorentz transformations, by lowering the indices show that in this case $\varepsilon_{0123} = -1$. Check this formula, and after that write down and prove (similar to what can be found in the first part of the book in $3D$) the matrix representation for the product $\varepsilon_{\alpha\beta\mu\nu}\varepsilon^{\rho\sigma\kappa\omega}$. After that, derive the following contractions:

$$\varepsilon_{\alpha\beta\mu\nu}\varepsilon^{\alpha\sigma\kappa\omega}, \quad \varepsilon_{\alpha\beta\mu\nu}\varepsilon^{\alpha\beta\kappa\omega}, \quad \varepsilon_{\alpha\beta\mu\nu}\varepsilon^{\alpha\beta\mu\omega} \text{ and } \varepsilon_{\alpha\beta\mu\nu}\varepsilon^{\alpha\beta\mu\nu} = -24. \quad (12.52)$$

Explain how the five relations can be obtained starting from the last one in (12.52).

Exercise 13 It proves useful to define the dual tensor

$$\tilde{F}^{\alpha\beta} = \frac{1}{2}\varepsilon^{\alpha\beta\mu\nu} F_{\mu\nu}. \quad (12.53)$$

Show that

$$\tilde{F}_{\alpha\beta} = \begin{pmatrix} 0 & H_1 & H_2 & H_3 \\ -H_1 & 0 & E_3 & -E_2 \\ -H_2 & -E_3 & 0 & E_1 \\ -H_3 & E_2 & -E_1 & 0 \end{pmatrix} = (\mathbf{H}, \mathbf{E}) \quad (12.54)$$

and that $\tilde{F}^{\alpha\beta} = (-\mathbf{H}, \mathbf{E})$.

Solution. Let us remember that $\varepsilon^{0123} = 1$. Consider

$$\tilde{F}^{01} = \frac{1}{2}\left(\varepsilon^{0123} F_{23} + \varepsilon^{0132} F_{32}\right) = \varepsilon^{0123} F_{23} = F_{23} = H_1, \quad (12.55)$$

where we used $\varepsilon^{0123} = -\varepsilon^{0132}$ and $F_{23} = -F_{32}$. Other coefficients can be obtained in a similar way.

Exercise 14 Explore the invariance of Eqs. (12.49) and (12.50) under parity transformation (also called P-transformation),

$$x \to -x, \quad y \to -y, \quad z \to -z, \quad \text{or simply } \mathbf{r} \to -\mathbf{r}. \quad (12.56)$$

Namely, show that the quantities that enter these equations transform as

$$\mathbf{p} \to -\mathbf{p}, \quad \varphi \to \varphi, \quad \mathbf{A} \to -\mathbf{A}, \quad \mathbf{E} \to -\mathbf{E}, \quad \mathbf{H} \to \mathbf{H}. \quad (12.57)$$

The last relation means that $\mathbf{H}$ is an axial vector. It is necessary to check out that (12.49) and (12.50) do not change.

Exercise 15 Explore the invariance of Eqs. (12.49) and (12.50) under time inversion, or T-transformation: $t \to -t$; and also under charge inversion, or C-transformation: $e \to -e$. Show that the same equations do not change when C, P, and T transformations are applied simultaneously.

Observation The last symmetry, called CPT, is likely to hold at the fundamental level in particle physics, while P-symmetry alone is violated in the weak interactions sector of the Standard Model. Indeed, the CPT-symmetry can be violated by the presence of some (scalar, vector, or even tensor) field at the astrophysical scale. There are serious experimental and theoretical efforts working to establish upper bounds on CPT-breaking fields. According to the present-day data, if such fields exist, they are extremely weak.

Exercise 16 As a follow-up of the previous exercise, describe different types of extending Eq. (12.50) that violate CPT-symmetry.

12.3.1 Gauge Invariance of the Particle Action

The most important feature of Eq. (12.41) is that it possesses a symmetry called gauge invariance. In order to see this symmetry, we note that the electromagnetic potential A_μ enters these equations only through the combination (12.42). And this combination does not change under *gauge transformation*

$$A_\mu \to A'_\mu = A_\mu + \partial_\mu f, \qquad f = f(x) = f(\mathbf{r}, t). \tag{12.58}$$

It is easy to check by direct substitution that the tensor of electromagnetic field does not change,

$$\begin{aligned} F'_{\mu\nu} &= \partial_\mu A'_\nu - \partial_\nu A'_\mu = \partial_\mu A_\nu + \partial_\mu \partial_\nu f - \partial_\nu A_\mu - \partial_\nu \partial_\mu f \\ &= \partial_\mu A_\nu - \partial_\nu A_\mu = F_{\mu\nu}. \end{aligned} \tag{12.59}$$

The immediate consequence of this is that Eq. (12.41) does not change under the transformation (12.58), and hence the motion of a charged particle cannot distinguish between the potentials A'_μ and A_μ. In other words, at least one out of the four degrees of freedom of A_μ is *not* physical, as it does not affect the motion of a test particle. Another manifestation of the same fact is that electric and magnetic fields, $\mathbf{E}$ and $\mathbf{H}$, which form the tensor $F_{\mu\nu}$, according to Eq. (12.43), also do not change under the action of (12.58). We leave it as an exercise to the reader to check this symmetry property using the $3D$ definitions (11.12) and (11.11). It is useful to start by rewriting the gauge transformation (12.58) in the $3D$ form,

$$\mathbf{A} \to \mathbf{A}' = \mathbf{A} - \text{grad } f, \tag{12.60}$$

$$\varphi \to \varphi' = \varphi + \frac{1}{c}\frac{\partial f}{\partial t}. \tag{12.61}$$

The last thing we have to do is to see how the gauge transformation (12.58) affects the action of the particle. Obviously, we need to worry only about the interaction term (12.32) since the free action of the particles does not depend on A_μ. Transforming (12.32), we get

$$
\begin{aligned}
S_{\text{int}} \to S'_{\text{int}} &= -\frac{e}{c} \int_{s_1}^{s_2} A'_\mu \, dx^\mu = -\frac{e}{c} \int_{s_1}^{s_2} \left(A_\mu + \partial_\mu f \right) dx^\mu \\
&= S_{\text{int}} - \frac{e}{c} \int_{s_1}^{s_2} ds \, \frac{\partial f}{\partial x^\mu} \frac{dx^\mu}{ds} = S_{\text{int}} - \frac{e}{c} \int_{s_1}^{s_2} ds \, \frac{df}{ds}. \quad (12.62)
\end{aligned}
$$

The last term is the integral of the total derivative and hence does not represent an essential change of the action.

Exercise 17 Verify that (12.60) and (12.61) together are equivalent to (12.58). After that, prove the gauge invariance of (11.12) and (11.11) using relations (12.60) and (12.61).

Exercise 18 Consider the case of a constant and homogeneous magnetic field $\mathbf{H} = \hat{\mathbf{k}} H_0$. Write down the general form of the vector potential $\mathbf{A}$, which has only linear dependence on x, y, z and corresponds to such an $\mathbf{H}$. After that, show that it is possible to find a linear function $f(x, y, z)$ that after the corresponding gauge transformation $A_y = A_z = 0$.

Exercise 19 Repeat the consideration of the previous exercise for the case of a constant and homogeneous electric field $\mathbf{E} = \hat{\mathbf{k}} E_0$. In fact, this case is more complicated, because the components of $\mathbf{A}$ and φ can depend on time. However, by means of the gauge transformation this dependence can be removed without changing the electric field.

12.3.2 Continuous Description and Electric Current

The next step in our consideration is to introduce the notion of electric current. To this end, we consider many point-like charges e_a, assuming that a is the index that numbers the particles. We assume that these particles create electric $\mathbf{E}$ and magnetic $\mathbf{H}$ fields, also there may be external electric and magnetic fields. Then each of the particles interacts with the field created by all the particles and by an external field, if the last is present. The interaction part of the action has the form

$$
S_{\text{int}} = -\sum_a \frac{e_a}{c} \int A_\mu(x_a) \, dx_a^\mu . \quad (12.63)
$$

Let us now use continuous description and introduce the notation $\rho(t, \mathbf{r})$ for the density of charge. The average density in the volume V is defined as $\rho = Q/V$, where $Q = \sum_b e_b$ is the total charge since the last sum is taken over all the charges inside the volume V. It is important to note that $\rho(t, \mathbf{r})$ is *not* a scalar. In order to understand this, consider the same volume at rest, V_0, and in the inertial reference frame that moves with the constant speed $\mathbf{v}$. According to Eq. (11.36), the volume in this frame is $V = V_0\sqrt{1 - \beta^2}$. Thus, the charge density transforms as

$$\rho = \frac{\rho_0}{\sqrt{1 - \beta^2}}, \tag{12.64}$$

which is not the scalar rule. As a result, one can define the four-current

$$j^\mu = \rho\frac{dx^\mu}{dt} = \rho\frac{ds}{dt}\frac{dx^\mu}{ds} = c\rho\sqrt{1 - \beta^2}\frac{dx^\mu}{ds}, \tag{12.65}$$

and it is easy to see that it transforms as a four-vector with respect to the Lorentz transformations. The time and space components of j^μ form scalar and $3D$ vector with respect to the purely spacial change of coordinates,

$$j^\mu = (c\rho, \mathbf{j}), \qquad \mathbf{j}(t, \mathbf{r}) = \rho(t, \mathbf{r})\mathbf{v}. \tag{12.66}$$

The comparison between (12.63) and (12.65) shows that the former can be rewritten in the continuous notations as

$$S_{\text{int}} = -\int_{t_1}^{t_2} d^3x\, dt\, A_\mu\frac{dx^\mu}{cdt}\rho(t, \mathbf{r}) = \frac{1}{c^2}\int d^4x\, A_\mu\, j^\mu. \tag{12.67}$$

Now the total action can be cast into the form

$$S = -\sum\int mc\, ds - \frac{1}{c^2}\int d^4x\, A_\mu\, j^\mu, \tag{12.68}$$

with the sum over all particles of the system.

Now we are in a position to understand the fundamental role of the gauge symmetry. Let us perform the gauge transformation (12.58) in the action (12.68). The variation of the action is

$$\delta S = -\frac{1}{c^2}\int d^4x\, j^\mu\, \partial_\mu f. \tag{12.69}$$

Integrating by parts, we can ignore the surface term just assuming that the current j^μ vanishes on the boundary of the volume where the charges are moving. Then the requirement of vanishing δS for any choice of the gauge parameter f leads to the identity

$$\partial_\mu j^\mu = 0 \qquad \Longrightarrow \qquad \frac{\partial\rho}{\partial t} + \text{div}\,\mathbf{j} = 0. \tag{12.70}$$

The last relation is the continuity equation for the charge. The physical sense of this relation becomes clear if we consider the time dynamics of the overall charge Q inside a compact figure (V) with the volume V and boundary surface (∂V),

$$\frac{dQ}{dt} = \frac{d}{dt} \iiint\limits_{(V)} \rho \, dV = \iiint\limits_{(V)} \frac{\partial \rho}{\partial t} \, dV . \qquad (12.71)$$

Using Eq. (12.70) and the Gauss theorem, this becomes

$$\frac{dQ}{dt} = - \oiint\limits_{(\partial V)} \mathbf{j} \cdot d\mathbf{S}, \qquad (12.72)$$

where $d\mathbf{S}$ is the oriented element of integration in the surface integral of the second type. The r.h.s. of the last equation represents the flux of the current $\mathbf{j}$ through the boundary of the volume. From the physics side, this means that the charge inside (V) can change only due to the flux of the current through the surface (∂V). In other words, electric charge cannot be created inside the volume.

Finally, we arrive at the direct relation between the gauge symmetry and the conservation of electric charge. This relation certainly shows that the gauge invariance is a fundamental principle of electrodynamics. In the next sections, we shall see how this principle will help us to establish the action of the electromagnetic field and finally reformulate Maxwell equations (11.2) and (11.3) in the explicitly relativistic form.

References

1. O.-M.P. Bilaniuk, V.K. Deshpande, E.C.G. Sudarshan, Am. J. Phys. **30**, 718 (1962)
2. O.-M. Bilaniuk, E.C.G. Sudarshan, Phys. Today **22**, 43 (1969)

Chapter 13
Maxwell Equations in Relativistic Form

13.1 Action and Relativistic Equations for A^μ

In what follows we shall deal a lot with fields, hence it is important to formulate Lagrange equations for relativistic fields on a regular basis. After doing this, we go on to construct the action and derive the dynamical equations for electromagnetic potential in relativistic form.

13.1.1 Minimal Action Principle for Fields

Consider the field $\varphi(x)$, where $x = x^\mu = (x^0, x^1, x^2, x^3)$. This field $\varphi(x)$ can be multicomponent, $\varphi = \varphi_i$, it can be scalar, vector, tensor, or have another nature.[1] Since these details are not relevant in what follows, for the sake of simplicity we use the condensed notation $\varphi(x)$.

The action of the field $\varphi = \varphi(x)$ has the form of the $4D$ integral

$$S = \int_{\mathcal{M}} d^4x \, \mathcal{L}(\varphi, \partial_\mu \varphi), \qquad (13.1)$$

where $\mathcal{L}$ is the density of the Lagrange function, also called Lagrangian. In order to formulate the least action principle, we assume that the dynamics of $\varphi(x)$ occurs in the space–time volume $\mathcal{M}$, surrounded by the surface $\partial \mathcal{M}$. The action (13.1) depends only on the field and its first partial derivatives. This is the simplest case, and we mainly consider it because in general, higher derivatives in the action lead to instabilities and other complications. Some examples of such theories will be

---

[1] In quantum mechanics and particle physics, we need to deal with fields called spinors, for instance. But their consideration is beyond the scope of this book.

© Springer Nature Switzerland AG 2019
I. L. Shapiro, *A Primer in Tensor Analysis and Relativity*, Undergraduate
Lecture Notes in Physics, https://doi.org/10.1007/978-3-030-26895-4_13

suggested as exercises and also considered in the third part of the book, when we will be discussing extensions of general relativity.

Let's take an arbitrary variation of the field $\delta\varphi(x)$ in the bulk $\mathcal{M}$, but require that it vanish on the boundary,

$$\delta\varphi(x)\Big|_{\partial\mathcal{M}} \equiv 0. \tag{13.2}$$

The variation $\delta\varphi(x)$ is not related to the variation of independent coordinates x, hence

$$\delta\left(\frac{\partial\varphi}{\partial x^\mu}\right) = \delta\partial_\mu\varphi(x) = \partial_\mu\delta\varphi(x) = \frac{\partial(\delta\varphi)}{\partial x^\mu}. \tag{13.3}$$

From this point, we always assume this relation and also avoid writing the space–time argument x of the field φ and its derivatives, for the sake of compactness. As usual in the variational principle, we require that the first variation of S vanishes, hence

$$\delta S = \int_\mathcal{M} d^4x \left\{\frac{\partial\mathcal{L}}{\partial\varphi}\delta\varphi + \frac{\partial\mathcal{L}}{\partial_\mu\varphi}\delta(\partial_\mu\varphi)\right\}. \tag{13.4}$$

Integrating by parts in the last term, one can cast (13.4) in the form

$$\delta S = \int_\mathcal{M} d^4x \left\{\frac{\partial\mathcal{L}}{\partial\varphi} - \frac{\partial}{\partial x^\mu}\left(\frac{\partial\mathcal{L}}{\partial_\mu\varphi}\right)\right\}\delta\varphi. \tag{13.5}$$

Here we took into account the fact that the surface term equals to zero because of the boundary condition for variations (13.2),

$$\int_\mathcal{M} d^4x \frac{\partial}{\partial x^\mu}\left\{\frac{\partial\mathcal{L}}{\partial_\mu\varphi}\delta\varphi\right\} = \int_{\partial\mathcal{M}} d\Sigma_\mu \frac{\partial\mathcal{L}}{\partial_\mu\varphi}\delta\varphi = 0, \tag{13.6}$$

where $d\Sigma_\mu$ is the oriented element of integration on the surface $\partial\mathcal{M}$, and we used the $4D$ version of the Stokes theorem. In the first part of the book, we proved only the $3D$ version of this theorem. The reader can find a complete formulation and proof in many books, e.g., in [1].

Starting from Eq. (13.5), the requirement of $\delta S = 0$ for an arbitrary variation $\delta\varphi(x)$ provides the Lagrange equation for the field,

$$\frac{\partial\mathcal{L}}{\partial\varphi} = \frac{\partial}{\partial x^\mu}\left[\frac{\partial\mathcal{L}}{\partial(\partial_\mu\varphi)}\right]. \tag{13.7}$$

Exercise 1 Generalize the consideration leading to Lagrange equation (13.7) for the multicomponent field φ_i, where $i = 1, \ldots, N$.

Exercise 2 Consider the action for the multicomponent scalar field

$$S_{sc} = \int d^4x \left\{ \frac{1}{2} \eta^{\mu\nu} \partial_\mu \varphi_i \, \partial_\nu \varphi_i - \frac{1}{2} m^2 \varphi_i \varphi_i \right\}, \tag{13.8}$$

where the sum over i is assumed and the units with $c = 1$ and Planck constant $\hbar = 1$ are used.[2] (i) Using dimensional arguments, recover the powers of c and $\hbar$ in the standard units. (ii) Derive the dynamical equations for φ_i using (13.7) and also directly, by taking the variational derivative of the action. Write down the result (called Klein–Gordon equation)

$$\left(\Box + \frac{c^2 m^2}{\hbar^2} \right) \varphi_i = 0 \tag{13.9}$$

in the $3D$ notations.

Exercise 3   Consider the solution of Eq. (13.9) in the form of the monochromatic plane wave propagating in the direction $\hat{n}$,

$$\varphi_i(t, \mathbf{r}) = f_i(\mathbf{k}) \, e^{i\omega_k t - \frac{i}{c} \mathbf{r} \cdot \mathbf{k}} + f_i^*(\mathbf{k}) \, e^{-i\omega_k t + \frac{i}{c} \mathbf{r} \cdot \mathbf{k}}, \tag{13.10}$$

where $\mathbf{k} = k\hat{n}$. Find the dispersion relation between ω_k and $k = |\mathbf{k}|$. Compare it to the similar relation for the relativistic particle (12.28) and discuss the interpretation of ω_k and $\mathbf{k}$ as energy and momentum of the wave.

Exercise 4   Consider the higher derivative action

$$S_{hd} = \int_{\mathcal{M}} d^4x \, \mathcal{L}(\varphi, \partial_\mu \varphi, \partial_\mu \partial_\nu \varphi). \tag{13.11}$$

Formulate the principle of minimal action and the variational problem, including the modified form of the condition (13.2). Try to make further generalizations, including derivatives up to the order k. Explain why the Lagrange equations (13.7) in the theory with the standard action (13.1) have only second order in partial derivatives. What is the order of generalized Lagrange equations in the model with k derivatives in the action, e.g., with $k = 2$ for the action (13.11)?

Solution.   The boundary conditions for the variation in the case of (13.11) are $\varphi(\partial\mathcal{M}) = \partial_\alpha \varphi(\partial\mathcal{M}) = 0$ and the generalized Lagrange equations have the form

$$\frac{\partial \mathcal{L}}{\partial \varphi} - \frac{\partial}{\partial x^\mu} \left[\frac{\partial \mathcal{L}}{\partial(\partial_\mu \varphi)} \right] + \frac{\partial^2}{\partial x^\mu \partial x^\nu} \left[\frac{\partial^2 \mathcal{L}}{\partial(\partial_\mu \varphi) \, \partial(\partial_\nu \varphi)} \right] = 0. \tag{13.12}$$

Exercise 5   Write down the action of the free action of the type (13.11) with a single scalar field. The word "free" here means that there are no interactions. This can be provided if the action is quadratic (bilinear) in the field and its derivatives. Calculate

---

[2]The units with $c = 1$ and $\hbar = 1$ have dimension of mass inverse to length. The defining relation for $\hbar$ is $E = \hbar\nu$, where E is the energy of the photon and ν is its frequency.

explicitly Eq.(13.12) in this case and consider the plane wave solution (13.10). Analyze the dispersion relation in this case and show that there are solutions with (i) negative ω_k (energy), which are called higher derivative ghosts, with (ii) negative m^2 (tachyons), or with (iii) tachyonic ghosts with negative ω_k and also negative m^2. White down the second-order actions for the free field, which has each of these three cases as plane wave solutions.

13.2 Maxwell Equations in Relativistic Form

As we have seen in the previous sections, the fundamental equations of electrodynamics should be invariant under two distinct symmetries: Lorentz covariance and gauge invariance. The first symmetry means that the action of the electromagnetic field is supposed to be a scalar with respect to space transformation and Lorentz boosts. The second symmetry is equally important since it guarantees the conservation of electric charge.

13.2.1 Invariant Action for the Electromagnetic Field

We have seen that the action of particles interacting with external electromagnetic potential A^μ satisfies both conditions listed above. As a result, we obtained the Lorentz force equation formulated is the explicitly Lorentz covariant form (12.41). Now we shall see whether it is possible to achieve the same invariance for the action of the electromagnetic field itself. As far as we demand Lorentz invariance, the action should be written in terms of the 4D vector $A^\mu = (\mathbf{A}, \varphi)$.

The list of our requirements for the action of the electromagnetic field should be supplemented by extra conditions. Let us forbid the introduction of the terms with higher derivatives of the potential A^μ. As we have already mentioned above, if equations of motion have derivatives higher than the second ones, the theory is expected to have serious problems with instabilities of solutions and hence serious problems with physical interpretation. Moreover, our final target is to reproduce the Maxwell equations (11.2) and (11.3), and those are of the second order in derivatives of the potentials $\mathbf{A}$ and φ. Furthermore, the desired equations are linear in $\mathbf{E}$ and $\mathbf{H}$, and therefore linear in $\mathbf{A}$ and φ. Thus, we can expect that the action of A^μ should be quadratic (bilinear) in the potential.

All in all, we arrive at the general form of the candidate action

$$S_{\text{gen}} = \int d^4x \left\{ a F_{\mu\nu} F^{\mu\nu} + b(\partial_\mu A^\mu)^2 + c A_\mu A^\mu \right\}, \qquad (13.13)$$

where a, b, and c are arbitrary constants. The expression (13.13) is relativistic 4D scalar and is bilinear in A^μ. It is easy to check that the last two terms $(\partial_\mu A^\mu)^2$ and $A_\mu A^\mu$ are not invariant under gauge transformation (12.58). Therefore, one has to set $b = c = 0$, and now we only need to define the sign and the magnitude of the coefficient a.

The definition of the sign can be performed as follows. Due to the definition (12.43), the time derivatives enter the expression $F_{\mu\nu} F^{\mu\nu}$ in the combination $-2(\partial_0 A_k)^2$, where we used the relation $A_k = -A^k$. As far as we want a positively defined kinetic term of the potential, the sign of a should be negative. The magnitude of the coefficient a can be established on the basis of tradition, or convenience. The standard definition leads to the action[3]

$$S_{\text{em}} = -\frac{1}{16\pi c} \int d^4x \, F_{\mu\nu} F^{\mu\nu}. \tag{13.14}$$

Exercise 6 Analyze whether the action (13.14) is really the most general one that is Lorentz invariant and bilinear in the field A^μ.

Exercise 7 Consider the change of variables $A_*^\mu = k A^\mu$, with $k = const$, in the action (13.14) in the units with $c = 1$. Derive such values of k that trade the coefficient $(16\pi c)^{-1}$ of the action to (i) the value $\frac{1}{4}$, or (ii) the value $\frac{1}{4e^2}$. Find the form of the interacting term (12.67) in both of these cases. Discuss the technical disadvantages of the choice (ii), taking into account that the sign of the charge e can be either positive or negative. Discuss whether the sign issue can be taken into account in this case?

13.2.2   Relativistic Form of Maxwell Equations

Before using the least action principle to derive the equations of motion from the expression (13.14), let us explore in more detail the definition (12.43) of the tensor of electromagnetic field $F_{\mu\nu}$. It is easy to check that this antisymmetric tensor satisfies the identity

$$\partial_\alpha F_{\mu\nu} + \partial_\nu F_{\alpha\mu} + \partial_\mu F_{\nu\alpha} = 0. \tag{13.15}$$

After contracting (13.15) with the Levi-Civita symbol $\varepsilon^{\alpha\beta\mu\nu}$, one can rewrite the same relation in the equivalent form as

$$\varepsilon^{\alpha\beta\mu\nu} \left(\partial_\alpha F_{\mu\nu} + \partial_\nu F_{\alpha\mu} + \partial_\mu F_{\nu\alpha} \right) = 0 \quad \Longrightarrow \quad \partial_\alpha \tilde{F}^{\alpha\beta} = 0. \tag{13.16}$$

On the other hand, (13.15) can be restored from (13.16) by multiplying by the Levi-Civita symbol $\varepsilon_{\lambda\beta\mu\nu}$.

Let us consider the $3D$ representation of Eq. (13.16). Consider the case $\beta = 0$. Taking into account that the terms with $\alpha = \beta$ vanish and using Eq. (12.54), we arrive at the equation

---

[3]In quantum field theory, it is more common to use the coefficient $-1/4$ instead of $-1/(16\pi c)$ and the units with $c = 1$.

$$\partial_k \tilde{F}^{k0} = 0 \quad \Longrightarrow \quad \text{div } \mathbf{H} = 0. \tag{13.17}$$

Indeed, this is the first one of the homogeneous Maxwell equations (11.3). In its physical sense, there are no point-like sources for magnetic field (monopoles).

The theory with magnetic monopoles is beyond the scope of our book, but it is a theoretically interesting issue. The interested reader can start learning about this subject from the classical textbook [2], which refers to the original papers by Dirac.

In the case $\beta = i$, the equation that results from (13.16) can be easily elaborated using (12.54), to give

$$\partial_o \tilde{F}^{ko} + \partial_k \tilde{F}^{ki} = 0 \quad \Longrightarrow \quad \frac{1}{c}\frac{\partial \mathbf{H}}{\partial t} = -\text{rot } \mathbf{E}, \tag{13.18}$$

that is exactly the second equation of the second couple of Maxwell equations (11.3). Equation (13.18) is the local form of the Faraday's law of induction. The integral form of this law links the time derivative of the magnetic flux through the surface (Σ) and the circulation of the electric field over the boundary of this surface $(\partial \Sigma)$,

$$\frac{d}{dt}\iint_{(\Sigma)} \mathbf{H} \cdot d\Sigma = -\oint_{(\partial \Sigma)} \mathbf{E} \cdot d\mathbf{r}. \tag{13.19}$$

At this point, it is worth making a general comment. Indeed, it is remarkable that the homogeneous part of Maxwell equations can be obtained on the basis of definition (12.42) and identification of electric and magnetic fields (11.12) and (11.11), which combine into the tensor (12.43). Let us remember that we arrived at these relations from the general arguments based on Lorentz covariance and gauge invariance. On the other hand, these general arguments came from the desire to construct new space–time properties that make Maxwell equations compatible with the laws of mechanics. Now, if we can identify the covariant form of the remaining two equations, (11.2), our consideration will gain a logically closed form.

There is a curious detail about Eqs. (13.16) and (13.18) from one side and Eq. (13.15) from another side. The last is a trivial algebraic identity that immediately follows from the definition (12.43). At the same time, it is completely equivalent to (13.16) and (13.18), which both have deep physical sense and great importance. One can ask how it can be that something trivial is mathematically equivalent to something so important. The answer is that (13.15) is the equation in terms of A_μ which has four components. At the same time, (13.16) and (13.18) are written in terms of the two vectors $\mathbf{E}$ and $\mathbf{H}$, which have, taken together, six components. Thus, the secret is in the identification of potentials (11.12) and (11.11), as we have already mentioned above.

In order to arrive at the last couple of Maxwell equations, consider the action of electromagnetic field (13.14) together with the term that corresponds to the interaction with current from (12.68). The first term in (12.68) does not need to be included since it does not depend on A_μ and therefore cannot affect the corresponding variation. Thus, the relevant part of the total action is

$$S_t = \int d^4x \left\{ -\frac{1}{16\pi c} F_{\mu\nu} F^{\mu\nu} - \frac{1}{c^2} A_\mu j^\mu \right\}. \tag{13.20}$$

To obtain the dynamical equations we use variational principle. Let us perform an arbitrary variation $\delta A_\mu(x)$ of the potential and require that $\delta A_\mu(x)\big|_{(\partial M)} = 0$, where (∂M) is the boundary of the volume (M), where the dynamics of the fields occurs. For instance, (M) can be infinite and then (∂M) also lies at the space–time infinity. The variation of the action can be easily elaborated by using antisymmetry of $F_{\mu\nu}$ and integrating by parts,

$$\delta S_t = \int d^4x \left\{ \frac{1}{4\pi c} F^{\mu\nu} \cdot \partial_\nu \delta A_\mu - \frac{1}{c^2} j^\mu \cdot \delta A_\mu \right\}$$

$$= \int d^4x \, \delta A_\mu \left\{ -\frac{1}{4\pi c} \partial_\nu F^{\mu\nu} - \frac{1}{c^2} j^\mu \right\}. \tag{13.21}$$

Once again using antisymmetry of $F_{\mu\nu}$, we arrive at the nonhomogeneous equations

$$\partial_\mu F^{\mu\nu} = \frac{4\pi}{c} j^\nu. \tag{13.22}$$

Now we have to rewrite the last equation in Euclidean $3D$ notations. For $\nu = 0$, we can use (12.51) from the exercise and get

$$\partial_k F^{k0} = \text{div} \, \mathbf{E} = \frac{4\pi}{c} j^0 = 4\pi\rho, \tag{13.23}$$

which is exactly the first of the second couple of Maxwell equations (11.2).

And for $\nu = i$ the result is obtained by identifying $F^{01} = E_x$, $F^{21} = H_z$, $F^{31} = -H_y$, etc. After some algebra we arrive at

$$-\frac{1}{c} \frac{\partial \mathbf{E}}{\partial t} + \text{rot} \, \mathbf{H} = \frac{4\pi}{c} \mathbf{j}, \tag{13.24}$$

which is exactly the second equation of (11.2).

Now we can say that the logical chain that started from the Maxwell equations is closed. We started by observing that the speed of electromagnetic wave cannot depend on the observer (or reference frame) according to these equations. This resulted in the invariance of the interval and Lorentz transformations. Then we started to construct the equations of electromagnetic theory in the "top to bottom" style and found that the conservation of charge requires gauge invariance. Postulating this symmetry as a fundamental property, we arrived at the action of electromagnetic field coupled to charges and proved that the Maxwell equations are in fact relativistic equations (13.15) and (13.22). This means that we have found the general framework for relativistic mechanics that is compatible with the laws of Maxwell.

Exercise 8 Using symmetries of $F_{\mu\nu}$, calculate the number of independent equations in (13.15) and compare it with the number (four) of Eq. (13.16). Contracting with the

Levi-Civita symbol, show that (13.16) can be obtained from (13.15). Furthermore, go through the opposite way and confirm that the product $\partial_\alpha \tilde{F}^{\alpha\beta} \, \varepsilon_{\mu\nu\lambda\beta}$ gives Eq. (13.15). **Hint.** In the last part, use Eq. (12.54), the result of Exercise 6 of Sect. 12.3.

Exercise 9 Verify Eq. (13.15) and the consideration leading to Eq. (13.16). After that, recalculate Eqs. (13.17) and (13.18) using (12.43) and the first part of (13.16) in the form

$$\varepsilon^{\alpha\beta\mu\nu} \partial_\alpha F_{\mu\nu} = 0. \tag{13.25}$$

Exercise 10 Derive the Lagrange equations (13.7) for the action (13.20), identifying $\varphi(x) = A_\mu(x)$. Verify that the result is nothing else but (13.22).

Exercise 11 Consider the simplest generalizations of the action (13.14), which are relativistic and gauge invariant, but at the same time produce linear or nonlinear equations of motion. For instance, add the term $(F_{\mu\nu} F^{\mu\nu})^2$ with an arbitrary coefficients and derive the corresponding contribution to the equations of motion. This structure is termed by the names of Euler and Heisenberg, who derived it as a quantum electrodynamics correction to the action (13.14).

Exercise 12 Another possible addition to the electromagnetic field action is

$$F_{\mu\nu} f\!\left(\tfrac{\square}{M^2}\right) F^{\mu\nu}, \tag{13.26}$$

where $f(x)$ is an arbitrary function called *form factor*. A useful exercise is to obtain the dispersion relation in this theory, starting from the linear function $f(x)$. Try to find such a function $f(x)$ that the dispersion relation has no ghost-like or tachyonic solutions.

13.2.3   Lorentz Transformation of Electromagnetic Field

An important result of our identification of electric and magnetic fields as parts of the $4D$ tensor (12.43) in Minkowski space is that now we can calculate how these fields transform from one inertial reference frame to another.

Assuming that the reference frame K' moves with the velocity $v = \beta c$ in the direction OX with respect to the static frame K, we get

$$ct' = x^{0'} = \frac{x^0 - \beta x^1}{\sqrt{1-\beta^2}}, \quad x' = x^{1'} = \frac{x^1 - \beta x^0}{\sqrt{1-\beta^2}},$$
$$y' = x^{2'} = x^2, \quad z' = x^{3'} = x^3 \tag{13.27}$$

and for the inverse transformation

$$x^0 = \frac{x^{0'} + \beta x^{1'}}{\sqrt{1 - \beta^2}}, \quad x^1 = \frac{x^{1'} + \beta x^{0'}}{\sqrt{1 - \beta^2}}, \quad x^2 = x^{2'}, \quad x^3 = x^{3'}. \quad (13.28)$$

Let us show in detail the calculation of only the two coefficients and leave the rest as an exercise to the reader. Using (12.43) we get

$$E'_x = F'_{01} = \frac{\partial x^\alpha}{\partial x^{0'}} \frac{\partial x^\beta}{\partial x^{1'}} F_{\alpha\beta} = \frac{\partial x^0}{\partial x^{0'}} \frac{\partial x^1}{\partial x^{1'}} F_{01} + \frac{\partial x^1}{\partial x^{0'}} \frac{\partial x^0}{\partial x^{1'}} F_{10},$$

while all other terms in the sum vanish. Using (13.28) we obtain

$$E'_x = F'_{01} = \frac{1}{\sqrt{1 - \beta^2}} \frac{1}{\sqrt{1 - \beta^2}} E_x + \frac{\beta}{\sqrt{1 - \beta^2}} \frac{\beta}{\sqrt{1 - \beta^2}} (-E_x) = E_x. (13.29)$$

Similarly,

$$E'_y = F'_{02} = \frac{\partial x^\alpha}{\partial x^{0'}} \frac{\partial x^\beta}{\partial x^{2'}} F_{\alpha\beta} = \frac{\partial x^0}{\partial x^{0'}} \frac{\partial x^2}{\partial x^{2'}} F_{02} + \frac{\partial x^1}{\partial x^{0'}} \frac{\partial x^2}{\partial x^{2'}} F_{12}$$
$$= \frac{E_y - \beta H_z}{\sqrt{1 - \beta^2}}. \quad (13.30)$$

Other transformations can be obtained in a similar way,

$$E'_z = \frac{E_z + \beta H_y}{\sqrt{1 - \beta^2}}, \quad H'_x = H_x, \quad H'_y = \frac{H_y + \beta E_z}{\sqrt{1 - \beta^2}}, \quad H'_z = \frac{H_z - \beta E_y}{\sqrt{1 - \beta^2}}. \quad (13.31)$$

Exercise 13 (i) Derive the transformation rules (13.29)–(13.31) using other tensors, e.g., (12.51) and (12.54). (ii) Derive the parity transformation for the components $E_{x,y,z}$ and $H_{x,y,z}$ using (12.43) and (12.54).

Exercise 14 Consider the transformation law for the potentials φ and $\mathbf{A}$, which form a covariant vector $A^\mu = (\varphi, \mathbf{A})$. Using the definitions of electric (11.12) and magnetic (11.11) fields, verify the transformation laws for $\mathbf{E}$ and $\mathbf{H}$. Compare with the calculation based on the identification (12.43).
Answer:

$$\varphi = \frac{\varphi' + \beta A'_x}{\sqrt{1 - \beta^2}}, \quad A_x = \frac{A'_x + \beta \varphi'}{\sqrt{1 - \beta^2}}, \quad A_{y,z} = A'_{y,z}, \quad (13.32)$$

where $A_{x,y,z} = A^{1,2,3} = -A_{1,2,3}$ as a rule of transition from Minkowski to Euclidean notations.

Exercise 15 After verification of Eq. (13.31), derive the transformations inverse to (13.29)–(13.31) directly and explain why it corresponds to the flip of sign $\beta \to -\beta$.

Exercise 16 Verify that

$$F_{\mu\nu}\tilde{F}^{\mu\nu} = -8\,\mathbf{E}\cdot\mathbf{H} \quad \text{and} \quad F_{\mu\nu}F^{\mu\nu} = 2(\mathbf{H}^2 - \mathbf{E}^2). \qquad (13.33)$$

Discuss why these two expressions transform as scalars under Lorentz transformations, using both $4D$ and $3D$ representations. Using general arguments, explain why in this particular case the scalar transformation law means that the two quantities are invariant, meaning they do not change.

Exercise 17 Using the previous exercise prove that the following is true for inertial reference frames:
(i) If $\mathbf{E} \perp \mathbf{H}$ in one inertial reference frame, the same is true in any other such frame.
(ii) If $E = |\mathbf{E}| = |\mathbf{H}| = H$ in one frame, the same is true in any other such frame.
(iii) If $E > H$ (or $v.v.$) in one inertial reference frame, the same is true in another frame.
(iv) If the angle α between vectors $\mathbf{E}$ and $\mathbf{H}$ satisfies $0 < \alpha < \frac{\pi}{2}$ in one inertial reference frame, the same is true in any inertial frame.

Exercise 18 Use the Lorentz transformation rules for electric and magnetic fields and the definition of four-force (12.29) with the properties (12.31) to make a consistent analysis of the change or reference frame in the problem of two point-like charges described at the beginning of this chapter, e.g., in Fig. 11.1. Verify that the transformation of fields plus Lorentz force (11.1) gives the same result as the transformation of the vector of four-force in (12.41). Discuss what was wrong in Fig. 11.1.

Exercise 19 Repeat the previous exercise for the case of two infinite, homogeneously charged parallel lines. Namely, derive the force per unit of length in both static and uniformly moving along the lines reference frames, and show that the transformations (13.29)–(13.31) provide the consistency of description between both frames.

13.3 Energy–Momentum Tensor

The important aspects of relativistic theory are the notions of energy, momentum, and angular momentum. Until now we learned, e.g., how to write the energy of free relativistic particle (12.24) and for a particle in an external electromagnetic field (12.35). However, this is not sufficient if we consider the dynamics of the closed system, which includes both particles and field. In this case, we need to formulate the notion of energy and momentum for the field itself and then understand how the exchange of energy and momentum occurs between particles and fields. This consideration will make the physical sense of the formulas (12.49) and (12.50) more apparent.

Indeed, the approach of the modern physics assumes that charged particles do not interact directly between themselves. Each of them exchanges energy and momentum with the field, and thus field plays the role of intermediator in the interaction between particles. From this perspective, the formulation of energy and momentum of the field has fundamental importance. In what follows we will introduce these notions in two different ways.

13.3.1 Energy Density of the Electromagnetic Field in 3D Notations

Let us start with the formula (12.49), which describes how the energy of the particle changes under the action of electromagnetic field. As we have mentioned above, charged particles do not interact with each other directly, but rather through the intermediating field. If we consider the closed system of particles and fields, then the kinetic energy lost by a particle should go to the field. Hence, we have to construct such a combination of **E** and **H** that would satisfy the energy conservation law for the complete system.

Let us take the two of the Maxwell equations, namely, the second ones of (11.2) and (11.3), and combine them as follows:

$$\frac{1}{c}\left(\mathbf{E}\cdot\frac{\partial \mathbf{E}}{\partial t}+\mathbf{H}\cdot\frac{\partial \mathbf{H}}{\partial t}\right)=-\frac{4\pi}{c}\,\mathbf{j}\cdot\mathbf{E}-\left(\mathbf{H}\cdot\operatorname{rot}\mathbf{E}-\mathbf{E}\cdot\operatorname{rot}\mathbf{H}\right). \quad (13.34)$$

The last parenthesis can be rewritten using the result of Exercise 4 from Chap. 1.6,

$$\mathbf{H}\cdot\operatorname{rot}\mathbf{E}-\mathbf{E}\cdot\operatorname{rot}\mathbf{H}=\operatorname{div}\,[\mathbf{E},\mathbf{H}]\,,$$

to give

$$\frac{1}{2c}\frac{\partial}{\partial t}\left(E^2+H^2\right)=-\frac{4\pi}{c}\,\mathbf{j}\cdot\mathbf{E}-\operatorname{div}\,[\mathbf{E},\mathbf{H}]\,, \quad (13.35)$$

or, finally,

$$\frac{\partial}{\partial t}\left(\frac{E^2+H^2}{8\pi}\right)=-\,\mathbf{j}\cdot\mathbf{E}-\operatorname{div}\mathbf{S}, \quad (13.36)$$

where

$$\mathbf{S}=\frac{c}{4\pi}\,\mathbf{E}\times\mathbf{H} \quad (13.37)$$

is a new object called *Poynting vector*.

In order to give a physical interpretation to the relation (13.36), let us rewrite the first term in the *r.h.s.* in the discrete formulation, and then use Eq. (12.49). The result is

$$- \mathbf{j} \cdot \mathbf{E} = - \sum e\mathbf{v} \cdot \mathbf{E} = - \frac{d}{dt} \sum \varepsilon_{\text{kin}}, \qquad (13.38)$$

where the sums are taken over all charged particles of the system. Consider a simple space volume (V) with the boundary (∂V) and write down the integrated form of (13.36),

$$\frac{d}{dt} \left\{ \iiint\limits_{(V)} \frac{E^2 + H^2}{8\pi} + \sum \varepsilon_{\text{kin}} \right\} = \oiint\limits_{(\partial V)} \mathbf{S} \cdot d\Sigma. \qquad (13.39)$$

The standard physical interpretation of this equation is that $\rho_{\text{r}} = \frac{E^2 + H^2}{8\pi}$ is the energy density of the electromagnetic field, and hence the volume integral of this expression is the total energy of the field inside (V). Together with the sum of the kinetic energies of the particle, the whole volume integral is a total energy in (V) in the *l.h.s.*. The energy has a nonzero time derivative because there is a flux of energy through the surface of the volume (∂V) in the *r.h.s.*. Thus, the physical interpretation of the Poynting vector is that it is the space density of the vector flux of energy. One can guess that this vector is related to the density of momentum of the field.

13.3.2 Conservation of the Energy–Momentum Tensor

Consider a $4D$ approach to the energy and momentum of the electromagnetic field. This approach is closer to the one of modern theoretical physics and will provide us some notions that will turn out to be very useful in the next part of the book, when we deal with the relativistic theory of gravity, that is, Einstein's general relativity.

It proves useful to start with the general formalism for the field φ described in Sect. 13.1. It can be a multicomponent field, but we avoid writing the index on φ_i, trying to make formulas more compact and simple.

Consider a closed system of fields φ, without the action of external forces. This means that the density of the Lagrange function cannot depend on x^μ explicitly, but only through the fields and their partial derivatives, $\mathcal{L} = \mathcal{L}(\varphi, \partial\varphi)$. Then

$$\frac{\partial \mathcal{L}}{\partial x^\mu} = \frac{\partial \mathcal{L}}{\partial \varphi} \frac{\partial \varphi}{\partial x^\mu} + \frac{\partial \mathcal{L}}{\partial (\partial_\alpha \varphi)} \frac{\partial (\partial_\alpha \varphi)}{\partial x^\mu}. \qquad (13.40)$$

Now we use the Lagrange equations (13.7) in the first factor in the first term in the *l.h.s.* and the commutativity of partial derivatives $\partial_\mu \partial_\alpha \varphi = \partial_\alpha \partial_\mu \varphi$. Thus, we arrive at the relation

$$\frac{\partial \mathcal{L}}{\partial x^\mu} = \frac{\partial}{\partial x^\alpha}\left(\frac{\partial \mathcal{L}}{\partial(\partial_\alpha \varphi)}\right)\frac{\partial \varphi}{\partial x^\mu} + \frac{\partial \mathcal{L}}{\partial(\partial_\alpha \varphi)}\frac{\partial}{\partial x^\alpha}\left(\frac{\partial \varphi}{\partial x^\mu}\right)$$
$$= \frac{\partial}{\partial x^\alpha}\left(\frac{\partial \mathcal{L}}{\partial(\partial_\alpha \varphi)}\frac{\partial \varphi}{\partial x^\mu}\right). \tag{13.41}$$

Using Kronecker delta symbol, we cast this relation into the form

$$\partial_\alpha \theta^{\alpha}_{\cdot \mu} = 0, \tag{13.42}$$

where

$$\theta^{\alpha}_{\cdot \mu} = \frac{\partial \mathcal{L}}{\partial(\partial_\alpha \varphi)}\partial_\mu \varphi - \delta^{\alpha}_{\mu}\mathcal{L} \tag{13.43}$$

is called energy–momentum tensor of the field φ. The relation (13.42) is called conservation law. As we have seen, it comes from the requirement that the system is closed, meaning there are no external forces, and as a result we have the initial relation (13.40). One can see the same relation from another viewpoint. If the fundamental object such as the Lagrangian depends on x^μ explicitly, this means that the equations of motion can be different in different points of space–time regions. For example, the dependence on x^0 means the non-invariance with respect to time translations. Thus, the main condition (13.40) means the homogeneity of space and time, and the conservation law (13.42) reflects this fundamental property of space and time.

It is possible and useful to use a modified form of the energy–momentum tensor (13.42), which also satisfies the same conservation law (13.43). The reason is that one can simplify the expression (13.42) without violating the conservation law. Consider the same tensor with two contravariant indices,

$$\theta^{\alpha\beta} = \theta^{\alpha}_{\cdot \mu}\eta^{\mu\beta}. \tag{13.44}$$

In general, tensor $\theta^{\alpha\beta}$ may not be symmetric, but one can modify it by adding a new term

$$\theta^{\alpha\beta} \longrightarrow T^{\alpha\beta} = \theta^{\alpha\beta} + \partial_\gamma \Psi^{\alpha\beta\gamma}, \quad \text{where} \quad \Psi^{\alpha\beta\gamma} = -\Psi^{\gamma\beta\alpha}. \tag{13.45}$$

It is easy to see that $T^{\alpha\beta}$ satisfies the same conservation law, because $\partial_\alpha\partial_\gamma \Psi^{\alpha\beta\gamma} = 0$. Thus, we arrive at the new version of the energy–momentum tensor, which is symmetric $T^{\beta\alpha} = T^{\alpha\beta}$ and satisfies the conservation law

$$\partial_\alpha T^{\alpha\beta} = 0, \tag{13.46}$$

which reflects the same fundamental symmetry, that is, homogeneity of the space–time.

An important example of the energy–momentum tensor for the of electromagnetic field with the action (13.14) and Lagrangian

$$\mathcal{L} = -\frac{1}{16\pi} F_{\mu\nu}^2 = -\frac{1}{8\pi} \partial_\mu A_\nu \left(\partial^\mu A^\nu - \partial^\nu A^\mu\right). \tag{13.47}$$

In this formula, we used condensed notations $F_{\mu\nu}^2 = F_{\mu\nu}F^{\mu\nu}$ and $\partial^\mu = \eta^{\mu\lambda}\partial_\lambda$. Taking the required derivatives, we get

$$\frac{\partial\mathcal{L}}{\partial(\partial_\rho A_\sigma)} = -\frac{1}{4\pi}\delta_\mu^\rho \delta_\nu^\sigma \left(\partial^\mu A^\nu - \partial^\nu A^\mu\right) = -\frac{1}{4\pi} F^{\rho\sigma}$$

$$\text{and} \quad \frac{\partial\mathcal{L}}{\partial A_\rho} = 0. \tag{13.48}$$

Then

$$\theta_{\cdot\mu}^\rho = \frac{\partial\mathcal{L}}{\partial(\partial_\rho A_\sigma)} \cdot \partial_\mu A_\sigma - \delta_\mu^\rho \mathcal{L} = \frac{1}{4\pi}\left(-F^{\rho\sigma}\partial_\mu A_\sigma + \frac{1}{4}\delta_\mu^\rho F_{\alpha\beta}^2\right). \tag{13.49}$$

It is easy to note that $\theta^{\rho\sigma}$ is not symmetric. It is a relatively easy exercise (the second one at the end of this section) to construct the symmetric version of the energy–momentum tensor for the electromagnetic field

$$T_{\alpha\beta} = \frac{1}{4\pi}\left(-F_\alpha^{\ \rho} F_{\beta\rho} + \frac{1}{4} g_{\alpha\beta} F_{\rho\sigma}^2\right). \tag{13.50}$$

Taking into account the Maxwell equations in vacuum, $\partial_\mu F^{\mu\nu} = 0$, it is easy to show that $T_{\alpha\beta}$ satisfies the conservation law (13.46). Another important property of this tensor is that its trace is zero, $T_\alpha^\alpha = T_{\alpha\beta}g^{\alpha\beta} = 0$. Later on, we shall discuss the physical significance of this condition.

Using the definition of the electromagnetic field tensor (12.43), we can derive the components of $T_{\alpha\beta}$, namely,

$$T_{00} = \frac{E^2 + H^2}{8\pi}, \tag{13.51}$$

$$T_{0k} = -\frac{1}{c} S^k, \tag{13.52}$$

$$T_{ik} = \sigma_{ik} = \frac{1}{4\pi}\left\{-E_i E_k - H_i H_k + \frac{1}{2}\delta_{ik}(E^2 + H^2)\right\}. \tag{13.53}$$

The first one of these expressions is the density of energy of the field that appears in the conservation law (13.39). The second one represents the components of the Poynting vector (13.37), and the last part σ_{ik} defined in (13.53) is called Maxwell's stress tensor in vacuum. Sometimes in the literature, the whole $T_{\alpha\beta}$ is also called stress tensor.

The relations (13.51), (13.52), and (13.53) hint that the tensor $T_{\alpha\beta}$ defines density of energy and momenta of the field. In the rest of this section, we shall explore this feature in more detail.

Consider the $3D$ space-like surface S in the $4D$ Minkowski space. In the simplest case, S is just a space volume taken at the fixed instant of time. However, taking into account the possibility of Lorentz transformation, it is better to consider an arbitrary space-like $3D$ surface S with the normal time-like vector and the differential element of surface (space volume in the fixed-time case) df_α. By definition, the total energy of the system inside S is

$$P^\mu = \frac{1}{c} \int T^{\mu\alpha} \, df_\alpha, \tag{13.54}$$

which boils down to

$$P^\mu = \frac{1}{c} \int T^{\mu 0} \, d^3x, \tag{13.55}$$

in the mentioned simplest case. It is clear that for P^0 this gives a total energy, if we accept the identification of (13.51) as the energy density of the electromagnetic field and if there are no particles inside S.

To better understand the physical sense of other components of $T^{\mu\alpha}$, let us use the conservation law $\partial_\mu T^{\mu\alpha} = 0$ in the integral (13.54). In this way, we get

$$\partial_0 P^0 + \operatorname{div} \mathbf{P} = \partial_\mu P^\mu = \frac{1}{c} \int \partial_\mu T^{\mu 0} \, d^3x = 0. \tag{13.56}$$

From the last relation follows that the density of momentum of the field corresponds to the Poynting vector, which is also the density of the energy transfer in the direction of the vector $\mathbf{P}$.

Similarly, we can write for the space components

$$\frac{1}{c} \frac{\partial}{\partial t} \int_V T^{i0} \, d^3x = - \oint_{\partial V} T^{ik} \, df_k, \tag{13.57}$$

where df_k is the infinitesimal element of integration over the oriented surface of the boundary (∂V) of the volume (V). The volume integral on the left is the time derivative of the total momentum of the field in the volume (V). Then it is clear that the integral of the r.h.s. is the flux over the surface (∂V) of the component i of momentum in the direction k. Since the relation (13.57) is valid for any volume (V), one can conclude that the component $\sigma_{ik} = T_{ik}$ of the stress tensor describes the transfer of the component i of the density of momentum in the direction k. The symmetry $\sigma_{ik} = \sigma_{ki}$ means that it is equal to the transfer of the component k in the direction i.

Now, since we learned the energy and momentum densities of the field, let us remember that our target is the system of charged particles coupled to the field. Therefore, consider the energy–momentum tensor of a system of particles, such that later on we can implement their interaction with the field. The starting point will be the volume density of the mass[4] of the set (we can call it gas) of particles numbered by the index a,

$$\mu(\mathbf{r}) = \sum_a m_a \, \delta(\mathbf{r} - \mathbf{r}_a). \tag{13.58}$$

In the last formula $\mathbf{r}_a = \mathbf{r}_a(t)$. The next step is to rewrite last formula in the relativistic form. To this end, we write the density of four-momentum of the gas of particles,

$$p^\alpha = \mu(x^\beta) c \, u^\alpha, \tag{13.59}$$

where u^α is the four-velocity of the particle in the given point of space.

Analogous to the energy–momentum tensor of electromagnetic tensor, we assume that $p^\alpha = \frac{1}{c} T_m^{0\alpha}$. Now, in order to obtain the full $T_m^{\alpha\beta}$, we note that $\mu(\mathbf{r})$ is nothing else but the time component of the four-vector $\frac{\mu}{c}\frac{dx^\alpha}{dt}$. Thus, we arrive at the expression for the density of the energy–momentum tensor,

$$T_m^{\alpha\beta} = \mu c \, u^\alpha \frac{dx^\beta}{dt} = \mu c \, u^\alpha \, u^\beta. \tag{13.60}$$

Our method of deriving Eq. (13.60) may look a little bit empirical; therefore, let us verify it in a direct way, by checking that the total energy–momentum tensor of particles and electromagnetic field satisfies the conservation law

$$\partial_\alpha \left(T_m^{\alpha\beta} + T_{field}^{\alpha\beta} \right) = 0. \tag{13.61}$$

If confirmed, this relation will mean that our identification of $T_m^{\alpha\beta}$ is correct.

For the field part $T_{field}^{\alpha\beta} = T^{\alpha\beta}$ from (13.50) we have

$$\partial_\alpha T^\alpha_{.\beta} = \frac{1}{4\pi} \left(\frac{1}{2} F_{\rho\sigma} \partial_\beta F^{\rho\sigma} - \partial_\alpha F^\alpha_{.\rho} F_\beta^{.\rho} - F^\alpha_{.\rho} \partial_\alpha F_\beta^{.\rho} \right). \tag{13.62}$$

Using Maxwell equations

$$\partial_\beta F_{\rho\sigma} = -\partial_\rho F_{\sigma\beta} - \partial_\sigma F_{\beta\rho} \qquad \text{and} \qquad \partial_\alpha F^\alpha_{.\rho} = \frac{4\pi}{c} j_\rho,$$

after short calculation we obtain

$$\partial_\alpha T^\alpha_{.\beta} = \frac{1}{4\pi} \left(\frac{1}{2} F^{\rho\sigma} \partial_\rho F_{\beta\sigma} + \frac{1}{2} F^{\rho\sigma} \partial_\rho F_{\beta\sigma} - F^{\alpha\rho} \partial_\alpha F_{\beta\rho} - \frac{4\pi}{c} j_\rho F_\beta^{.\rho} \right). \tag{13.63}$$

It is easy to see that the first three terms in the parentheses cancel, and hence for the field part

---

[4] Here and throughout the rest of this book we assume that the "mass" is the "rest mass", that is, an intrinsic property of the particle that does not depend on its motion.

$$\partial_\alpha T^\alpha{}_\beta = \frac{1}{c} j^\alpha F_{\alpha\beta}. \tag{13.64}$$

On the other hand, for the counterpart of the particles (13.60) we have

$$\partial_\alpha T_m^{\alpha\beta} = c\, u^\beta \frac{\partial}{\partial x^\alpha} \left(\mu \frac{dx^\alpha}{dt} \right) + \mu c \frac{dx^\alpha}{dt} \frac{\partial u^\beta}{\partial x^\alpha}. \tag{13.65}$$

Here $\mu \frac{dx^\alpha}{dt}$ is the mass current and hence its four-divergence vanish due to the conservation of mass,

$$\frac{1}{c} \left(\mu \frac{dx^0}{dt} \right) + \operatorname{div} \left(\mu \, \mathbf{v} \right) = 0. \tag{13.66}$$

For the last term, we remember the equation for the charged particle (12.41),

$$mc \frac{du^\beta}{ds} = \frac{e}{c} F_{\beta\alpha} u^\alpha. \tag{13.67}$$

This equation is valid for each particle of the gas, but in the continuous description we need to use the following relation, which is valid for the gas of particles with equal masses and charges,

$$\frac{\mu}{m} = \frac{\rho}{e}. \tag{13.68}$$

Replacing this formula into Eq. (13.67) we arrive at

$$\mu c \frac{du^\beta}{ds} = \frac{\rho}{c} F_{\beta\alpha} u^\alpha \tag{13.69}$$

and finally from (13.67) we have

$$\mu c \frac{du^\beta}{dt} = \frac{\rho}{c} F_{\beta\alpha} u^\alpha \frac{ds}{dt} = \frac{1}{c} F_{\beta\alpha} \rho \frac{dx^\alpha}{dt} = \frac{1}{c} F_{\beta\alpha} j^\alpha, \tag{13.70}$$

that precisely cancels with (13.64), confirming the validity of (13.61).

All in all, we have confirmed the expressions for the energy–momentum tensor of the electromagnetic field (13.50) and of the gas of free massive particles (13.60). There is an interesting similarity between the two expressions, which is worth special discussion.

Let us consider what happens with the energy–momentum tensor of the gas of free particles (13.60) in the low-energy (IR) and high-energy (UV) limits. In the first case, the four-velocities of all particles approximately equal to $u^\alpha = (1, 0, 0, 0)$, such that (13.60) becomes (we use the system of units with $c = 1$ here)

$$T_m^{\alpha\beta} = \mu c\, u^\alpha u^\beta \approx \operatorname{diag} \left(\mu, 0, 0, 0 \right) \quad \text{and} \quad (T_m)_\alpha^\alpha \approx \mu. \tag{13.71}$$

This result means that in the IR we have a gas of almost static particles, where the energy density strongly dominates over the density of momenta and over other elements of the stress tensor.

In the UV limit, one can get all information from the trace. In order to do so, remember that $u^\alpha u_\alpha = 1$ and hence

$$(T_m)_\alpha^{\ \alpha} \approx \mu = nm, \tag{13.72}$$

where n is the concentration of the particles with the mass m. On the other hand, the energy density of these particles is

$$\rho = \frac{nm}{\sqrt{1 - \beta^2}}, \tag{13.73}$$

where we assume that the speeds of all particles are approximately equal. In the UV, when this speed is close to that of light, $\beta \approx 1$, we obviously have

$$\rho \gg nm = (T_m)_\alpha^{\ \alpha}. \tag{13.74}$$

In the extreme UV limit the trace is zero. As we know from the first chapter of the book, there is always a reference frame in which the symmetric tensor $T_m^{\alpha\beta}$ is diagonal,

$$T_m^{\alpha\beta} \approx \mathrm{diag}\left(\rho,\, p,\, p,\, p\right), \tag{13.75}$$

where p is called the pressure of the gas.[5] Obviously, the vanishing trace means that the *equation of state* of the gas is $p \approx \rho/3$. We can see that the vanishing trace of the $T^{\alpha\beta}$ is a common feature of the UV limit of the gas of massive particles and of the electromagnetic field, where the relations $T_\alpha^\alpha = 0$ and $p_r = \rho_r/3$ are exact. Indeed, this result for the electromagnetic field can be obtained in several different ways, including from the Boltzmann approach in statistical mechanics, based on the Bose–Einstein distribution for photons. In one of the next sections, we consider a very simple example of how this equation of state can be obtained as a massless limit for the ideal gas of massive relativistic particles.

Exercise 20 Using the definition (13.43), derive θ_μ^α for the scalar field with the action (13.8). Identify the energy density of the field. Consider the generalization, subtracting from the free Lagrangian an arbitrary potential term $V(\varphi)$. Compare the result for the energy density with the expression for energy of a particle in classical mechanics.

Exercise 21 Verify that $T_{\alpha\beta} = \theta_{\alpha\beta} + \Delta T_{\alpha\beta}$, where

---

[5]This identification can be confirmed from the nonrelativistic limit in hydrodynamics, see e.g. [3].

$$\Delta T_{\alpha\beta} = \frac{1}{4\pi} F_{\alpha}{}^{\rho} \partial_{\rho} A_{\beta}.$$

Using the vacuum version of Maxwell equations, verify that all three tensors $\theta_{\alpha\beta}$, $\Delta T_{\alpha\beta}$ and $T_{\alpha\beta}$ satisfy the conservation law (13.46).

Exercise 22 Using the definition (12.43) of $F_{\mu\nu}$ verify the relations (13.51), (13.52), and (13.53). Write down the matrix form of $T_{\alpha\beta}$, $T_{\alpha}{}^{\nu}$, $T^{\mu}{}_{\beta}$, and $T^{\mu\nu}$. Discuss the difference between these four expressions in view of the symmetry $T_{\alpha\beta} = T_{\beta\alpha}$.

Exercise 23 Verify the tensor form of the expression (13.59). For this end, it is necessary to verify the factors of $\sqrt{1 - \beta^2}$ in the Lorentz transformation of the density $\mu = \frac{mass}{volume}$ and $\frac{1}{c} \frac{ds}{dt}$ cancel out in this expression. The second part of the exercise is to establish the Lorentz transformation of the delta functions in Eq. (13.58).

Exercise 24 Write down the matrix representation of the energy–momentum tensor of the gas of particles (13.60). Consider the nonrelativistic limit of this expression, assuming that the distribution of velocities is isotropic. Show that in this limit we have the energy–momentum tensor of the form $T_{\alpha}^{\beta} = \text{diag}\left(\mu c^2, 0, 0, 0\right)$. Consider the ultrarelativistic limit, when the speed of particles is very close to c, and show that in this case $T_{\alpha}^{\beta} = \text{diag}\left(\varepsilon, p, p, p\right)$, with $p = \frac{1}{3}\varepsilon$, and that the trace of the energy–momentum tensor vanishes in this limit.

Exercise 25 Discuss a generalization of Eq. (13.70) for the system with two kinds of particles with different relations between mass and charge, of the form (13.68).

13.4 Electromagnetic Waves

In this section, we consider various aspects of the electromagnetic wave solutions of the Maxwell equation. As we already know, the term *wave* means the solution without sources. And as we have seen in Sect. 11.2, the equation for the electric field in this case is (11.9), also the equations for scalar and vector parts of electromagnetic potential have the same form. At the beginning of the chapter, we have found that the equations for any of these quantities have plane wave solutions of the form (11.16), which correspond to the universal speed c of the electromagnetic wave. In the present section, we are going to consider the propagation of the electromagnetic wave in more detail, and also discuss some aspects of emission of wave emissions.

It is now the right time to solve one of the Exercises of Sect. 11.2. Our starting point will be the gauge transformation (12.58), which can be rewritten in the 3D notations as (12.60) and (12.61). One can always choose an arbitrary function $f(x)$ in such a way that the scalar potential is $\varphi(x) \equiv 0$. In what follows we assume this choice, such that the definition of electric and magnetic fields (11.12) and (11.11) boil down to

$$\mathbf{H} = \text{rot} \, \mathbf{A} \quad \text{and} \quad \mathbf{E} = -\frac{1}{c} \frac{\partial \mathbf{A}}{\partial t}. \tag{13.76}$$

Replacing these expressions into (11.6) and (11.7) we obtain

$$\text{rot rot } \mathbf{A} = \frac{1}{c}\frac{\partial}{\partial t}\left(-\frac{1}{c}\frac{\partial \mathbf{A}}{\partial t}\right) = -\frac{1}{c^2}\frac{\partial^2}{\partial t^2}. \tag{13.77}$$

Using the formula for rot rot $\mathbf{A}$ from Chap. 1, after some algebra we arrive at

$$\Box \mathbf{A} = \left(\frac{1}{c}\frac{\partial}{\partial t} - \Delta\right)\mathbf{A} = -\text{ grad div } \mathbf{A}. \tag{13.78}$$

Can we use the gauge ambiguity (12.60) to provide div $\mathbf{A} \equiv 0$? This would be nice because we already know that the gauge transformation affects neither electric nor magnetic fields, such that we can perform it at our convenience. On the other hand, we already used the part (12.61) to guarantee $\varphi(x) \equiv 0$. Since the function f is unique, how can we use it another time? The point is that we did not need to use this function completely to eliminate the scalar part of the potential. In (12.61), there is only the time derivative in the *r.h.s.*, while in (12.60) there are only space derivatives. Thus, in the first case, the solution is

$$f(t, \mathbf{r}) = f(t_0, \mathbf{r}) + \int_{t_0}^{t} c\varphi(t', \mathbf{r})dt' \implies \varphi' = \varphi - \frac{1}{c}\frac{\partial f}{\partial t} \equiv 0, \tag{13.79}$$

while the second solution starts with $\Delta f = \text{div } \mathbf{A} - \text{div } \mathbf{A}'$ and has the form, e.g., or retarded potential[6]

$$\text{div } \mathbf{A}'(t, \mathbf{r}) = \int d^3x' \frac{\text{div } \mathbf{A}(\mathbf{r}', t - R/c)}{R}, \quad \text{where} \quad R = |\mathbf{r}' - \mathbf{r}|. \tag{13.80}$$

This solution provides $\Delta f \equiv \text{div } \mathbf{A}$ and hence automatically div $\mathbf{A}' = \text{div } \mathbf{A} - \Delta f \equiv 0$. Thus, the freedom to perform a gauge transformation is sufficient to provide $\varphi' \equiv 0$ and div $\mathbf{A}' \equiv 0$ at the same time. And moreover, some of the freedom in the choice of the gauge transformation still remains, because one can still change the initial condition $f(t_0, \mathbf{r})$ in (13.79).

The special choice which is done in (13.80) is called the Coulomb gauge fixing (or just *Coulomb gauge*) div $\mathbf{A} = 0$. From this point, we will not write primes over $\mathbf{A}$ and φ and just assume that the electromagnetic potentials satisfy the Coulomb gauge and $\varphi = 0$. Alternatively, one can use the *Lorentz gauge* $\partial_\mu A^\mu = 0$, which is in fact more useful for the electromagnetic waves.

Exercise 26 Consider the gauge fixing that makes zero and nonzero scalar potential $\varphi = A^0$. Find the gauge transformation between Coulomb and Lorentz gauge fixing conditions and explore its ambiguity.

---

[6]For the reader who is not familiar with this solution, it is explained in the next section.

13.4.1 Plane Wave Solution

As a first step, let us consider the plane electromagnetic wave solutions of the Maxwell equation. Let us note that constant electric and magnetic fields must be considered separately, and this part of electrodynamics lies beyond the scope of this book. As far as Maxwell equations are linear, one can always consider any nonconstant configuration of electromagnetic field as a superposition of the plane electromagnetic waves. In this sense, the problem we will discuss here is a rather general one.

The individual plane wave can be described by its wave vector $\mathbf{k} = k\hat{\mathbf{n}}$ and the frequency $\omega = k^0$, which form a four-vector $k^\mu = (k^0, \mathbf{k})$. Such a wave can be presented in the form

$$A_\mu(x) = \varepsilon_\mu(k) e^{ik_\mu x^\mu} + \varepsilon_\mu^*(k) e^{-ik_\mu x^\mu}. \tag{13.81}$$

Here $\varepsilon_\mu(k)$ is the complex vector that depends on $\mathbf{k}$ and is the Fourier image of $A_\mu(x)$. Vector $\varepsilon_\mu(k)$ has four components. In what follows we explore how the number of independent components of $\varepsilon_\mu(k)$ changes when we take into account the gauge ambiguity.

In the Lorentz gauge $\partial_\mu A^\mu = 0$, the vacuum field Eq. (13.22) become

$$\partial_\mu F^{\mu\nu} = \partial_\mu(\partial^\mu A^\nu - \partial^\nu A^\mu) = \Box A^\nu - \partial^\nu \partial_\mu A^\mu = \Box A^\nu = 0, \tag{13.82}$$

where we used commutativity of partial derivatives. Replacing plane wave representation (13.81) into Eq. (13.82) and into the gauge condition, we get two relations for the vector $\varepsilon_\mu(k)$ and the four-vector of the wave

$$k^\mu \varepsilon_\mu = 0 \tag{13.83}$$

and

$$k_\mu k^\mu = 0. \tag{13.84}$$

The last condition is a constraint that shows that out of the four components of $\varepsilon_\mu(k)$, at most three are independent, because the four components satisfy this constraint. However, there is a further reduction of independent components, which are called physical degrees of freedom. Consider the gauge transformation (12.58), with the parameter $f(x)$ satisfying the same equation of motion as the vector field itself,

$$A_\mu \rightarrow A_\mu' = A_\mu + \partial_\mu f, \qquad \Box f(x) = 0. \tag{13.85}$$

Consider the Fourier mode of the gauge parameter $f(x)$ with the same wave vector,

$$f(x) = i\xi(k) e^{ik_\mu x^\mu} - i\xi^*(k) e^{-ik_\mu x^\mu}, \tag{13.86}$$

then the gauge transformation (13.85) becomes

$$\varepsilon_\mu \rightarrow \varepsilon'_\mu = \varepsilon_\mu + k_\mu \xi, \quad k^2 = 0. \tag{13.87}$$

It is easy to see that the condition (13.83) is *not* fixing the gauge ambiguity completely. The point is that this condition defines the $3D$ surface in the space of the fields, and even with $k^\mu \varepsilon_\mu = 0$ satisfied we can still make gauge transformations in the direction along this surface.

Without loss of generality, consider the wave propagating along the OZ-axis. Then

$$k^\mu = (k, 0, 0, k), \quad \text{with} \quad k^\mu k_\mu = k^\mu k^\nu \eta_{\mu\nu} = 0, \tag{13.88}$$

such that the transformation (13.87) becomes

$$\varepsilon'_0 = \varepsilon_0 + k\xi, \quad \varepsilon'_3 = \varepsilon_3 - k\xi. \tag{13.89}$$

The condition (13.83) becomes simply $\varepsilon_0 = -\varepsilon_3$ and according to (13.89) we also have $\varepsilon'_0 = -\varepsilon'_3$. The gauge fixing condition (13.83) is not violated by this particular form of the gauge transformation, and hence we can eliminate both ε_0 and ε_3 by a single transformation (13.89) with $\xi(k) = \frac{\varepsilon_3(k)}{k}$. We can then conclude that the two degrees of freedom ε_0 and ε_3 are completely dependent on the gauge transformation. Since gauge-dependent quantities do not affect electric and magnetic fields, ε_0 and ε_3 are physically irrelevant. Therefore, the physical degrees of freedom of the wave are only transverse components ε_1 and ε_2.

The previous consideration has shown that the plane electromagnetic wave is transverse, such that $\mathbf{A} \cdot \mathbf{k} = 0$. For the wave propagating along the OZ-axis, this means that only A_x and A_y components are relevant, while A_0 and A_z can be set to zero. Furthermore, we know from the generic wave solution (11.15) that $\mathbf{A}$ is a function of a single combination of space and time coordinates, which may be either $\eta = t - \frac{z}{c}$ or $\zeta = t + \frac{z}{c}$ for the waves propagating into the positive or negative side of OZ. Taking the first case we get

$$\mathbf{A} = \hat{\mathbf{i}} A_x + \hat{\mathbf{j}} A_y = \mathbf{A}(\eta). \tag{13.90}$$

Using this formula, we can easily calculate the electric and magnetic fields,

$$\mathbf{E} = -\frac{1}{c} \frac{\partial \eta}{\partial t} \mathbf{A}' = -\frac{1}{c} \mathbf{A}', \quad \text{where} \quad \mathbf{A}' = \frac{\partial \mathbf{A}}{\partial \eta}; \tag{13.91}$$

$$\mathbf{H} = \text{rot}\, \mathbf{A} = -\frac{1}{c} \hat{\mathbf{k}} \times \mathbf{A}'. \tag{13.92}$$

An obvious conclusion is that in the plane wave $\mathbf{E} \perp \mathbf{k}$, $\mathbf{H} \perp \mathbf{k}$, and $\mathbf{H} \perp \mathbf{E}$. For the wave propagating in an arbitrary direction $\hat{\mathbf{n}}$ we arrive at the relations

$$H = \hat{n} \times E \quad \Longrightarrow \quad |H| = |E|. \tag{13.93}$$

For the energy density and the energy density flux (Poynting vector, or density of momentum) we get, after some algebra,

$$\rho_r = \frac{H^2 + E^2}{8\pi}, \tag{13.94}$$

$$S = \frac{c}{4\pi} E \times H = \frac{c}{4\pi} H^2 \hat{n} = \frac{c}{4\pi} E^2 \hat{n} = c\rho_r \hat{n}. \tag{13.95}$$

Finally, for the components of the stress tensor (13.53), the result is that only one of them is different from zero,

$$\sigma_{zz} = \rho_r, \qquad \sigma_{xi} = \sigma_{yi} = 0. \tag{13.96}$$

Exercise 27 Verify formulas (13.91)–(13.96).

13.4.2 Emission of Electromagnetic Waves

Consider the emission of electromagnetic waves. For this, we need the complete version of Maxwell equation with source (11.2) and (11.3). Alternatively, we can use the relativistic version (13.22). In the Lorentz gauge $\partial_\mu A^\mu = 0$ the equation is

$$\Box A^\mu = \frac{4\pi}{c} j^\mu, \tag{13.97}$$

which in the $3D$ notations gives

$$\Box \varphi = 4\pi \rho, \qquad \Box A = \frac{4\pi}{c} j. \tag{13.98}$$

Since the four equations are independent and have the same structure, we can solve any of them and will choose the first one in (13.98).

Our equation of interest is linear, and hence we can divide the space into small, compact, infinitesimal volumes and find contribution of each such volume to the potential φ. After that the total potential by the whole volume will be the sum over all the contributions of the small volumes. Thus, we take the time-dependent, point-like charge in the point with the radius-vector r', then

$$\rho = de(t) \cdot \delta^3(r') \tag{13.99}$$

and ignore the charges in the rest of the space. Now we have to solve the first of Eq. (13.98), that is, to find the scalar potential φ in the point $r = r' + R$. The variable **R** is certainly the most useful one, because in this case the charge $de(t)$ is at the origin of the coordinates and one can use spherical symmetry to simplify the equations.

Our equation of interest can be written in the form

$$\Delta\varphi - \frac{1}{c^2}\frac{\partial^2\varphi}{\partial t^2} = -4\pi\, de(t) \cdot \delta^3(0), \qquad (13.100)$$

where Δ is defined with the components of the vector **R**. Using spherical coordinates, the function φ can depend only on R. For the Laplace operator, which we know from the first chapter of the book, we rewrite Eq. (13.100) for **R** $\neq 0$ in the form

$$\frac{1}{R^2}\frac{\partial}{\partial R}\left(R^2\frac{\partial\varphi}{\partial R}\right) - \frac{1}{c^2}\frac{\partial^2\varphi}{\partial t^2} = 0. \qquad (13.101)$$

Thus, our problem reduced to the solution for the spherical wave in the vacuum, plus the boundary condition in the point **R** $= 0$.

It proves useful to introduce the new field variable $\chi(R,t) = R\varphi$. After some short algebra we get Eq. (13.101) in the form

$$\frac{\partial^2\chi}{\partial R^2} - \frac{1}{c^2}\frac{\partial^2\chi}{\partial t^2} = 0. \qquad (13.102)$$

This is exactly the same wave equation that we already solved in Sect. 11.2, and the solution is a sum of retarded and advanced components, but this time in the radial direction,

$$\chi = f_1\left(t - \frac{R}{c}\right) + f_2\left(t + \frac{R}{c}\right). \qquad (13.103)$$

In our case, it is sufficient to have only one solution, hence we set $f_2 = 0$, and hence the solution we need is

$$\varphi(R,t) = \frac{\chi\left(t - \frac{R}{c}\right)}{R}, \qquad (13.104)$$

where $\chi(x)$ is an arbitrary function.

The next step is to fit the boundary condition from (13.100) at the point **R** $= 0$. Since $\chi(x)$ is a differentiable function, (13.104) tends to infinity at **R** $\to 0$, together with all its derivatives. Therefore, we can ignore the time derivatives and simply solve the space part of Eq. (13.100), $\Delta\varphi = 4\pi\rho$, with the point-like distribution $\rho = de \cdot \delta^3(0)$. The solution is well known, and recovering the time dependence from (13.104) we get

$$\varphi(R,t) = \frac{de\left(t - \frac{R}{c}\right)}{R}. \qquad (13.105)$$

Integrating over $\mathbf{r}' = \mathbf{R} + \mathbf{r}$ we arrive at the solutions of Eqs. (13.98) in the form

$$\varphi(\mathbf{r}, t) = \int \frac{\rho\left(\mathbf{r}', t - \frac{R}{c}\right)}{|\mathbf{r}' - \mathbf{r}|} d^3 x', \tag{13.106}$$

$$\mathbf{A}(\mathbf{r}, t) = \frac{1}{c} \int \frac{\mathbf{j}\left(\mathbf{r}', t - \frac{R}{c}\right)}{|\mathbf{r}' - \mathbf{r}|} d^3 x'. \tag{13.107}$$

13.5 Summary: Symmetries of Maxwell Equations

As we have learned in the previous sections, the equations of classical electrodynamics possess several symmetries. Let us list those we can discuss now. There are two more symmetries that can be revealed only in the framework of general relativity, and hence we postpone them for the last part of the book and mention them in the course of discussion of the diffeomorphism invariance (also called general covariance) and local conformal symmetry.

To start with, Maxwell equations are Lorentz invariant, reflecting the fundamental property of the space and time. From the mathematical side, this invariance can be seen in that the action of both the electromagnetic field (13.14) and the massive charged particles (12.68) is scalars under Lorentz transformations, space rotations, and inversion of coordinates (parity transformation). The dynamical field $A^\mu(x)$ is a vector under the same transformations, such that the equations of motion (11.2) and (11.3) can be cast into tensor form (13.15) and (13.22) with respect to the mentioned transformations. The same is true for the dynamical equations for the massive charges (11.1), which can be written in the tensor relativistic form (12.41). In the third part of the book, we consider how to generalize this symmetry to include arbitrary infinitesimal transformations of coordinates. This is called general covariance of the Maxwell equations, which we mentioned above.

Another fundamental property of the actions (13.14) and (12.68) symmetry is the gauge symmetry (12.58), which implies the conservation of electric charge. Taken together, Lorentz symmetry, gauge invariance, absence of higher derivatives, and the linearity of dynamical equations (superposition principle) enable one to establish the action of the electromagnetic field in a unique way. Thus, the two fundamental symmetries form the basis of electromagnetic theory, including the possibility of the Lorentz mixing of electric and magnetic fields described by Eqs. (13.29)–(13.31). Let us mention that the importance of gauge symmetry goes beyond the electromagnetic interactions. Indeed, all fundamental forces are described by the gauge theories. At high energies, electromagnetic force should be considered together with the weak interaction, forming what is called the electroweak (EW) sector of the Standard Model of particle physics. Another part of the Standard Model is the theory of strong interactions or quantum chromodynamics (QCD). Both weak and strong interactions are transmitted by vector potential fields, which are W and Z bosons for EW sector and gluons for QCD. In both cases, these are gauge fields that are described by the theories invariant under the transformations, similar but essentially more complicated than (12.58). The main difference with the photon described by A^μ is that the equations for W and Z bosons and gluons are nonlinear and include self-interactions of these fields. Furthermore, as we shall see in the last part of the book, relativistic gravity is also a gauge theory. Thus, gauge symmetry can be seen as the most universal and fundamental property, which is common for all known fundamental interactions.

Along with the main two symmetries, there are two others that emerge only for the pure field system, without charges. In the absence of sources, Maxwell equations possess two more symmetries. The first is the electromagnetic duality, which we deal with briefly in Exercise 2 in (11.2). The mixing described by (11.10) is qualitatively different from the Lorentz mixing (13.29)–(13.31). The electromagnetic duality found important applications in string theory, which is supposed to be the unique theory of all four fundamental interactions.

Another symmetry of the action of a free electromagnetic field (13.14) is the global rescaling of coordinates and potentials

$$x^\mu \longrightarrow x'^\mu = x^\mu e^{-\lambda}, \qquad A^\mu \longrightarrow A'^\mu = A^\mu e^\lambda, \qquad (13.108)$$

where λ is an arbitrary constant. It is easy to see that this transformation leaves the action invariant. As we shall see in the last part of the book, this symmetry is related to the local (or global) conformal symmetry that takes place in curved space–time, typical for general relativity.

Exercise 28 Try to find the four-dimensional (Minkowski space) version of the transformation of electromagnetic duality (11.10).
Hint. Start from the particular case of $\alpha = \pi/2$. In general, this is not a simple exercise.

Exercise 29 Explore whether the Lorentz transformations (13.29)–(13.31) are compatible with the transformation (11.10).

Exercise 30 Write the modified expression for the Lorentz force (12.41) after the transformation (11.10).

13.6   Ideal Relativistic Gas of Massive Particles

The equation of state for different types of fluids is interesting and especially useful for the cosmological models that we will consider at the end of the next chapter. Consider an especially simple case of the ideal gas of identical, classical massive particles. From the consideration in Sect. 13.3, we know that when such a gas is ultrarelativistic (meaning that the average speed of the particles in the units of c is approaching one), the energy–momentum tensor of such a gas should have vanishing trace. In the co-moving frame, when the energy–momentum tensor is diagonal, this means that $T^\alpha_\beta = \text{diag}(\rho, -p, -p, -p)$, where $p = \rho/3$. At the other limit, in the IR, the pressure satisfies the relation $p \ll \rho$ and this type of gas is called *dust*. Just to clarify things, let us note that the pressure of the air in the room where the reader is now can be very well approximated by zero. The reason is that here ρ includes the mass part, nmc^2 (n is concentration of particles). Since $c = 3 \times 10^8$ m/s and the typical speed of the molecule of air is about $v_a = 500$ m/s, the ratio p/ρ is of the order of $(v_a/c)^2 \approx 3 \cdot 10^{-13}$. Certainly, taking $p = 0$ if compared to *relativistic density of energy* is a very good approximation in this case.

Let us consider the general case of the particles that have an arbitrary average speed, and derive the relation between p and ρ. The problem can be solved on the basis of the relativistic version of Maxwell distribution. This solution was obtained in the paper [4] (see also the book [5]). One can also arrive at the same result from the relativistic version of the Boltzmann's H-theorem (see, e.g., [6]). In what follows we first review the Jüttner model [4] and then consider the simplified version introduced by A.D. Sakharov in the seminal work [7], where he used this simple equation of state to describe what is nowadays called baryon acoustic oscillations.

The most simple calculation of the Jüttner formula requires basic knowledge of statistical mechanics. We present this solution and leave a slightly more complicated elementary one as an exercise. The starting formula is the relation between kinetic energy and momenta of the particle,

$$\varepsilon = \sqrt{c^2 \mathbf{p}^2 + c^4 m^2}. \tag{13.109}$$

Then the statistical integral for a single particle is given by the expression

$$Z = \int e^{-\varepsilon/kT} d^3 p = 4\pi m^2 c \cdot K_2\left(\frac{mc^2}{kT}\right), \tag{13.110}$$

where $K_\nu(x)$ is a modified Bessel function of index ν. After that, the equation of state for the gas of N particles can be derived in a standard way

$$PV = kTN \left(\frac{\partial \ln Z}{\partial \ln V}\right)_T = NkT, \tag{13.111}$$

while the average energy of the particle is

$$\bar{\varepsilon} = \frac{1}{Z} \int e^{-\varepsilon/kT} \varepsilon \, d^3 p = mc^2 \frac{K_3(mc^2/kT)}{K_2(mc^2/kT)} - kT. \tag{13.112}$$

The energy density $\rho = \frac{N\bar{\varepsilon}}{V}$ and pressure p are related by an implicit functional dependence that consists of Eqs. (13.111) and (13.112).

The Jüttner formulas enable one to get the equation of state for the ideal relativistic gas, but (13.111) and (13.112) are not very useful for practical purposes. In the last part of the book, we will see how the almost equivalent equation of state can be effectively used to construct a simple cosmological model, and now we shall derive this equation.

Consider a single relativistic particle with the rest mass m in a $3D$ volume V. The difference with the Jüttner model is that now we do not assume the Maxwell distribution of the velocities. Instead we take all velocities to have the same absolute values, and later on compare the equation of state that follows from this assumption with the one that emerges from (13.111) and (13.112).

For the sake of simplicity, we assume that the volume V is a cuboid of $a \times b \times c$, and first consider a single particle that moves in the volume $V = abc$. If the scattering of this particle with the walls is always elastic, the component v_x does not change with time. Then the time between two collisions of the particle with one of the rectangle walls with the area bc is $\Delta t_a = 2a/v_x$. The momentum of the particle is given by Eq. (12.23) and hence each collision transfers to the wall the quantity of momentum

$$\mathbf{p} = \frac{m\mathbf{v}}{\sqrt{1 - v^2/c^2}} \quad \Longrightarrow \quad \Delta p_x = \frac{2mv_x}{\sqrt{1 - v^2/c^2}}, \tag{13.113}$$

such that the average pressure acting on the wall is

$$p_a = \frac{\Delta p_x}{\Delta t_a \cdot bc} = \frac{1}{\sqrt{1 - v^2/c^2}} \frac{mv_x^2}{abc}. \tag{13.114}$$

Now consider the gas of N such particles. Due to isotropy, the average of the squares of the three components of velocities is equal,

$$\langle v_x^2 \rangle = \langle v_y^2 \rangle = \langle v_z^2 \rangle = \frac{1}{3} \langle v^2 \rangle = \frac{1}{3} v^2, \tag{13.115}$$

where at the last step we took into account that the speeds of all particles is the same. Thus, the pressure of the gas with N such particles (13.114) is

$$p = \frac{N}{3\sqrt{1 - v^2/c^2}} \frac{mv^2}{abc} = \frac{N}{3\sqrt{1 - v^2/c^2}} \frac{mv^2}{V} = \frac{n\,mv^2}{3\sqrt{1 - v^2/c^2}}, \tag{13.116}$$

where $n = N/V$ is the concentration of the particles in the volume V.

It is an elementary exercise to show that the pressure (13.116) and the energy density

$$\rho = n\varepsilon = \frac{n\,mc^2}{\sqrt{1 - v^2/c^2}} \tag{13.117}$$

satisfy the relation (equation of state)

$$p = \frac{\rho}{3} \cdot \left[1 - \left(\frac{mc^2}{\varepsilon} \right)^2 \right]. \tag{13.118}$$

The expression (13.118) is remarkable for at least three reasons. The first reason is that Eq. (13.118) nicely interpolates between the equations of state of radiation $p = \rho/3$ and dust $p = 0$. It is easy to see that the ratio $\omega = p/\rho$ tends to $1/3$ in the ultrarelativistic limit $\varepsilon \to \infty$ and becomes zero in the nonrelativistic limit $\varepsilon \to mc^2$. It proves useful to introduce the special notation for the density of the rest energy of the particles $\rho_d = Nmc^2/V$. With this notation one can cast Eq. (13.118) into the form

$$P = \frac{\rho}{3} \cdot \left[1 - \frac{\rho_d^2}{\rho^2} \right]. \tag{13.119}$$

The second reason to say that (13.118) is remarkable is that it is extremely simple, for instance, if being compared to the Jüttner formulas (13.111) and (13.112). Regardless of its simplicity, (13.118) nicely reproduce the equation of state of the Jüttner model, with the relative deviation never exceeding 2.5% [8].

The third reason to regard this model as remarkable is its historic importance, related to the important work of A.D. Sakharov [7], whom we mentioned before. This seminal paper can be seen as an impressive proof that one can calculate from the first principles the details of what took place in our universe at the early stage of its existence. In his analysis, Sakharov used the equation of state that is equivalent to (13.118), but did not provide details of its derivation. In the last part of the book, we consider the cosmological model based on this equation of state, which interpolates between the universe dominated by radiation and matter.

Exercise 31 Using available software, compare (13.118) with the equation of state given by (13.111) and (13.112) and verify the corresponding plot of Ref. [8]. Explain why the difference between the two formulas vanishes in both IR and UV limits.
Hint. The last part requires basic knowledge of statistical mechanics.

References

1. A.T. Fomenko, S.P. Novikov, B.A. Dubrovin, *Modern Geometry-Methods and Applications, Part I: The Geometry of Surfaces, Transformation Groups, and Fields* (Springer, Berlin, 1992)
2. J.D. Jackson, *Classical Electrodynamics*, 3rd edn. (Wiley, Hoboken, 1998)
3. L.D. Landau, E.M. Lifshits, *Fluid Mechanics - Course of Theoretical Physics*, vol. 6, 2nd edn. (Elsevier, Amsterdam, 2013)
4. F. Jüttner, *Das Maxwellsche Gesetz der Geschwindigkeitsverteilung in der Relativtheorie*, Ann. der Phys. Bd **116** (1911) S. 145
5. W. Pauli, *Theory of Relativity* (Dover, United States, 1981)
6. C. Cercignani, G.M. Kremer, *The Relativistic Boltzmann Equation: Theory and Applications* (Birkhäuser, Basel, 2002)
7. A.D. Sakharov, The initial stage of an expanding Universe and the appearance of a nonuniform distribution of matter. Zh. Eksp. Teor. Fiz. **49**(1), 345; Sov. Phys. JETP **22**, 241 (1966)
8. G. de Berredo-Peixoto, I.L. Shapiro, F. Sobreira, Simple cosmological model with relativistic gas. Mod. Phys. Lett. A **20**, 2723 (2005)

Part III
Applications to General Relativity

Historically, tensor analysis became incorporated in the physics curriculum after the creation of general relativity by Albert Einstein. The first papers on general relativity were partially written in collaboration with Marcel Grossmann, who should be credited a lot for the implementation of tensor methods in Physics. This is one reason why one of the main meetings on general relativity bears his name. Nowadays, tensor calculus represents a relevant part of mathematical education, especially for physicists. For those who know tensors, the general relativity is becoming much simpler to learn. Thus, it is quite natural that the book on tensors has a part devoted to general relativity.

The purpose of this part of the book is to give a brief technical introduction to general relativity. The exposition is designed in such a way that the reading should be smooth for those who are familiar with the contents of the first and second parts of the book. At the same time, we do not pretend to replace the existing books on general relativity, many of which offer a comprehensive coverage of the subject— something that cannot be done in an introductory textbook such as the present one. We expect, however, that this chapter can be useful for a first reading and will make more accessible the study of more complete and detailed books.

Let me give a short list for possible further reading. Note that this is not a complete list of the literature or of the textbooks on general relativity, which is very extensive and rapidly growing. The books listed below are selected on the basis of personal preference.

The reader who is looking for a more comprehensive introduction to relativity should perhaps start from the standard classical textbooks such as [1, 2, 3] (see also more recent textbooks [4] and [5]) and the book of problems [6]. The books approach general relativity from the mathematical side include [7, 8, 9] and [10] (the black hole part). The recommended books on the black holes include the comprehensive monograph [11] and the more textbook style [12]. The books on cosmology recommended for the first and subsequent readings are [13, 14, 15]. There are many other excellent books on cosmology and gravity in general (some are cited in the text), but the mentioned ones look most pedagogical and sufficiently complete.

References

1. L.D. Landau, and E.M. Lifshits, The Classical Theory of Fields-Course of Theoretical Physics, Vol. 2, (Butterworth-Heinemann, 1987).
2. S. Weinberg, Gravitation and Cosmology. (John Wiley and Sons. Inc., 1972).
3. R. D'Inverno, Introducing Einstein's relativity. (Oxford University Press, 1992-1998).
4. T. Padmanabhan, Gravitation: Foundations and Frontiers, (Cambridge University Press, 2010).
5. M.P. Hobson, G.P. Efstathiou, A.N. Lasenby, General Relativity: An Introduction for Physicists, (Cambridge University Press, 2006).
6. A.P. Lightman, W.H. Press, R.H. Price, and S.A. Teukolsky, Problem book in Relativity and Gravitation, (Princeton University Press, 1975).
7. S.W. Hawking and G.F.R. Ellis, The large scale structure of space-time, (Cambridge University Press, 1973).
8. A.Z.Petrov, Einstein Spaces. (Pergamon Press, Oxford, 1969).
9. A.T. Fomenko, S.P. Novikov, and B. A. Dubrovin, Modern Geometry-Methods and Applications, Part I: The Geometry of Surfaces, Transformation Groups, and Fields, (Springer, 1992).
10. S. Chandrasekhar, The Mathematical Theory of Black Holes, (Clarendon Press, Oxford, 1998).
11. V.P.Frolov and I.D.Novikov, Black Hole Physics-Basic Concepts and New Developments. (Kluwer Academic Publishers, 1989).
12. V.P. Frolov, and A. Zelnikov, It Introduction to Black Hole Physics. (Oxford University Press, 2015).
13. S. Dodelson, Modern Cosmology. (Academic Press, 2003).
14. V. Mukhanov, Physical Foundations of Cosmology. (Cambridge University Press, 2005).
15. D.S. Gorbunov, and V.A. Rubakov, Introduction to the Theory of the Early Universe: Vol. I. Hot Big Bang Theory. (World Scientific Publishing, 2011); Vol. II. Cosmological Perturbations and Inflationary Theory. (ibid, 2011).

Chapter 14
Equivalence Principle, Covariance, and Curvature Tensor

14.1 Classical Gravity and Relativity

The previous part of the book started from the Maxwell equations, one of which is the Gauss law. We already know that this law of electrostatics is nicely embedded into relativistic framework, being part of the full set of equations, which are invariant under the Lorentz transformations. At the same time, there is one more case when we meet the same Gauss equation. Classical Newton's law of gravitational attraction implies that

$$\Delta \varphi = 4\pi G\rho\,, \tag{14.1}$$

where φ is the gravitational potential, G is Newton's gravitational constant, and ρ the mass density, meaning the quantity of mass per volume of the $3D$ space. Equation (14.1) is perfectly compatible with the Galilean transformation rule, especially because both φ and ρ do not depend on time.

At this moment one discovers a contradiction. Remember that considerations in Part 2 have shown that the Galilean transformation between inertial reference frames is an approximation to the more general Lorentz transformation rule. Indeed, the Gauss equation (14.1) alone is not compatible with the Lorentz transformations.

Exercise 1 Prove this incompatibility, and try to do it in several different ways. After that, construct the simplest relativistic generalization of Eq. (14.1) by replacing $\Delta \to -\Box$ and the mass density by the trace of the energy–momentum tensor, $\frac{1}{c^2} T^{\mu}_{\mu}$. Verify that the nonrelativistic limit of such an equation is (14.1).

Observation The notion of energy–momentum tensor was introduced in Part 2, and will be discussed again from another perspective in Sect. 15.1.1.

This incompatibility signals a real contradiction between gravity and relativity, but the Lorentz invariance of the Maxwell equations already taught us how it can be resolved. It is natural to suppose that Eq. (14.1) represents just a part of the more

© Springer Nature Switzerland AG 2019

I. L. Shapiro, *A Primer in Tensor Analysis and Relativity*, Undergraduate
Lecture Notes in Physics, https://doi.org/10.1007/978-3-030-26895-4_14

extensive system of equations, and that this part is what remains from the whole system in the static case. Starting from this hypothesis, we first have to find the relativistic equations which produce the Gauss equation as the lowest order, nonrelativistic approximation. Second, it is necessary to find some potentially observable consequences of the new equations. And third, these consequences should be tested in experiments and observations.

All these points are correct, with one relevant detail. Indeed, the simple relativistic extension of the Gauss equation that is compatible with existing observations and experiments does not exist. The construction of relativistic gravitational theory requires introducing a qualitatively new physical concept, and this is one of the most important issues in Einstein's general relativity.

Here we must stress the critical importance of three steps described above. As we already saw in Exercise 1, there is a theoretically nice and relatively simple realization of the first part, based on the scalar field. One can certainly construct another approach based on vectors, as it is done in electrodynamics. However, in both cases, there is no experimental/observational confirmation of the corresponding theories. An interested reader can come back to this statement after reading Sect. 15.3 below and check it by calculations.

14.2   Space–Time Metric and Equivalence Principle

The main idea of general relativity is that the gravitational field is described by the metric of the four-dimensional space–time. In special relativity, the space–time is always the same, independent of the dynamics of particles and fields that may be present in this space–time. In general relativity, the space–time itself becomes dynamical. The variable that describes gravity is metric, but as we shall see in what follows the gravitational force acting on the particle depends on metric derivatives.

In order to understand how the force acting on the particle emerges, we have to introduce the main physical assumption of general relativity: the equivalence principle. This principle states that the point-like test particle cannot distinguish whether it is a subject of the force of gravity or of inertia. In other words, locally, gravity can be always compensated by inertia. In order to understand the sense of this postulate[1] let us imagine an observer confined in a very small box in space. Suppose that the observer feels no gravity, so according to the equivalence principle he can not know whether his box is far from all bodies that are producing a gravitational field, or it is in fact close to sources of gravity, but the box is in the state of a free fall.

A simple example is when an observer feels a constant gravitational acceleration **g**. This means that the force $m\mathbf{g}$ acts on any test body inside the box. The observer cannot distinguish whether this force is produced by gravity, or the box is moving with a constant acceleration $-\mathbf{g}$ somewhere far from all gravitating bodies.

---

[1] We will not distinguish between different versions of the equivalence principle; the interested reader can look, e.g., into the books [1, 2].

Fig. 14.1 The trajectories of the free fall of the points B and C are straight lines, which are not exactly parallel to that of the central point A. The effect becomes negligible for a very small-sized box. For a point-like box, the effect of gravity can be completely compensated by the force of inertia

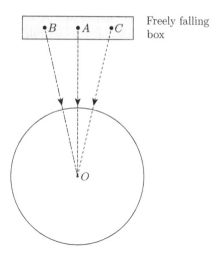

It is immediately clear that the equivalence between gravity and inertia can hold only locally. This means that an equivalence cannot be achieved in the finite volume of space, if we do not neglect the size of the volume. Imagine, for example, the box that is in the free fall to the Earth moving along the radial line directed to the center of the planet (see Fig. 14.1 for the illustration of this situation). As far as we take the size of the box into account, the real situation is that the center of mass of the box, point A, is moving along this line. Other points inside the box, which do not belong to the same radial line, are in fact moving along parallel straight lines. However, for these points, the radial line directed to the center of the planet is not exactly parallel to the trajectory of point A. The gravitational and inertial forces for these points will have slightly different directions, and the compensation between gravity and inertia is not precise. The observer inside a finite-sized box can measure a mutual acceleration of test bodies; as a manifestation of that, the equivalence principle is local and does not hold in a whole box. The violation also takes place for the test particles that belong to the same radial line. In the case of homogeneous gravity $\mathbf{g} = const$, we do not observe the effect. But in all real situations $\mathbf{g} = const$ is only an approximation.

Exercise 2 Calculate the maximal mutual acceleration of the points which are separated by 1 m in the box which is freely falling to Earth along the radial line at the altitude 6.4 thousand kilometers over the surface of the planet. Consider different positions of the test particles with respect to the radial line. Imagine we can use precise interferometers to detect the accelerations of 10^{-8} cm/s^2. How large should the box be to provide detectable mutual acceleration?

Let us now put the equivalence principle into formulas. The starting point is that the gravity is described by the metric of space–time. The equivalence principle tells us we can make the metric to be precisely Minkowski one $\eta_{\alpha\beta}$ in a given point. For this, it is sufficient to relate the reference frame with one of the freely falling particles. We can call the coordinates in this locally inertial frame ξ^α. The laboratory reference frame is not freely falling and the coordinates are x^μ. Here both $\alpha, \mu = 0, 1, 2, 3$. According to the standard tensor rule of transformation, the metric in the laboratory coordinates is

$$g_{\mu\nu}(x) = \frac{\partial \xi^\alpha}{\partial x^\mu} \frac{\partial \xi^\beta}{\partial x^\nu} \eta_{\alpha\beta}. \tag{14.2}$$

In the falling frame, the equation of motion of a free particle is certainly the equation for the particle without external forces, that is,

$$\frac{d^2\xi^\alpha}{ds^2} = 0. \tag{14.3}$$

Now, all we need is to rewrite this equation in terms of external laboratory coordinates x^μ. Tensor analysis tells us immediately how this can be done. We assume that in the vicinity of a given point both sets $\xi^\alpha(x^\mu)$ and $x^\mu(\xi^\alpha)$ are at least twice differentiable functions, then

$$0 = \frac{d^2\xi^\alpha}{ds^2} = \frac{d}{ds}\left(\frac{\partial \xi^\alpha}{\partial x^\mu} \frac{dx^\mu}{ds}\right) = \frac{\partial \xi^\alpha}{\partial x^\mu} \frac{d^2 x^\mu}{ds^2} + \frac{\partial^2 \xi^\alpha}{\partial x^\mu \partial x^\nu} \frac{dx^\mu}{ds} \frac{dx^\nu}{ds}. \tag{14.4}$$

This equation can be recast into the form

$$\frac{d^2 x^\lambda}{ds^2} + \Gamma^\lambda_{\mu\nu} \frac{dx^\mu}{ds} \frac{dx^\nu}{ds} = 0, \tag{14.5}$$

where

$$\Gamma^\lambda_{\mu\nu} = \frac{\partial x^\lambda}{\partial \xi^\alpha} \frac{\partial^2 \xi^\alpha}{\partial x^\mu \partial x^\nu} = -\frac{\partial \xi^\alpha}{\partial x^\mu} \frac{\partial \xi^\beta}{\partial x^\nu} \frac{\partial^2 x^\lambda}{\partial \xi^\alpha \partial \xi^\beta} \tag{14.6}$$

are the coefficients of affine connection, which are familiar to us from the first part of the book. As we know, this object is used to construct the covariant derivatives, but we leave the discussion of this subject to the next section.

Exercise 3 *(a)* Verify Eqs. (14.5) and (14.6). As we have seen in Part 1, this requires the relations

$$\frac{\partial x^\rho}{\partial \xi^\alpha} \frac{\partial \xi^\alpha}{\partial x^\sigma} = \delta^\rho_\sigma \implies \frac{\partial^2 x^\rho}{\partial \xi^\alpha \partial \xi^\beta} = -\frac{\partial^2 \xi^\mu}{\partial x^\lambda \partial x^\tau} \frac{\partial x^\rho}{\partial \xi^\mu} \frac{\partial x^\lambda}{\partial \xi^\alpha} \frac{\partial x^\tau}{\partial \xi^\beta}.$$

(b) Using definition (14.6), calculate the law of transformation of affine connection from the coordinates (four-dimensional reference frame) x^μ to another frame x'^ρ.

Solution.

$$\Gamma'^\tau_{\rho\sigma} = \frac{\partial x'^\tau}{\partial x^\lambda} \frac{\partial x^\mu}{\partial x'^\rho} \frac{\partial x^\nu}{\partial x'^\sigma} \Gamma^\lambda_{\mu\nu} - \frac{\partial x^\mu}{\partial x'^\rho} \frac{\partial x^\nu}{\partial x'^\sigma} \frac{\partial^2 x'^\tau}{\partial x^\mu \partial x^\nu}. \tag{14.7}$$

Coming back to the definition of covariant derivative in Part 1 shows that with the usual expression for affine connection (14.6) can be perfectly well used in case of curved space. The covariant derivative of a tensor is a tensor, exactly as in the globally Euclidean (flat) space.

Exercise 4 Verify whether one can use the expression for the metric (14.2), to rewrite the affine connection (14.6) in the form (compare to the same transformation in the Part 1)

$$\Gamma^\lambda_{\alpha\beta} = \frac{1}{2} g^{\lambda\tau} \left(\partial_\alpha g_{\beta\tau} + \partial_\beta g_{\alpha\tau} - \partial_\tau g_{\alpha\beta} \right) = \left\{ {}^{\ \lambda}_{\alpha\beta} \right\}. \tag{14.8}$$

Let us make the first conclusions. We postulated that the gravitational field is described by the metric of space–time, and also introduced the equivalence principle between gravity and inertia. These two things together led us to the equation for the free particle in the laboratory reference frame, Eq. (14.5), while the equation for the same particle in the frame that freely falls together with the particle is (14.3). The next step should be to find the classical nonrelativistic limit of (14.5) and compare it to the Gauss law of Newtonian gravity. However, it proves useful to first introduce some mathematical notions, which will put all our considerations on the more solid basis. Without this we cannot be sure of the difference between the case of a curved space, which we are dealing with, and the pseudo-Euclidean space in the curvilinear coordinates, where formulas look quite similar.

Therefore, we postpone the comparison with Newtonian gravity to Sect. 14.7 and in the next few sections engage with more involved math.

14.3 Manifolds and Riemann Spaces

After reading the previous section, the most natural questions are as follows: (1) How we can describe the curved space, which is supposed to be our space, with the gravitational field describing the deviation from the flat space? (2) How we can know that the given expression for the metric describes flat or curved space? (3) Why we say that it is the metric that describes gravity, while the equation for the free particle is (14.5), which has no metric, but only affine connection. Moreover, we know that this connection is coordinate (reference frame) dependent, and in fact it can be eliminated in (14.3) describing the special frame. So, what is the real role of the metric?

Let us start from the definitions of the space, which is the geometric description of gravity in general relativity. Mathematically, curved space is a structure called manifold. For the sake of generality, we consider D-dimensional manifolds and describe both Euclidean and pseudo-Euclidean spaces. In case of general relativity, our main interest is the pseudo-Euclidean $3 + 1$-dimensional case, but in many situations it is useful to have a general definition.

Fig. 14.2 Stereographic
projection on the plane. All
points P of the sphere are
projected into points P′ on
the plane except the point O,
which is opposite to the point
O′, where the tangent plane
touches the sphere. The
sphere cannot be covered by
a single map

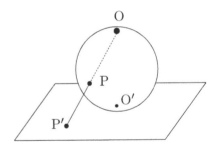

Def. 1 The set of points M is called differentiable manifold, if it possesses the
following structure:

(i) M can be considered as a set of maps U_q that means $M = \bigcup_q U_q$. Here q can be
finite or infinite. The set of all maps is called atlas.

(ii) Each of the maps U_q can be mapped to an open region in $\mathbb{R}^D$. Such a mapping
provides each point of the map by the coordinates x^α, where $\alpha = 1, 2, \ldots, D$. Let
us remember that $\mathbb{R}^D$ is defined as a set of all $x^1, x^2, \ldots, x^D$. It can be Euclidean or
pseudo-Euclidean space, or not have any notion of distance.

(iii) If the two maps intersect, in the common area one meets two sets of coordinates
of the same point (element of the manifold), say x^α and y^μ. It is required that the
functions $x^\alpha(y^\mu)$ and $y^\mu(x^\alpha)$ exist and are differentiable. If these functions are r
times differentiable, the manifold is called r-differentiable. Correspondingly, there
can also be ∞-differentiable manifold.

Let's look at two simple examples.

Example 1 Consider simple, flat Euclidean $2D$ surface. It can be mapped to $\mathbb{R}^2$
in infinitely many ways, e.g., by introducing different Cartesian or polar (or other)
coordinates. If we require that these coordinates satisfy the condition *(iii)*, this is
a two-dimensional Euclidean manifold. Of course, this particular manifold can be
covered by a single map.

Example 2 Consider a $2D$ Euclidean sphere. It can be mapped into $\mathbb{R}^2$ in infinitely
many ways by introducing spherical coordinates with different orientations of the
axes, or by means of stereographic projection, as shown in Fig. 14.2. However, each
of these mappings does not provide the existence of differentiable transformation of
coordinates for *all* points of the sphere. There will be always at least one point on
the sphere that cannot be mapped in a way consistent with the condition *(iii)* of the
Definition. Therefore, the sphere can be regarded as a manifold that requires at least
two maps.

Observation The existence of differentiable functions $x^\alpha(y^\mu)$ and $y^\mu(x^\alpha)$ means
that

$$J = \frac{D(x^\alpha)}{D(y^\mu)} \neq 0 \quad \text{and} \quad J^{-1} = \frac{D(y^\mu)}{D(x^\alpha)} \neq 0. \tag{14.9}$$

Def. 2 The differentiable manifold is called oriented, if the Jacobians from Eq. (14.9) are positive in all intersections of all maps.

Observation 1 For example, the flat $2D$ surface and sphere are examples of oriented manifolds, and the Möbius strip is an example of a non-oriented manifold. In what follows, we will assume that our spaces of interest are oriented.

Observation 2 Let us remember that in the first part of the book we considered examples of the change of coordinates that violate the condition of Definition 2. These examples show that some maps should be excluded in order to have an oriented manifold.

Def. 3 A contravariant vector field on the manifold M in a system of coordinates $x = x^\alpha$ is given by the set of functions V^μ. It is required that the transformation rule to another map with coordinates $x' = x^{\beta'}$ is

$$V^{\rho'}(x') = \frac{\partial x^{\rho'}}{\partial x^\mu} V^\mu(x). \tag{14.10}$$

Observation It is easy to see that this definition is exactly the same as the one we discussed in Part 1 for the contravariant vector in Euclidean space in curvilinear coordinates. The definitions of a scalar, covariant vector, and all types of tensors are also the same. Therefore, we leave these definitions as an **Exercise 5** to the reader and only write one particular relation for the mixed (m, n)-type tensor,

$$T^{\rho'_1 \ldots \rho'_m}{}_{\sigma'_1 \ldots \sigma'_n}(x') = \frac{\partial x^{\rho'_1}}{\partial x^{\mu_1}} \cdots \frac{\partial x^{\rho'_m}}{\partial x^{\mu_m}} \frac{\partial x^{\nu_1}}{\partial x^{\sigma'_1}} \cdots \frac{\partial x^{\nu_n}}{\partial x^{\sigma'_n}} T^{\mu_1 \ldots \mu_m}{}_{\nu_1 \ldots \nu_n}(x). \tag{14.11}$$

Exercise 6 Verify that the partial derivative of a scalar transforms as a covariant vector. Discuss whether the partial derivative of any other tensor is a tensor.

Hint. If you need to think more than a minute about the last question, it may be a good idea to go back to Part 1 and check out whether you remember well the sections concerning covariant derivatives.

Def. 4 A Riemannian (pseudo-Riemannian, in case of a curved version of Minkowski space) metric $g_{\mu\nu} = g_{\mu\nu}(x)$ is a bilinear form defined on the vectors at any point of the manifold, and smoothly dependent on the coordinates. For a map U_q with coordinates x^α we have, for the two vectors ξ^α and η^β, the scalar product

$$\langle \xi | \eta \rangle = g_{\mu\nu} \xi^\mu \eta^\nu, \tag{14.12}$$

such that $|\xi^2| = g_{\mu\nu} \xi^\mu \xi^\nu$.

The number of positive and negative eigenvalues of the metric defines its *signature*. For example, the signature of Euclidean $4D$ space is $+4$ and the signature of Minkowski $1 + 3$ space (one time and three space coordinates, $\eta_{\mu\nu} = \text{diag}(1, -1, -1, -1)$ is -2.

Observation In the literature, one can find another definition of Minkowski metric, $\eta_{\mu\nu} = \text{diag}(-1, 1, 1, 1)$, and we usually use one or another depending on convenience. The same concerns some other definitions, typical only for a curved space. The unfortunate consequence of this situation is that one needs to take special care when using formulas from the books or papers.

Exercise 7 (a) Verify that the scalar product of two vectors depends on coordinates as a scalar field, if the metric is a second-rank covariant tensor. (b) Verify that this feature holds for the metric defined in (14.2). (c) Discuss whether the relation (14.2) is the most general one in the situation when the equivalence principle holds, and to what extent the equivalence principle defines this relation.

Def. 5 The manifold provided by the metric is called Riemannian space, if there are no other fields that describe the geometry. We will discuss briefly the non-Riemannian spaces in the next section, devoted to affine connection and covariant derivative.

Def. 6 An absolutely antisymmetric contravariant tensor of a maximal rank in the $4D$ space with the signature -2 is defined by the relations

$$\varepsilon^{\alpha\beta\mu\nu} = \frac{1}{\sqrt{-g}} E^{\alpha\beta\mu\nu}, \qquad E^{0123} = 1. \tag{14.13}$$

Here $E^{\alpha\beta\mu\nu}$ is an absolutely antisymmetric Levi-Civita symbol with constant coefficients, which is a tensor density. The definition (14.13) provides the tensor properties of $\varepsilon^{\alpha\beta\mu\nu}$, which does not require a special proof, since it is completely analogous to the definition of a Levi-Civita tensor in Euclidean space in curvilinear coordinates, which we know from Part 1.

Observation The maximally antisymmetric covariant Levi-Civita tensor in the $4D$ Minkowski space can be defined by the relations

$$\varepsilon_{\alpha\beta\mu\nu} = \sqrt{-g}\, E_{\alpha\beta\mu\nu}, \qquad E_{0123} = -1. \tag{14.14}$$

Exercise 8 (a) Verify that the antisymmetric tensors (14.13) and (14.14) can be obtained one from another by lowering or raising indices with the metric $g_{\mu\nu}$ and its inverse $g^{\mu\nu}$. Generalize the two definitions (14.13) and (14.14) to the case of D-dimensional space with an arbitrary signature k.

Def. 7 The invariant volume element in the Riemann space of dimension D is defined as

$$d^D x \sqrt{|g|} = \frac{1}{D!} \varepsilon_{\alpha_1\alpha_2\ldots\alpha_D} dx^{\alpha_1} dx^{\alpha_2} \ldots dx^{\alpha_D}. \tag{14.15}$$

This expression enables one to perform covariant integration of tensor quantities on the manifold. In this respect, it is completely analogous to the similar one which we know from the curvilinear coordinates in flat space.

The Gauss and Stokes theorems can be obtained by mapping the corresponding formulas for the D-dimensional volume (V) restricted by the ($D - 1$)-dimensional surface (∂V) onto the corresponding expressions in the Euclidean space. In particular, the Gauss theorem has the form

$$\int_V dV\,\nabla_\mu A^\mu = \int_{(\partial V)} \sqrt{-g}\,A^\mu\,dS_\mu\,. \tag{14.16}$$

14.3.1 Affine Connections

Due to the similarity between curved-space manifold and the description of a flat (e.g., Minkowski in the pseudo-Euclidean $4D$ case) space in curvilinear coordinates, it is natural to ask whether one can define the notion of a covariant derivative, that is, a generalization of partial derivative that is capable of producing tensors when taking derivative of other tensors. It is easy to see that the considerations from the first part of the book work perfectly well for this purpose. The reason is that the form of the tensor transformations (14.11) is the same as we have discussed in Chap. 1 of Part 1. Thus, the considerations from Chap. 5 of Part 1 regarding transformation of derivatives and the derivation of compensating term do not change when we consider the issue in a Riemann space instead of the flat (Euclidean or pseudo-Euclidean Minkowski) space. As a result, we arrive at the conclusion that the covariant derivative of a tensor is[2]

$$
\begin{aligned}
\nabla_\lambda T^{\rho_1\ldots\rho_m}{}_{\sigma_1\ldots\sigma_n} &= T^{\rho_1\ldots\rho_m}{}_{\sigma_1\ldots\sigma_n;\lambda} \\
&= \partial_\lambda T^{\rho_1\ldots\rho_m}{}_{\sigma_1\ldots\sigma_n} + \Gamma^{\rho_1}_{\tau\lambda} T^{\tau\ldots\rho_m}{}_{\sigma_1\ldots\sigma_n} + \ldots + \Gamma^{\rho_m}_{\tau\lambda} T^{\rho_1\ldots\tau}{}_{\sigma_1\ldots\sigma_n} \\
&\quad - \Gamma^\tau_{\sigma_1\lambda} T^{\rho_1\ldots\rho_m}{}_{\tau\ldots\sigma_n} - \ldots - \Gamma^\tau_{\sigma_n\lambda} T^{\rho_1\ldots\rho_m}{}_{\sigma_1\ldots\tau}\,.
\end{aligned} \tag{14.17}
$$

This is a tensor if the object of affine connection transforms according to (14.7). A pertinent observation based only on the transformation law (14.7) is that the difference between two affine connections (constructed with two different metrics or in any other way) is a tensor. This fact proves relevant in many aspects.

Consider the same question from another perspective. One can ask whether it is possible to establish the unique form of $\Gamma^\tau_{\mu\nu}$ using the transformation rule (14.7). The answer is negative, which has important physical implications. To better understand this issue, let us consider again Exercise 4 from Sect. 14.2.

First of all, one can easily arrive at the same expression (14.8) just assuming that the gravitational field is described only by the metric, without other fields. Technically, this requirement means two things. First, in the globally flat space, with the metric $g_{\mu\nu} \equiv \eta_{\mu\nu}$, we know that $\partial_\lambda \eta_{\mu\nu} \equiv 0$ and therefore $\nabla_\lambda g_{\mu\nu} \equiv 0$. This

---

[2]Note two alternative notations for covariant derivative. Similar notations are also used for partial derivative, e.g., $\partial_\alpha \mathbf{T} = \mathbf{T}_{,\alpha}$.

is called the metricity condition of covariant derivative.[3] If there are other elements of the gravitational field, this could not be true, because even if the metric is globally Minkowski, other gravitational fields do not switch off. However, as far as we assume that the geometry is described by the metric and that there are no other geometric fields, then the metricity condition should be satisfied. Second, if there are no other fields, the affine connection is defined by the expression (14.6) and hence it is symmetric in the lower indices. These two conditions can be used in the following way: Let us write the metricity condition as

$$\partial_\lambda g_{\mu\nu} = \partial_\lambda g_{\mu\nu} - \Gamma^\tau_{\mu\lambda} g_{\tau\nu} - \Gamma^\tau_{\nu\lambda} g_{\mu\tau} = 0. \tag{14.18}$$

Making permutation of indices we get the three equations,

$$\partial_\lambda g_{\mu\nu} = \Gamma^\tau_{\mu\lambda} g_{\tau\nu} + \Gamma^\tau_{\nu\lambda} g_{\mu\tau} ,$$
$$\partial_\mu g_{\lambda\nu} = \Gamma^\tau_{\lambda\mu} g_{\tau\nu} + \Gamma^\tau_{\nu\mu} g_{\lambda\tau} ,$$
$$\partial_\nu g_{\lambda\mu} = \Gamma^\tau_{\lambda\nu} g_{\tau\mu} + \Gamma^\tau_{\nu\mu} g_{\lambda\tau} . \tag{14.19}$$

Summing up the last two equations and subtracting the first one, we arrive at (14.8). This formula is the same in curved space and in the flat space in curvilinear coordinates, as we know from Part 1.

It is instructive to explore how these equations modify in case the geometry is described by some extra fields, along with the metric. The solution is as follows. The expressions (14.8) (also called the Christoffel symbol and denoted by the symbol $\left\{ {\lambda \atop \alpha\beta} \right\}$) and (14.6) provide the tensor nature of covariant derivatives of vectors and tensors. However, it is easy to see that this solution for the covariance is not unique. In fact, one can write, instead of (14.8), a more general expression

$$\Gamma^\lambda_{\alpha\beta} = \left\{ {\lambda \atop \alpha\beta} \right\} + C^\lambda_{\alpha\beta} , \tag{14.20}$$

where $C^\lambda_{\alpha\beta}$ is an *arbitrary* tensor, and easily check that the tensor nature of the covariant derivative, and the Leibnitz rule, is guaranteed. In particular, one can violate the metricity condition,

$$\nabla_\lambda g_{\mu\nu} = \Omega_{\lambda,\mu\nu} , \tag{14.21}$$

or/and introduce torsion

$$\Gamma^\lambda_{\alpha\beta} - \Gamma^\lambda_{\beta\alpha} = T^\lambda_{\alpha\beta} = -T^\lambda_{\beta\alpha}, \tag{14.22}$$

---

[3]We also met this condition in Part 1, but then it was somehow trivial, since in flat space it immediately follows from the existence of a global orthonormal baisis. In the case of a general curved manifold this condition is completely nontrivial, as we shall see briefly.

and the covariant derivative of the tensor will still be perfectly working to make tensors out of other tensors. However, in the case of a globally Minkowski space, this new version of covariant derivative will not become a partial derivative. The reason is that the new fields, namely, nonmetricity $\Omega_{\lambda, \mu\nu}$ and torsion $T^\lambda_{\alpha\beta}$, are tensors and cannot be eliminated by transformations of coordinates to the global M_4 basis.

Exercise 9 Derive the explicit form of the tensor $C^\lambda_{\alpha\beta}$ in Eq. (14.20) that corresponds to the nonzero torsion and to the nonzero nonmetricity. Then answer the following questions:

(i) Can this task can be completed for torsion and nonmetricity separately, or they should be taken into account at the same time?

(ii) Is the solution for $C^\lambda_{\alpha\beta}$ with both torsion and nonmetricity the most general one?

(iii) Divide the torsion tensor into three irreducible components, namely, vector $T_\alpha = T^\beta_{\alpha\beta}$, axial vector $S^\alpha = \varepsilon^{\alpha\beta\mu\nu} T_{\beta\mu\nu}$, and the field $q^\lambda_{\alpha\beta}$, which satisfied the two conditions, $q^\beta_{\alpha\beta} = 0$ and $\varepsilon^{\alpha\beta\mu\nu} q_{\beta\mu\nu} = 0$. Find an explicit expression of a general torsion being a linear combination of these three fields.

(iv) Construct a similar representation in terms of irreducible components for nonmetricity.

(v) Is it possible that torsion and nonmetricity can replace each other? In order to answer the last two questions, one has to start by evaluating the number of independent components of $C^\lambda_{\alpha\beta}$, of torsion, and of nonmetricity.

Observation The full detailed answer to the last question requires real scientific research.

14.3.2 Parallel Transport

At this point, we can consider an important notion of the parallel transport of a tensor along the given curve in curved space. To this end, we consider a vector field $A^\mu(\tau)$ defined along a curve $x^\alpha = x^\alpha(\tau)$, where τ is a parameter along the curve and all functions $x^\alpha(\tau)$ are smooth. Indeed, the consideration can be easily generalized for an arbitrary tensor, but we start from the vector case for the sake of simplicity. It is useful to assume that $d\tau$ does not change under the change of coordinates. For instance, if the tangent vectors to the curve, $v^\alpha = dx^\alpha/d\tau$, are timelike, one can take $cd\tau = ds$, where s is the interval along the curve.

The first question to answer is what we mean by saying that the vectors A^μ taken at the different points of the curve τ_1 and τ_2 are parallel to each other. Geometrically it is not immediately clear. In order to understand the problem, it is sufficient, e.g., to try different paths of transporting the vector on a $2D$ sphere. In general, there may be different possibilities, in particular, because (as we shall see below) the result of the parallel transport between two points depends also on the choice of the curve that links these points. The same example of the $2D$ sphere shows this pretty well. It is clear that we need to provide a new definition of what the parallel transport is, such that it could make sense in an arbitrary (flat or curved) space.

In the flat space, one can use the existence of the global (e.g., Cartesian) basis and consider that the transport of the vector between the two different points is parallel if the components of the vector remain the same. Then in Cartesian coordinates we obtain the simple criterion of the parallel transport,

$$\frac{dA^\beta}{d\tau} = \frac{dX^\alpha}{d\tau}\frac{\partial A^\beta}{\partial X^\alpha} = 0. \tag{14.23}$$

In an arbitrary curvilinear coordinates, the last relation involves the covariant derivative and can be cast into the form

$$\frac{DA^\mu}{d\tau} = \frac{dx^\lambda}{d\tau}\nabla_\lambda A^\mu = 0, \tag{14.24}$$

where we introduced a special notation for the covariant derivative $D/d\tau$ along the curve. The last expression can be presented in such a way that the difference between $D/d\tau$ and $d/d\tau$ becomes explicit

$$\frac{DA^\mu}{d\tau} = \frac{dA^\mu}{d\tau} + \Gamma^\mu_{\lambda\sigma}\frac{dx^\lambda}{d\tau}A^\sigma = 0. \tag{14.25}$$

It is clear that in curvilinear coordinates in the $3D$ Euclidean flat space, the relation $\frac{d\mathbf{A}}{d\tau} = 0$ does not correspond to our geometric understanding of the parallel transport while the relation $\frac{D\mathbf{A}}{d\tau} = 0$ does, for it corresponds to the same components of the transported vector *in the special Cartesian coordinates*. Therefore, the relation (14.25) makes the definition of the geometric notion of parallel transport coordinate independent, and hence correct.

Now we come back to the situation in curved space. Since in curved space there is no global Cartesian coordinate system (or, equivalently, no global orthonormal vector basis) we cannot rely on the scheme presented above and must introduce the notion of parallel transport via an independent definition. This definition is constructed by analogy with the flat-space relation (14.24).

Def. 8 The parallel transport of the vector A^α along the curve $x^\alpha(\tau)$ in curved space is defined by the relation

$$\frac{DA^\alpha}{d\tau} = \frac{\partial x^\mu}{\partial \tau}\nabla_\mu A^\alpha = \frac{dA^\alpha}{d\tau} + \Gamma^\alpha_{\mu\nu}\frac{dx^\mu}{d\tau}A^\nu = 0, \tag{14.26}$$

where $\frac{dA^\alpha}{d\tau} = \frac{\partial A^\alpha}{\partial x^\mu}\frac{dx^\mu}{d\tau}$. In a similar way, we can define the parallel transport of an arbitrary tensor. For example, for the mixed $(1, 1)$-type tensor we have, by definition, the relation

$$\frac{DT^\alpha_\beta}{d\tau} = \frac{dx^\mu}{d\tau}\nabla_\mu T^\alpha_\beta = 0, \tag{14.27}$$

or, in a more detailed form,

$$\frac{DT_\mu^\nu}{d\tau} = \frac{dT_\mu^\nu}{d\tau} + \Gamma_{\alpha\lambda}^\nu \frac{dx^\alpha}{d\tau} T_\mu^\lambda - \Gamma_{\mu\alpha}^\lambda \frac{dx^\alpha}{d\tau} T_\lambda^\nu = 0. \tag{14.28}$$

14.4 Curvature Tensor

Suppose we have some metric and need to know whether this metric corresponds to the curved space or to the flat space in the curvilinear coordinates. In order to solve this problem, one can try to find the transformation of coordinates that reduces the metric to the flat form. This procedure is technically complicated and not too efficient. It would be nice to have some simple test for the flatness of the Riemann space. As we shall see below, this criterion is possible, and its construction provides many advantages for the applications of tensor analysis on manifolds.

Let us remember Exercise 4 at the end of Sect. 5 of Part 1. Now we can resolve it in a more general form. Consider the commutator of the two covariant derivatives applied to an arbitrary tensor in the flat space, but in arbitrary curvilinear coordinates

$$\left[\nabla_i, \nabla_j\right] T^{k_1 k_2 \cdots}{}_{l_1 l_2 \cdots}, \quad \text{where} \quad \left[\nabla_i, \nabla_j\right] = \nabla_i \nabla_j - \nabla_j \nabla_i. \tag{14.29}$$

Exercise requires us to prove that this commutator is zero. Since the space is flat, there are global Cartesian coordinates. In these coordinates X^a we have $\nabla_a = \partial_a$. The relation (14.29) is tensor equality and therefore it can be easily transformed into Cartesian coordinates

$$\left[\partial_a, \partial_b\right] T^{c_1 c_2 \cdots}{}_{d_1 d_2 \cdots}. \tag{14.30}$$

The last expression is zero because the partial derivatives commute. Then, making tensor transformation to the original coordinates x^i, we arrive at the required relation

$$\left[\nabla_i, \nabla_j\right] T^{k_1 k_2 \cdots}{}_{l_1 l_2 \cdots} \equiv 0. \tag{14.31}$$

It is easy to see that this solution is completely based on the existence of the global Cartesian coordinates. However, they exist only in flat spaces. Therefore, we can expect that in a curved space the commutator

$$\left[\nabla_\alpha, \nabla_\beta\right] T^{\mu_1 \mu_2 \cdots}{}_{\nu_1 \nu_2 \cdots} \tag{14.32}$$

may be nonzero. Moreover, it is possible to obtain the useful criterion of flatness using this fact.

Before going on in the derivation of the commutator (14.32), let us notice its close relation to the parallel transport of a tensor. Consider the parallel transport of the tensor $\mathbf{T}$ from the initial point $\mathbf{x} = x_0^\mu$ to the infinitesimal vector e^μ, that is,

$x_0^\mu \rightarrow x_1^\mu = x_0^\mu + e^\mu$. Up to the second order in e^μ we can write

$$\mathbf{T}(\mathbf{x}_0 + \mathbf{e}) = \mathbf{T}(\mathbf{x}_0) + e^\mu \nabla_\mu \mathbf{T}(\mathbf{x}_0) + \frac{1}{2} e^\mu e^\nu \nabla_\mu \nabla_\nu \mathbf{T}(\mathbf{x}_0) + \mathcal{O}(e^3). \quad (14.33)$$

After the consequent parallel transport to the infinitesimal vector f^ν we arrive at

$$\mathbf{T}(\mathbf{x}_0; \mathbf{e}, \mathbf{f}) = \mathbf{T}(\mathbf{x}_0) + (f^\mu + e^\mu) \nabla_\mu \mathbf{T}(\mathbf{x}_0)$$
$$+ \left(\frac{1}{2} f^\mu f^\nu + e^\mu f^\nu + \frac{1}{2} e^\mu e^\nu \right) \nabla_\mu \nabla_\nu \mathbf{T}(\mathbf{x}_0) + \mathcal{O}(f^3, e^3). \quad (14.34)$$

If we now perform the parallel transport to the same vectors $\mathbf{f}$ and $\mathbf{e}$ but in the opposite order, and then subtract the result from (14.34), we arrive at amazing result: the two parallel transports do not commute if the commutator of covariant derivatives is nonzero. And we already have an experience that shows that the two parallel transports do not commute in the case of the $2D$ sphere. Therefore, at least in this case

$$\mathbf{T}(\mathbf{x}_0; \mathbf{e}, \mathbf{f}) - \mathbf{T}(\mathbf{x}_0; \mathbf{f}, \mathbf{e}) = e^\mu f^\nu [\nabla_\mu, \nabla_\nu] + \mathcal{O}(f^3, e^3) \quad (14.35)$$

(and in fact in many others) the commutator of covariant derivatives is nonzero, violating the statement which we discussed in Part 1. In the general case of a curved space, one can expect to meet a noncommutative covariant derivatives.

As we shall see in what follows, the commutator $[\nabla_\mu, \nabla_\nu]$ acting on any tensor can always be expressed through the same unique object, called Riemann tensor, or curvature tensor, or simply curvature. Thus, this tensor possesses high degree of universality. Let us see why this is so. Consider first the commutator acting on the contravariant vector

$$[\nabla_\mu, \nabla_\nu] T^\alpha = \nabla_\mu \nabla_\nu T^\alpha - \nabla_\nu \nabla_\mu T^\alpha. \quad (14.36)$$

By direct calculations we find

$$\nabla_\mu \nabla_\nu T^\alpha = \partial_\mu (\nabla_\nu T^\alpha) - \Gamma_{\nu\mu}^\lambda (\nabla_\lambda T^\alpha) + \Gamma_{\lambda\mu}^\alpha (\nabla_\nu T^\lambda)$$
$$= \partial_\mu \partial_\nu T^\alpha + T^\lambda \partial_\mu \Gamma_{\lambda\nu}^\alpha + \Gamma_{\lambda\nu}^\alpha \partial_\mu T^\lambda - \Gamma_{\nu\mu}^\lambda \partial_\lambda T^\alpha - \Gamma_{\nu\mu}^\lambda \Gamma_{\tau\lambda}^\alpha T^\tau$$
$$+ \Gamma_{\lambda\mu}^\alpha \partial_\nu T^\lambda + \Gamma_{\lambda\mu}^\alpha \Gamma_{\tau\nu}^\lambda T^\tau.$$

Taking into account the commutativity of partial derivatives $[\partial_\mu, \partial_\nu] T^\alpha = 0$ we obtain

$$[\nabla_\mu, \nabla_\nu] T^\alpha = T^\lambda \left(\partial_\mu \Gamma_{\lambda\nu}^\alpha - \partial_\nu \Gamma_{\lambda\mu}^\alpha \right) + T^\tau \left(\Gamma_{\lambda\mu}^\alpha \Gamma_{\tau\nu}^\lambda - \Gamma_{\lambda\nu}^\alpha \Gamma_{\tau\mu}^\lambda \right)$$
$$= -T^\lambda \cdot R^\alpha_{\cdot\lambda\nu\mu}, \quad (14.37)$$

where

$$R^\alpha_{\cdot\lambda\nu\mu} = \partial_\nu \Gamma^\alpha_{\lambda\mu} - \partial_\mu \Gamma^\alpha_{\lambda\nu} + \Gamma^\tau_{\lambda\mu} \Gamma^\alpha_{\tau\nu} - \Gamma^\tau_{\lambda\nu} \Gamma^\alpha_{\tau\mu} \tag{14.38}$$

is the main object of our present discussion, the *curvature tensor*. The universality of the Riemann curvature can be seen, in particular, in the following two examples.

Exercise 10 Using direct calculus similar to the one for the contravariant vector, check the relation between the commutator of covariant derivatives and the curvature tensor for the covariant vector and for the mixed tensor

$$[\nabla_\mu, \nabla_\nu] Z_\beta = R^\lambda_{\cdot\beta\nu\mu} Z_\lambda , \tag{14.39}$$

$$[\nabla_\mu, \nabla_\nu] W^\alpha_\beta = R^\lambda_{\cdot\beta\nu\mu} W^\alpha_\lambda - R^\alpha_{\cdot\lambda\nu\mu} W^\lambda_\beta . \tag{14.40}$$

The result of this exercise demonstrates that the commutators of covariant derivatives demonstrate universality property, for they are always expressed in terms of the curvature tensor. The next step is to explain this fact and also derive the general rule for the commutator of an arbitrary tensor.

Let us start from the first relation (14.39). It is easy to see that the commutator acting on an arbitrary scalar field φ is zero

$$[\nabla_\mu, \nabla_\nu] \varphi = \partial_\mu \partial_\nu \varphi - \Gamma^\lambda_{\mu\nu} \partial_\lambda \varphi - \partial_\nu \partial_\mu \varphi + \Gamma^\lambda_{\nu\mu} \partial_\lambda \varphi = 0. \tag{14.41}$$

In order to find the relation between formulas (14.37) and (14.39), consider the commutator acting on the scalar $Z_\alpha \cdot T^\alpha$. Obviously such a commutator is zero and we obtain

$$0 = [\nabla_\mu, \nabla_\nu] (Z_\alpha T^\alpha) = Z_\alpha [\nabla_\mu, \nabla_\nu] T^\alpha + T^\alpha [\nabla_\mu, \nabla_\nu] Z_\alpha$$
$$+ \nabla_\mu Z_\alpha \cdot \nabla_\nu T^\alpha + \nabla_\nu Z_\alpha \cdot \nabla_\mu T^\alpha - \nabla_\nu Z_\alpha \cdot \nabla_\mu T^\alpha - \nabla_\mu Z_\alpha \cdot \nabla_\nu T^\alpha. \tag{14.42}$$

It is easy to see that the terms in the second line of the last formula cancel out. Then, using (14.37), we arrive at the following identity:

$$T^\alpha [\nabla_\mu, \nabla_\nu] Z_\alpha = Z_\alpha T^\lambda R^\alpha_{\cdot\lambda\nu\mu} = Z_\lambda T^\alpha R^\lambda_{\cdot\alpha\nu\mu} . \tag{14.43}$$

Since T^α is an arbitrary vector, the only solution of (14.43) is Eq. (14.39).

In a similar way, one can express the commutator of covariant derivatives acting on any tensor, just by contracting it with some amount of vectors to achieve a scalar product. It is clear that the rule for the general (k, l)-type tensor $T^{\alpha_1 \alpha_2 \ldots \alpha_k}_{\beta_1 \beta_2 \ldots \beta_l}$ can be obtained from the consideration of the scalar

$$\phi = T^{\alpha_1 \alpha_2 \ldots \alpha_k}_{\beta_1 \beta_2 \ldots \beta_l} Z_{\alpha_1} Z_{\alpha_2} \ldots Z_{\alpha_k} T^{\beta_1} T^{\beta_2} \ldots T^{\beta_l}$$

and has the form

$$\left[\nabla_\mu, \nabla_\nu\right]T^{\alpha_1 \alpha_2 \dots \alpha_k}{}_{\beta_1 \beta_2 \dots \beta_l} = R^{\alpha_1}{}_{\cdot \lambda \nu \mu} T^{\lambda \alpha_2 \dots \alpha_k}{}_{\beta_1 \beta_2 \dots \beta_l} + \cdots$$
$$+ R^{\alpha_k}{}_{\cdot \lambda \nu \mu} T^{\alpha_1 \alpha_2 \dots \lambda}{}_{\beta_1 \beta_2 \dots \beta_l} - R^{\lambda}{}_{\cdot \beta_1 \nu \mu} T^{\alpha_1 \alpha_2 \dots \alpha_k}{}_{\lambda \beta_2 \dots \beta_l} - \cdots$$
$$- R^{\lambda}{}_{\cdot \beta_l \nu \mu} T^{\alpha_1 \alpha_2 \dots \alpha_k}{}_{\lambda \beta_2 \dots \beta_l} . \tag{14.44}$$

For example, in order to obtain Eq. (14.40), one has to consider the scalar $W^\alpha_\beta Z_\alpha T^\beta$.

Exercise 11 Using the scalars $W^\alpha_\beta Z_\alpha T^\beta$ and $Y^{\alpha\beta} Z_\alpha Z_\beta$ obtain formula (14.40) and a similar formula for the covariant second-rank tensor

$$\left[\nabla_\mu, \nabla_\nu\right]Y^{\alpha\beta} = -R^{\alpha}{}_{\cdot \lambda \nu \mu} Y^{\lambda\beta} - R^{\beta}{}_{\cdot \lambda \nu \mu} Y^{\alpha\lambda} . \tag{14.45}$$

We have seen that the curvature tensor enables one to construct the universal expression for the commutator of the covariant derivatives acting on an arbitrary tensor. This commutator is a tensor, and thus the Riemann curvature also transforms as a tensor. Furthermore, we already know that the commutator is zero if and only if the space is flat, and hence the condition $R^{\alpha}{}_{\cdot \beta \nu \mu} \equiv 0$ is a necessary and sufficient criterion of the flatness.

The curvature tensor plays an important role in applications and especially in the relativistic theory of gravity. The rest of the section will be devoted to discussing its algebraic and other properties, and to efficient methods of its derivation for the particularly interesting metrics.

Consider first algebraic symmetries of the Riemann tensor. We can observe directly from its definition (14.38) that this tensor is antisymmetric in the last two indices

$$\text{(I)} \qquad R^{\alpha}{}_{\cdot \beta \mu \nu} = -R^{\alpha}{}_{\cdot \beta \nu \mu} . \tag{14.46}$$

In order to see other symmetries, one has to lower the first index

$$R_{\alpha\beta\mu\nu} = g_{\alpha\gamma} R^{\gamma}{}_{\cdot \beta \mu \nu} = g_{\alpha\gamma} \left(\partial_\mu \Gamma^\gamma_{\nu\beta} - \partial_\nu \Gamma^\gamma_{\mu\beta} + \Gamma^\lambda_{\nu\beta} \Gamma^\gamma_{\lambda\mu} - \Gamma^\lambda_{\mu\beta} \Gamma^\gamma_{\lambda\nu}\right). \tag{14.47}$$

At this point, we need the following Lemma, which was already mentioned earlier, when we discussed equivalence principle.

Lemma *At any point P in the Riemannian or pseudo-Riemannian space, one can choose such coordinates that, at the point P, the affine connection is zero $\Gamma^\lambda_{\mu\nu}(P) = 0$. Furthermore, we can choose such coordinates that the metric has the form of the flat space metric in the Cartesian coordinates.*

Proof Let us use the relation (14.7) between the coefficients of the affine connection in two different coordinates x^μ and x'^α,

$$\Gamma'^\lambda_{\mu\nu} = \frac{\partial x'^\lambda}{\partial x^\tau} \frac{\partial x^\alpha}{\partial x'^\mu} \frac{\partial x^\beta}{\partial x'^\nu} \Gamma^\tau_{\alpha\beta} - \frac{\partial x^\alpha}{\partial x'^\mu} \frac{\partial x^\beta}{\partial x'^\nu} \frac{\partial^2 x'^\lambda}{\partial x^\alpha \partial x^\beta} . \tag{14.48}$$

Our purpose is to prove the possibility of choosing new coordinates x'^α in such a way that $\Gamma'^\lambda_{\mu\nu} = 0$. This condition can be rewritten in the form

$$\Gamma^\tau_{\alpha\beta} = \frac{\partial x^\tau}{\partial x'^\lambda} \frac{\partial^2 x'^\lambda}{\partial x^\alpha \partial x^\beta}, \qquad \text{or}$$

$$\Gamma^\tau_{\alpha\beta} = \left(\Omega^{-1}\right)^\tau_{\lambda'} \, \partial_\alpha \Omega^{\lambda'}_\beta, \qquad \text{where} \qquad \Omega^{\lambda'}_\beta = \frac{\partial x'^\lambda}{\partial x^\beta}. \tag{14.49}$$

The last relation is the linear first-order differential equation for the elements of the matrix $\Omega^{\lambda'}_\beta$. According to the known theorems of the theory of differential equations, the solution of these equations exists, and therefore the object $\Gamma'^\lambda_{\mu\nu}$ can be made equal to zero by an appropriate choice of coordinates. Still we have a freedom to multiply the matrix $\Omega^{\lambda'}_\beta$ by the constant orthogonal matrix U^β_σ. Since $\partial_\alpha U^\beta_\sigma = 0$, this multiplication does not affect Eq. (14.49), and hence if $\Omega^{\lambda'}_\beta$ is a solution of Eq. (14.49), $(\Omega \cdot U)^{\lambda'}_\beta$ is also solution. By means of the orthogonal matrix, one can diagonalize the symmetric matrix $g_{\mu\nu}$ at any given point and then the metric can be transformed to the Minkowski one by rescaling coordinates.

An important observation is that the last Lemma concerns only the value of $\Gamma^\lambda_{\alpha\beta}$ at the point P, and not its partial derivatives $\partial_\tau \Gamma^\lambda_{\alpha\beta}$ at this point. An opposite situation would imply a contradiction with our previous statements. Imagine that one could, in a curved space, provide $\partial_\tau \Gamma^\lambda_{\alpha\beta} = 0$ simultaneously with $\Gamma^\lambda_{\alpha\beta} = 0$. Then one could eliminate the whole Riemann curvature tensor. As we already know, this is not possible in curved space. Let us note that the same issue will be discussed in much more detail in Sect. 17.3 devoted to Riemann normal coordinates. One can also find there a more elaborated proof of the same Lemma.

Another practically interesting aspect of the Lemma is that one can find *special coordinates* x'^α, where the Riemann tensor takes a simplified form

$$R^{\alpha'}_{\cdot \lambda' \nu' \mu'} = \partial_{\nu'} \Gamma^{\alpha'}_{\lambda' \mu'} - \partial_{\mu'} \Gamma^{\alpha'}_{\lambda' \nu'}, \tag{14.50}$$

while all $\mathcal{O}\left(\Gamma' \cdot \Gamma'\right)$ terms vanish. These coordinates prove useful in what follows.

Now we can see how the last Lemma provides a great help in establishing the symmetries of the Riemann tensor. The key point is that curvature is a tensor. Therefore, if we find that it has certain algebraic symmetry in some special coordinates, this means that the same algebraic symmetry holds in other coordinates. In particular, we can choose the coordinates where the affine connection is zero, like it is in Eq. (14.50). Using expression (14.8) for $\Gamma^\lambda_{\alpha\beta}$, we arrive at the following form of the curvature tensor:

$$R_{\mu\nu\alpha\beta} = \frac{1}{2}\left(\frac{\partial^2 g_{\mu\beta}}{\partial x^\nu \partial x^\alpha} - \frac{\partial^2 g_{\nu\beta}}{\partial x^\mu \partial x^\alpha} + \frac{\partial^2 g_{\nu\alpha}}{\partial x^\mu \partial x^\beta} - \frac{\partial^2 g_{\mu\alpha}}{\partial x^\nu \partial x^\beta}\right)$$
$$+ g_{\lambda\tau}\left(\Gamma^\lambda_{\nu\alpha} \Gamma^\tau_{\mu\beta} - \Gamma^\lambda_{\nu\beta} \Gamma^\tau_{\mu\alpha}\right). \tag{14.51}$$

Indeed, the last two terms are not relevant, because we can consider the curvature tensor in the x'^{α} coordinates, where $\Gamma'^{\lambda}_{\mu\nu} = 0$, and thus we get

$$R'_{\mu\nu\alpha\beta} = \frac{1}{2}\left(\partial'_{\nu\alpha}g'_{\mu\beta} - \partial'_{\mu\alpha}g'_{\nu\beta} + \partial'_{\mu\beta}g'_{\nu\alpha} - \partial'_{\nu\beta}g'_{\mu\alpha}\right). \tag{14.52}$$

Using the last formula, one can easily verify symmetry (14.46) and also obtain three other algebraic symmetries of the curvature tensor (number I was in Eq. (14.46))

$$\text{(II)} \qquad R_{\alpha\beta\mu\nu} = -R_{\beta\alpha\mu\nu}, \tag{14.53}$$

that is the antisymmetry of the first pair of indices.

$$\text{(III)} \qquad R_{\alpha\beta\mu\nu} = R_{\mu\nu\alpha\beta}, \tag{14.54}$$

that is, the symmetry between the pairs of the indices, and also the permutation formula

$$\text{(IV)} \qquad R_{\alpha\beta\mu\nu} + R_{\alpha\nu\beta\mu} + R_{\alpha\mu\nu\beta} = 0. \tag{14.55}$$

Exercise 12 Investigate the details of the last statements. Prove that the symmetries that hold in the special coordinates x'^{μ} also take place in general coordinates.

Exercise 13 Investigate whether from (I) and (III) follows (II).

Exercise 14 Obtain the identities (I)–(IV) directly from Eq. (14.51).

Let us now consider another symmetry of the curvature tensor, related to its first covariant derivatives. The relations, which we will prove, are called *Bianchi identities*. Once again, we shall use the advantage of the simplified expression (14.52).

The first Bianchi identity has the form

$$R_{\alpha\beta\mu\nu;\lambda} + R_{\alpha\beta\lambda\mu;\nu} + R_{\alpha\beta\nu\lambda;\mu} = 0. \tag{14.56}$$

Here $R_{\alpha\beta\mu\nu;\lambda} = \nabla_{\lambda}R_{\alpha\beta\mu\nu}$, as usual. In order to achieve (14.56), one can use the coordinates x'^{α} and (14.52). Since Eq. (14.56) has a tensor form, it can be proved in any coordinates. Taking the covariant derivative of (14.52) we meet the terms with the third partial derivatives of the metric, for which (14.56) can be checked directly. Furthermore, there are terms proportional to $\Gamma'^{\lambda}_{\mu\nu}$, but they vanish because $\Gamma'^{\lambda}_{\mu\nu} = 0$.

Exercise 15 Verify Eq. (14.56) by direct calculations in special coordinates x'^{α}.

Let us define the *Ricci* tensor $R_{\mu\alpha}$ as the contraction of the Riemann curvature

$$R_{\mu\alpha} = R^{\beta}_{\cdot\alpha\beta\mu} = g^{\nu\beta}R_{\nu\alpha\beta\mu}. \tag{14.57}$$

It is easy to see, using (14.54), that the Ricci tensor is symmetric, $R_{\mu\nu} = R_{\nu\mu}$. Another feature of the Ricci tensor with two covariant indices is that it can be constructed from affine connection, without explicit use of the metric,

$$R_{\mu\nu}(\Gamma) = \partial_\lambda \Gamma^\lambda_{\mu\nu} - \partial_\mu \Gamma^\lambda_{\nu\lambda} + \Gamma^\lambda_{\mu\nu} \Gamma^\tau_{\lambda\tau} - \Gamma^\lambda_{\mu\tau} \Gamma^\tau_{\lambda\nu}. \tag{14.58}$$

In the last exercises of Sect. 17.2, we suggest the reader develop the Palatini (first order) formalism in general relativity, which is based on this fact.

Further contraction gives us *scalar curvature*

$$R = g^{\mu\alpha} R_{\mu\alpha} = R^\alpha_\alpha. \tag{14.59}$$

Making a contraction of Eq. (14.56) with the metric $g^{\nu\alpha}$, we arrive at the reduced (second) Bianchi identity

$$\nabla_\mu R^\mu_{\cdot\beta\nu\lambda} = \nabla_\nu R_{\beta\lambda} - \nabla_\lambda R_{\beta\nu} . \tag{14.60}$$

Finally, making a contraction with $g^{\beta\nu}$ we arrive at the third Bianchi identity

$$\nabla_\mu R^\mu_\nu = \frac{1}{2} \nabla_\nu R . \tag{14.61}$$

As we shall see in the section devoted to Einstein equations, the last identity has special relevance in general relativity.

14.5 Number of Independent Components of the Curvature

The properties of the Riemann and Ricci tensors and of the scalar curvature that we obtained in the last section have general form and may be applied to any Riemann or pseudo-Riemann space, independent of the space–time dimension. At the same time, one may guess that in the spaces of lower dimensions these curvature tensors may be simpler than in the higher dimensional D cases. The extreme case is the $1D$-space, where all curvatures vanish identically, according to the definition (14.37).

Exercise 16 Prove the last statement in two distinct ways: using appropriate algebraic symmetries of the Riemann tensor and also via the commutations of the infinitesimal parallel transports.

Observation Since in $1D$ space $R_{\mu\nu\alpha\beta} \equiv 0$, it is certain that the Ricci tensor and scalar curvature also vanish identically $R_{\mu\nu} = R \equiv 0$. However, it is not easy to prove this geometrically, because the reduced forms of the curvature are not directly related to the commutations of the infinitesimal parallel transports.

To better understand the possible role of dimension D, let us calculate the number of independent components of the tensors $R_{\mu\nu\alpha\beta}$ and $R_{\mu\nu}$. There is no need to worry about other forms of the Riemann tensor (e.g., $R^{\mu}{}_{\nu\alpha\beta}$), because there is always a one-to-one correspondence with $R_{\mu\nu\alpha\beta}$.

Let us calculate the number of independent components of $R_{\mu\nu\alpha\beta}$. Due to the symmetries (14.46) and (14.53), the curvature tensor components with three equal indices vanish. Therefore, we have only three possible situations:

(i) Two couples of equal indices, that is, the construction like R_{0101}. It is easy to see that all other arrangements of these two indices give zero or reduce to R_{0101} via the symmetries (14.46) and (14.53). So, the problem is just combinatorial: how many distinct combinations like R_{0101} we can have in the D-dimensional space. Indeed it is

$$n_1 = C_D^2 = \frac{D(D-1)}{2}.$$

(ii) One couple of equal indices plus two distinct indices, that is, the construction like R_{0102}. The choice of the couple can be made in D distinct ways, and so for the remaining two indices we have C_{D-1}^2 possibilities. In total, we have

$$n_2 = D\,C_{D-1}^2 = \frac{D(D-1)(D-2)}{2}$$

options here.

(iii) Four distinct indices, that is, the construction like R_{0123}. It is easy to see that there are two different arrangements of the four indices here: R_{0123} and R_{0312}, which cannot be transformed into each other through the use of the algebraic symmetries (14.46), (14.53), and (14.54). The use of the last symmetry (14.55) does not reduce the number of linear independent structures to one, so two of them always remain independent. Then, we have to take the number of possibilities to choose four distinct indices and multiply it by two. This gives

$$n_3 = 2C_D^4 = \frac{D(D-1)(D-2)(D-3)}{12}.$$

Summing up the three expressions, we obtain the number of independent components of the Riemann tensor in D-dimensional space

$$N_D = n_1 + n_2 + n_3 = \frac{D^2(D^2-1)}{12}. \tag{14.62}$$

It is interesting that this expression is correct even for $2D$ and $3D$, where some of the three options do not exist.

Now we are in a position to consider the particular cases of lower dimensions. Let us start from the $2D$ case. Then $N_2 = 1$ and one can see that the number of independent components of the Riemann tensor equals to that of the scalar curvature. Since both of them are linear combinations of the second partial derivatives of the metric, the dependence between Riemann tensor and scalar curvature can be only linear. The only way to provide this and to respect the symmetries (I)–(IV) is by assuming

$$R_{\mu\nu\alpha\beta} = \tau R \left(g_{\mu\alpha} g_{\nu\beta} - g_{\mu\beta} g_{\nu\alpha} \right),$$

where τ is an unknown coefficient. Making a contraction, we arrive at

$$R_{\mu\alpha} = \tau R g_{\nu\beta} \quad \text{and} \quad R = 2\tau R. \tag{14.63}$$

Hence, $\tau = 1/2$ and we obtain the desirable $2D$ relations

$$R_{\mu\nu\alpha\beta} = \frac{1}{2} R \left(g_{\mu\alpha} g_{\nu\beta} - g_{\mu\beta} g_{\nu\alpha} \right), \quad R_{\mu\alpha} = \frac{1}{2} R g_{\nu\beta}. \tag{14.64}$$

Exercise 17 Calculate all components of $R_{\mu\nu\alpha\beta}$, $R_{\mu\alpha}$, and R for the following $2D$ surfaces:

(i) Sphere of radius a.

(ii) Straight cylinder if radius a.

(iii) Torus presented in Fig. 8.4 with the two radii a and b.

(iv) Check relations (14.64) for the two surfaces where it makes sense, that is, for the first and third examples.

The next example is the $3D$ space, where we have, according to (14.62), six independent components. It is clear that in the $3D$ case the Riemann tensor cannot be described in terms of the scalar curvature only. At the same time, six is exactly the number of independent components of the Ricci tensor $R_{\mu\nu}$. Therefore, the components of the Riemann tensor should be linear combinations of scalar curvature and the components of the Ricci tensor $R_{\mu\nu}$. Using the method that proved useful in the $2D$ case—that is, by taking into account all the algebraic symmetries—one can write

$$R_{\mu\nu\alpha\beta} = t R \left(g_{\mu\alpha} g_{\nu\beta} - g_{\mu\alpha} g_{\nu\beta} \right)$$
$$+ u \left(R_{\mu\alpha} g_{\nu\beta} - R_{\mu\beta} g_{\nu\alpha} + R_{\nu\beta} g_{\mu\alpha} - R_{\nu\alpha} g_{\mu\beta} \right), \tag{14.65}$$

where t, u are two unknown coefficients. Making the contraction with $g^{\nu\beta}$ we obtain

$$R_{\mu\alpha} = 2t R g_{\mu\alpha} + u \left(R_{\mu\alpha} + R g_{\mu\alpha} \right). \tag{14.66}$$

The solution is $u = 1$ and $t = -1/2$. Therefore, the relation for $3D$ is

$$R_{\mu\nu\alpha\beta} + R_{\mu\alpha} g_{\nu\beta} - R_{\mu\beta} g_{\nu\alpha} + R_{\nu\beta} g_{\mu\alpha} - R_{\nu\alpha} g_{\mu\beta}$$
$$- \frac{1}{2} R \left(g_{\mu\alpha} g_{\nu\beta} - g_{\mu\alpha} g_{\nu\beta} \right). \tag{14.67}$$

As far as the $3D$ curvature tensor satisfies the relation (14.67), it makes sense to look for a linear combination of the Riemann, Ricci tensors and of scalar curvature in an arbitrary D, which would have the same algebraic symmetries as the Riemann tensor and be identically zero in the special $3D$ case. Let us construct this new tensor $C_{\mu\nu\alpha\beta}$ by requesting that its contraction in any pair of indices is zero. Due to the algebraic symmetries (I)–(IV), it is sufficient to consider the contraction with respect to the indices $(\mu\alpha)$ only. Then we can start from (14.65), and the difference with (14.66) will only be due to the relation $g_{\mu\alpha}\, g^{\mu\alpha} = D$, which holds for an arbitrary D. In $3D$ we had $g_{\mu\alpha}\, g^{\mu\alpha} = 3$, as we have used in (14.66) to derive (14.67). The calculations for an arbitrary D give

$$
C_{\mu\nu\alpha\beta} = R_{\mu\nu\alpha\beta} - \frac{1}{D-2} \left(R_{\mu\alpha}\, g_{\nu\beta} - R_{\mu\beta}\, g_{\nu\alpha} + R_{\nu\beta}\, g_{\mu\alpha} - R_{\nu\alpha}\, g_{\mu\beta} \right)
$$

$$
+ \frac{1}{(D-1)(D-2)}\, R \left(g_{\mu\alpha}\, g_{\nu\beta} - g_{\mu\alpha}\, g_{\nu\beta} \right) . \tag{14.68}
$$

Def. 9 The object $C_{\mu\nu\alpha\beta}$ defined by Eq. (14.68) is called a *Weyl tensor*.

The algebraic symmetries of $C_{\mu\nu\alpha\beta}$ are the same as the Riemann tensor. Also, as we have just proved, the Weyl tensor has a vanishing contraction, $C^{\mu}{}_{\cdot\alpha\mu\beta} \equiv 0$. Later on, we can link this property to its conformal transformation rule. Furthermore, we already know that in $3D$ the Weyl tensor is zero.

Def. 10 The last sort of curvature that we have to define is called an *Einstein tensor*,

$$
G_{\mu\nu} = R_{\mu\nu} - \frac{1}{2}\, R\, g_{\mu\nu} . \tag{14.69}
$$

This tensor is symmetric and satisfies the Bianci identity $\nabla_\mu G^{\mu}{}_{\nu} = 0$ that follows directly from (14.61).

The Einstein tensor is constructed, exactly as Riemann, Ricci, Weyl tensors, and scalar curvature, from the metric and its derivatives. All these tensors are homogeneous functions of the second order in the derivatives. In the next sections, we present examples of calculating Riemann, Ricci, Weyl, and Einstein tensors for the metrics of special physical interest.

The metric that provides $R_{\mu\nu\alpha\beta} = 0$ is flat, as we already know. However, there are situations when $R_{\mu\nu\alpha\beta} \neq 0$, but some of the reduced versions of curvature vanish.

Def. 11 The metric space where the Ricci tensor vanishes is called *Ricci-flat*. The metric space where the Weyl tensor vanishes is called *Weyl-flat*.

Exercise 18 *(i)* Prove that the Riemann space that is Ricci-flat and Weyl-flat at the same time has a vanishing Riemann tensor and therefore is flat. *(ii)* Calculate the number of degrees of freedom of the Weyl tensor. *(iii)* Derive Bianci identities for the Weyl tensor.

As we already know, any $D = 3$ space is Weyl-flat. In this case, all information about the Riemann tensor is encoded into the Ricci tensor. The $2D$ space has the special property (14.64) and all information about the Riemann tensor is encoded into scalar curvature. So, we see that the structure of the Riemann tensor may be much simpler for smaller dimensions D.

According to (14.62), for $D \geq 4$, the Riemann tensor cannot be expressed through the Ricci tensor. This can be easily seen from the fact that the number of independent components of $R_{\mu\nu}$ equals $\frac{D(D-1)}{2} < N_D$. This is a general statement valid for an arbitrary metric. However, in many cases, we do not need an arbitrary metric but instead just a particular form of the metric that has certain symmetries and therefore less independent components than the general one. One can ask the question of whether it is possible to construct the $D \geq 4$ metric with a curvature tensor that could be expressed via Ricci tensor or scalar curvature. The answer to this question leads to the notion of symmetric spaces, which we will not consider here and just refer to the books [3, 4].

14.5.1 Factorized Spaces

Let us now explore some features of the factorized metrics that will prove to be useful in what follows.

Theorem *Consider the D-dimensional Riemann manifold with the coordinates $x^\mu = (x^a, x^i)$, where $a, b, c, \ldots = 1, 2, \ldots, n$ and $i, j, k, \ldots = n + 1, n + 2, \ldots, D$. Let us assume that the metric is factorized*

$$g_{\mu\nu} = \left\| \begin{array}{cc} g_{ab}(x^a) & 0 \\ 0 & g_{ij}(x^i) \end{array} \right\| . \tag{14.70}$$

For this metric, the elements with mixed indices g_{ai} are zero, and only the blocks with g_{ab} and g_{ij} don't vanish. Moreover, the metric elements in each block depend only on their "own" coordinates $g_{ab} = g_{ab}(x^a)$ and $g_{ij} = g_{ij}(x^i)$. The statement is that for the metric (14.70) the Riemann tensor is also factorized. In particular, this means that all components of the curvature with mixed indices (a and i) do vanish.

Proof First of all, it is clear that the inverse metric will also have the factorized block structure, that is,

$$g^{\mu\nu} = \left\| \begin{array}{cc} g^{ab}(x^a) & 0 \\ 0 & g^{ij}(x^i) \end{array} \right\| . \tag{14.71}$$

Consider the Christoffel symbol with mixed indices, e.g.,

$$\Gamma_{ib}^a = \frac{1}{2} g^{ac} \left(\partial_i g_{bc} + \partial_b g_{ic} - \partial_c g_{ib} \right) . \tag{14.72}$$

Taking into account the relations $g_{ic} = 0$ and $\partial_i g_{bc} = 0$, we obtain $\Gamma_{ib}^a = 0$. In the same way, one can prove that other mixed components vanish

$$\Gamma^i_{jb} = \Gamma^i_{ab} = \Gamma^a_{ij} = 0. \tag{14.73}$$

Thus, only "pure" components of the Christoffel symbol can be nonzero: $\Gamma^a_{bc}(x^a)$ and $\Gamma^k_{ij}(x^i)$.

Let us now consider the mixed components of a Riemann tensor. According to the formula (14.38), it is zero, because (i) the "mixed" derivatives (e.g., $\partial_i \Gamma^a_{bc}$) equal zero; (ii) the $\Gamma \times \Gamma$-type terms with mixed indices always include at least one factor of Γ with mixed indices and therefore all such products have to vanish.

Observation The last theorem enables one to simplify the analysis of the curvature tensor for the factorized metrics. For $4D$, the effect is especially strong in the $(2 + 2)$ case, when $a, b, c, \ldots = 1, 2$ and $i, j, k, \ldots = 3, 4$ or in the $(1 + 3)$ case $a, b, c, \ldots = 1$ and $i, j, k, \ldots = 2, 3, 4$. In the last case, the only components of the Riemann tensor that do not vanish are $R^i{}_{jkl}$, which can be expressed via the corresponding $3D$ Ricci tensor using Eq. (14.67). In the former case, we have two kinds of the nonzero components:

$$R_{abcd} = \frac{1}{2} \left(g_{ac}g_{bd} - g_{ad}g_{bc} \right) \cdot {}^{(2)}R(x^a),$$

$$R_{ijkl} = \frac{1}{2} \left(g_{ik}g_{jl} - g_{il}g_{jk} \right) \cdot {}^{(2)}R(x^i), \tag{14.74}$$

where ${}^{(2)}R(x^a)$ and ${}^{(2)}R(x^i)$ are corresponding $2D$ scalar curvatures.

Exercise 19 Prove the factorization theorem for the Ricci tensor. Does it also hold for the completely covariant tensor $R_{\alpha\beta\rho\sigma}$? Find an explicit expression for the D-dimensional scalar curvature in terms of the n-dimensional and $(D - n)$-dimensional scalar curvatures. Formulate the conditions of the factorization theorem for the Weyl tensor.

Exercise 20 Use the results of this section to reconsider the Exercise 17 (iii) of Sect. 14.5 and arrive at the result without calculations.

Exercise 21 Consider the $4D$ metric that is a product of the two $2D$ spheres with different radii A and B. Find explicit expressions for Riemann, Ricci, Weyl, and Einstein tensors, and for the scalar curvature in this case.

14.6 Conformal Transformations

In the present section[4], we develop a useful technique for deriving curvature tensor components and related quantities. This technique is based on the local conformal

---

[4]Based on the pedagogical paper [5].

transformation, which has a great importance by itself and especially due to its applications to the quantum theory of gravity and cosmology. In some cases, the work of deriving the components of the curvature tensor, the Ricci tensor, and especially quadratic contractions of these tensors can be done much faster by means of conformal transformations. Also, in many cases, the conformal symmetry enables one to achieve a better understanding of the geometrical or physical sense of the problem.

The local conformal transformation of the metric is defined as follows:

$$g_{\mu\nu}(x) = \bar{g}_{\mu\nu}(x) e^{2\sigma(x)} . \tag{14.75}$$

It is important to stress that (14.75) is not a coordinate transformation. Instead, it has to be seen as an alternative form of parameterizing the metric via the new variables $\bar{g}_{\mu\nu}(x)$ and $\sigma(x)$, without changing coordinates. Our purpose in this section will be to derive the relevant quantities in this new parameterization.

Since some formulas are bulky, we shall use the slightly condensed notations. All indices in the space with the metric $\bar{g}_{\mu\nu}(x)$ are lowered and raised with this metric and with its inverse, $\bar{g}^{\mu\alpha}(x)$, where $\bar{g}_{\mu\nu}(x)\,\bar{g}^{\mu\alpha}(x) = \delta_\nu^\alpha$. Further, we shall use notations $(\nabla\sigma)^2 = g^{\mu\nu}\nabla_\mu\sigma\nabla_\nu\sigma$ and $(\Box\sigma) = (\nabla^2\sigma) = g^{\mu\nu}\nabla_\mu\nabla_\nu\sigma$. The parentheses are used to bound the range of action of derivatives, e.g., $\nabla_\mu\sigma = (\nabla_\mu\sigma) + \sigma\nabla_\mu$. Finally, in most of the cases, we will not explicitly write the argument, such that $\sigma \equiv \sigma(x)$.

The transformation laws for the inverse metric and for the metric determinant have the form

$$g^{\mu\nu} = \bar{g}^{\mu\nu} e^{-2\sigma} , \qquad g = \bar{g}\, e^{2D\sigma} . \tag{14.76}$$

For the Christoffel symbols, we have

$$\Gamma_{\alpha\beta}^\lambda = \bar{\Gamma}_{\alpha\beta}^\lambda + \delta_\alpha^\lambda(\bar{\nabla}_\beta\sigma) + \delta_\beta^\lambda(\bar{\nabla}_\alpha\sigma) - \bar{g}_{\alpha\beta}(\bar{\nabla}^\lambda\sigma) , \tag{14.77}$$

and for the curvature tensors and scalar

$$\begin{aligned}
R^\alpha{}_{.\beta\mu\nu} &= \bar{R}^\alpha{}_{.\beta\mu\nu} + \delta_\nu^\alpha\,(\bar{\nabla}_\mu\bar{\nabla}_\beta\sigma) - \delta_\mu^\alpha(\bar{\nabla}_\nu\bar{\nabla}_\beta\sigma) + \bar{g}_{\mu\beta}(\bar{\nabla}_\nu\bar{\nabla}^\alpha\sigma) \\
&\quad - \bar{g}_{\nu\beta}(\bar{\nabla}_\mu\bar{\nabla}^\alpha\sigma) + \delta_\nu^\alpha\bar{g}_{\mu\beta}(\bar{\nabla}\sigma)^2 - \delta_\mu^\alpha\bar{g}_{\nu\beta}(\bar{\nabla}\sigma)^2 + \delta_\nu^\alpha(\bar{\nabla}_\mu\sigma)(\bar{\nabla}_\beta\sigma) \\
&\quad - \delta_\nu^\alpha(\bar{\nabla}_\mu\sigma)(\bar{\nabla}_\beta\sigma) + \bar{g}_{\nu\beta}(\bar{\nabla}_\mu\sigma)(\bar{\nabla}^\alpha\sigma) - \bar{g}_{\mu\beta}(\bar{\nabla}_\nu\sigma)(\bar{\nabla}^\alpha\sigma), \tag{14.78}
\end{aligned}$$

$$\begin{aligned}
R_{\mu\nu} &= \bar{R}_{\mu\nu} - \bar{g}_{\mu\nu}(\bar{\nabla}^2\sigma) \\
&\quad + (D-2)\left[(\bar{\nabla}_\mu\sigma)(\bar{\nabla}_\nu\sigma) - (\bar{\nabla}_\mu\bar{\nabla}_\nu\sigma) - \bar{g}_{\mu\nu}(\nabla\sigma)^2\right], \tag{14.79}
\end{aligned}$$

$$R = e^{-2\sigma}\left[\bar{R} - 2(D-1)(\bar{\Box}\sigma) - (D-1)(D-2)(\bar{\nabla}\sigma)^2\right]. \tag{14.80}$$

Exercise 22 Verify formulas (14.78), (14.79) and (14.80).

Exercise 23 Verify that the Weyl tensor (14.68) in D dimensions transforms trivially,

$$C^{\alpha}_{\cdot\,\beta\mu\nu} = \bar{C}^{\alpha}_{\cdot\,\beta\mu\nu}, \quad \text{i.e.} \quad C_{\alpha\beta\mu\nu} = e^{2\sigma}\,\bar{C}_{\alpha\beta\mu\nu}. \tag{14.81}$$

Exercise 24 Using expressions (14.79) and (14.80), derive the conformal transformation law for the Einstein tensor (14.69).

Exercise 25 *(i)* Using the result of Exercise 2, without special calculation, show that the expression

$$\sqrt{|g|}\,C_{\mu\nu\alpha\beta}\,C^{\mu\nu\alpha\beta} = \sqrt{|g|}\,C^2_{\mu\nu\alpha\beta} \tag{14.82}$$

is conformal invariant in $4D$ and only in $4D$.
(ii) Invent the expression constructed only from the Weyl tensor, which is conformal invariant in an arbitrary dimension D.

Observation This is an easy thing to do for any dimension, at least for one of the (depending on D) many possible solutions.

Exercise 26 Discuss the relation between the special conformal property (14.81) of the Weyl tensor and its tracelessness $C^{\alpha}_{\cdot\,\mu\alpha\nu} = 0$.

14.7   Motion of a Particle in Curved Space

In the previous sections, we already constructed the equation of motion for a free particle moving in the gravitational field.[5] The equation in an arbitrary coordinate has the form (14.5), which now can be immediately associated with the equation for the parallel transport (14.25). The same equation is also describing a geodesic line. As we can see from (14.25) and related consideration, this equation is covariant, meaning it transforms as a vector when we change from one reference frame to another. In fact, this is a particular manifestation of a general feature. The equivalence principle that led us to Eq. (14.5) can be mathematically formulated as a covariance principle, which requires that all physically relevant equations should be covariant. This should be seen as something natural, because the non-covariant, non-tensor equations depend on the choice of coordinates in the uncontrollable way.

14.7.1   Action for a Free Massive Particle

The natural question is whether we can obtain Eq. (14.5) from the action principle, with a generally covariant action. Let us choose the simplest option and consider the general covariant version of the action of a relativistic particle, which is already known,

---

[5]It is customary to call such a particle free, because gravity can be compensated by inertia in a special coordinate system.

$$S = -mc \int_{t_1}^{t_2} ds . \tag{14.83}$$

The difference with the Minkowski-space interval is that now there is a dependence on the coordinate-dependent metric $g_{\mu\nu} = g_{\mu\nu}(x)$,

$$ds = \sqrt{g_{\mu\nu}\, dx^\mu\, dx^\nu} = \sqrt{g_{00}\, (cdt)^2 + 2g_{0i}\, dx^i\, cdt + g_{ij}\, dx^i\, dx^j}. \tag{14.84}$$

According to the minimal action principle, consider the variation δx^μ that has to vanish at the ends of the interval of integration, t_1 and t_2, and is otherwise arbitrary. In the first order in variations we have

$$\delta S = -mc \int_{t_1}^{t_2} \frac{1}{2ds} \left\{ (\partial_\alpha g_{\mu\nu}) \delta x^\alpha\, dx^\mu\, dx^\nu + 2 g_{\mu\alpha}\, dx^\mu\, d\delta x^\alpha \right\}. \tag{14.85}$$

Using $d\delta x^\alpha = \delta dx^\alpha$, useful notation $u^\alpha = dx^\alpha / ds$, and integrating by parts, we arrive at

$$\delta S = -mc \int_{t_1}^{t_2} \delta x^\alpha \left\{ \frac{1}{2} u^\mu u^\nu (\partial_\alpha g_{\mu\nu}) - \frac{d}{ds}\left(g_{\mu\alpha} \frac{dx^\mu}{ds} \right) \right\} \tag{14.86}$$

$$= -mc \int_{t_1}^{t_2} \delta x^\alpha \left\{ -g_{\mu\alpha} \frac{d^2 x^\mu}{ds^2} - \frac{dx^\mu}{ds} \partial_\lambda g_{\mu\alpha} \frac{dx^\lambda}{ds} + \frac{1}{2} u^\mu u^\nu (\partial_\alpha g_{\mu\nu}) \right\}.$$

Thus, the variational equation can be written in the form

$$g_{\mu\alpha} \frac{dx^\mu}{ds} + u^\mu u^\nu \left(\partial_\nu g_{\mu\alpha} - \frac{1}{2} \partial_\alpha g_{\mu\nu} \right) = 0 . \tag{14.87}$$

After multiplying it by $g^{\alpha\beta}$, it is an easy exercise to show that this is nothing else but (14.5),

$$\frac{du^\mu}{ds} + \Gamma^\mu_{\alpha\beta} u^\alpha u^\beta = \frac{d^2 x^\mu}{ds^2} + \Gamma^\mu_{\alpha\beta} \frac{dx^\alpha}{ds} \frac{dx^\beta}{ds} = 0 , \tag{14.88}$$

describing the geodesic line.

A few questions remain to be answered. For instance,
(i) As we already discussed above, Eq. (14.88) does not have a metric, but only an affine connection. In which sense can we claim that the gravitational field is described by the metric?
(ii) Furthermore, is it possible to recover the Newtonian limit from this equation?

(iii) And what is the difference between flat and curved space–times? Can we see this in Eq. (14.88) for the geodesic line?

Let us answer these questions one by one. First of all, the fact that the equation has only connection is in a perfect correspondence with the equivalence principle, which states that locally, gravity can be eliminated by the choice of frame. So, the dependence on the metric should be seen when we deal with the nonlocal quantities, e.g., those, which describe the relative motion of at least two space–time points.

In order to address the issue *(ii)*, let us consider the nonrelativistic limit.

$$\left| \frac{dx^i}{dx^0} \right| = \left| \frac{dx^i}{cdt} \right| \ll 1. \tag{14.89}$$

It proves useful to consider the units with $c = 1$. Also, we have to assume that the gravitational field is weak, as otherwise the particle may easily become too fast and violate the previous restriction. Therefore, we can assume that

$$g_{\mu\nu} = \eta_{\mu\nu} + h_{\mu\nu}, \tag{14.90}$$

where $|h_{\mu\nu}| \ll 1$. From now on, all indices will be lowered and raised with the Minkowski metric.

It is easy to see that in this case, the expansion for the affine connection is (see more details in Exercise 28)

$$\Gamma^\alpha_{\mu\nu} = \frac{1}{2} \left(\eta^{\alpha\beta} - h^{\alpha\beta} + h^\alpha_\lambda h^{\lambda\beta} - h^\alpha_\lambda h^\lambda_\tau h^{\tau\beta} + \cdots \right) \\ \times \left(\partial_\mu h_{\beta\nu} + \partial_\nu h_{\beta\mu} - \partial_\beta h_{\mu\nu} \right), \tag{14.91}$$

where the first parenthesis represents the expansion of the inverse metric.

For a weak gravitational field, we can restrict consideration by the first approximation. Then

$$\Gamma^\alpha_{\mu\nu} = \Gamma^{\alpha\,(1)}_{\mu\nu} = \frac{1}{2} \eta^{\alpha\beta} \left(\partial_\mu h_{\beta\nu} + \partial_\nu h_{\beta\mu} - \partial_\beta h_{\mu\nu} \right). \tag{14.92}$$

Furthermore, the relation (14.89) means that the time component of the four-velocity of the particle dominates, $|u^i| \ll |u^0| = 1$. In this case, we can use the approximation

$$\Gamma^\alpha_{\mu\nu} u^\mu u^\nu \approx \Gamma^\alpha_{00} u^0 u^0 = \frac{1}{2} \eta^{\alpha\beta} \left(2\partial_0 h_{0\beta} - \partial_\beta h_{00} \right). \tag{14.93}$$

Let us remember that Newton's gravity law has no time dependence, and therefore we also have to take a static limit. Then the first term in the parenthesis in (14.93) vanishes and we arrive at

$$\Gamma^\alpha_{\mu\nu} u^\mu u^\nu = -\frac{1}{2} \delta^\alpha_i \eta^{ij} \partial_j h_{00}. \tag{14.94}$$

After replacing this relation to (14.88), the $\alpha = 0$ component gives $d^2t/d\tau^2 = 0$, and hence the dependence between proper time τ and physical time t is linear. Then the derivative $dt/d\tau$ is a constant coefficient, which drops out from the the space components of (14.94). In this way, using $\eta^{ij} = -\delta^{ij}$, we obtain

$$\frac{d^2x^i}{dt^2} = -\frac{1}{2}\partial_i h_{00}.\tag{14.95}$$

This equation can be easily associated to the Newton's law $\ddot{\mathbf{r}} = -\operatorname{grad}\varphi$, if we identify $h_{00} = 2\varphi$.

Thus, we have found a nice correspondence with the classical gravitational law in the limit of weak and static gravitational field and small velocities. Moreover, we learned that for the case of a weak gravitational field $g_{00} = 1 + 2\varphi$, where φ is the Newton's potential. For instance, in the case of a field created by a point-like source the Newtonian potential is

$$\varphi = -\frac{GM}{r}.\tag{14.96}$$

The reader probably noted that the Newtonian limit restricts only $g_{00} = 1 + 2\varphi$, while other metric components cannot be defined from the comparison with the nonrelativistic limit in the simple way we described here. Later on, we will see how this can be done on the basis of the Schwarzschild solution, which is derived in Sect. 15.2.

The last observation is that the correspondence $g_{00} = 1 + 2\varphi$ requires that the Newtonian potential φ be calibrated to zero at the space infinity. Thus, compared to the treatment of gravitational potential in mechanics we have lost the possibility to add an arbitrary constant to the gravitational Newtonian potential (14.96).

Exercise 27 The results presented above correspond to the units with $c = 1$. Using dimensional considerations, show that the expression for the Newtonian limit is

$$g_{00} = 1 + \frac{2\varphi}{c^2}.\tag{14.97}$$

Exercise 28 (i) Work out more general expansion of the metric. Consider the shift

$$g_{\mu\nu} \longrightarrow g'_{\mu\nu} = g_{\mu\nu} + h_{\mu\nu}\tag{14.98}$$

and prove that, in an arbitrary order in $h_{\mu\nu}$ we have

$$g'^{\alpha\beta} = g^{\alpha\beta} - h^{\alpha\beta} + h^\alpha_\lambda h^{\lambda\beta} - h^\alpha_\lambda h^\lambda_\tau h^{\tau\beta} + \dots,\tag{14.99}$$

$$\Gamma'^\alpha_{\mu\nu} = \frac{1}{2}g'^{\alpha\beta}\left(\nabla_\mu h_{\beta\nu} + \nabla_\nu h_{\beta\mu} - \nabla_\beta h_{\mu\nu}\right),\tag{14.100}$$

where the covariant derivative is based on the background metric $g_{\mu\nu}$.
(ii) Consider an alternative expansions of the metric, such as

$$g^{\mu\nu} \longrightarrow g'^{\mu\nu} = g^{\mu\nu} + \Phi^{\mu\nu} \qquad \text{and} \tag{14.101}$$

$$\sqrt{|g|}g^{\mu\nu} = \eta^{\mu\nu} + \gamma^{\mu\nu} \qquad \text{this one on the flat background.} \tag{14.102}$$

Find the relations between $h_{\mu\nu}$, $\Phi^{\mu\nu}$, and $\gamma^{\mu\nu}$ up to the second order in each of the perturbations on the corresponding background.

Observation and hint. From a practical point of view, this is an important exercise. In what follows, we will use the skills that are supposed to be developed in this exercise. The relation (14.99) can be verified by direct replacement into $g'_{\mu\nu}g'^{\nu\alpha} = \delta^\alpha_\mu$. If the derivation of the second formula (14.100) meets difficulties, try to start by solving the problem in the first order in $h_{\mu\nu}$ first.

Let us now consider the point *(iii)*. Since the equivalence principle tells us that locally, gravity can be eliminated by the choice of reference frame, it is natural to look for a nonlocal effects. Therefore, consider two close geodesic lines (14.88) and explore how the deviation between them changes with the parameter s along the curve. Our equations of interest are

$$\frac{d^2x^\mu}{ds^2} + \Gamma^\mu_{\alpha\beta}(x)\frac{dx^\alpha}{ds}\frac{dx^\beta}{ds} = \frac{du^\mu}{ds} = 0, \qquad \text{where} \qquad u^\alpha = \frac{dx^\alpha}{ds} \qquad \text{and}$$

$$\frac{d^2}{ds^2}\left(x^\mu + \delta x^\mu\right) + \Gamma^\mu_{\alpha\beta}(x + \delta x)\frac{d\left(x^\alpha + \delta x^\alpha\right)}{ds}\frac{d\left(x^\beta + \delta x^\beta\right)}{ds} = 0. \tag{14.103}$$

Subtracting the second equation in (14.103) from the first one, in the first order in δx we arrive at the equation

$$\frac{d^2\delta x^\mu}{ds^2} + \left(\partial_\nu\Gamma^\mu_{\alpha\beta}\right)\delta x^\nu\frac{dx^\alpha}{ds}\frac{dx^\beta}{ds} + 2\Gamma^\mu_{\alpha\beta}\frac{d\delta x^\alpha}{ds}\frac{dx^\beta}{ds} = 0, \tag{14.104}$$

where we used the symmetry $\Gamma^\mu_{\alpha\beta} = \Gamma^\mu_{\beta\alpha}$.

It is natural to suppose that Eq. (14.104) should be covariant, in particular, because it has a clear geometric sense. In order to check this hypothesis, consider the covariant generalization of the first term using the definition (14.26),

$$\frac{D^2\delta x^\mu}{ds^2} = \frac{D}{ds}\left(\frac{D\delta x^\mu}{ds}\right) = \frac{d}{ds}\left(\frac{D\delta x^\mu}{ds}\right) + \Gamma^\mu_{\alpha\beta}u^\mu\frac{D\delta x^\nu}{ds}. \tag{14.105}$$

Using once again the relation (14.26), after some calculations we arrive at the final form of Eq. (14.104),

$$\frac{D^2\delta x^\mu}{ds^2} = \delta x^\nu u^\alpha u^\beta\left(\partial_\alpha\Gamma^\mu_{\beta\nu} - \partial_\nu\Gamma^\mu_{\alpha\beta} + \Gamma^\mu_{\alpha\lambda}\Gamma^\lambda_{\beta\nu} - \Gamma^\mu_{\nu\lambda}\Gamma^\lambda_{\alpha\beta}\right)$$

$$= \delta x^\nu u^\alpha u^\beta R^\mu_{\cdot\alpha\nu\beta}. \tag{14.106}$$

The last relation clarifies the geometric sense of the Riemann curvature tensor, which is a measure of covariant deviation between two geodesic lines. From the dynamical viewpoint, this means mutual acceleration of the two close, independent particles moving in an external gravitational field.

Exercise 29 Verify Eq. (14.106) by making direct calculations. After that, discuss Eq. (14.106) in relation to the equivalence principle, especially to the difference between gravity and inertia at the nonlocal level. Show that the real gravity cannot be replaced everywhere at the same time by inertia because the curvature tensor of the curved space–time is nonzero.

14.7.2 Maxwell Equations in Curved Space–Time

As far as we introduced the action of particles, it is quite natural to ask whether the same can be done for fields. For this reason, we consider the formulation of Maxwell equations in curved space–time.

The simplest way to arrive at the Maxwell equations is to start from the action of the electromagnetic field and of the charged particles. Let us remember that in special relativity the full electromagnetic action must have two symmetries, namely, Lorentz and gauge invariance. In general relativity, Lorentz invariance is upgraded to general covariance. Thus, we need to care about covariance and gauge invariance.

The action of a charged particle can be obtained from the action of the free particle in exactly the same way as in the flat space. Thus

$$S = -mc \int \left(1 - A_\mu \cdot \frac{dx^\mu}{ds} \right) ds, \qquad (14.107)$$

where $A^\mu = A^\mu(x)$ is a covariant vector of electromagnetic field potential and x^μ is the $4D$ coordinate of the particle. Taking variational derivative with respect to x^μ we arrive at the equation, which is a covariant generalization of the equation which we already considered in flat space–time,

$$\frac{du^\mu}{ds} + \Gamma^\mu_{\alpha\beta} u^\mu u^\nu = \frac{e}{mc^2} F^{\mu\nu} u_\nu, \qquad (14.108)$$

where $F_{\mu\nu} = \nabla_\mu A_\nu - \nabla_\nu A_\mu$ is a covariant tensor of electromagnetic field. It is easy to see that

$$\begin{aligned} F_{\mu\nu} &= \nabla_\mu A_\nu - \nabla_\nu A_\mu = \partial_\mu A_\nu - \Gamma^\lambda_{\mu\nu} A_\lambda - \partial_\nu A_\mu + \Gamma^\lambda_{\nu\mu} A_\lambda \\ &= \partial_\mu A_\nu - \partial_\nu A_\mu. \end{aligned} \qquad (14.109)$$

As far as torsion is zero, $\Gamma^\lambda_{\mu\nu} = \Gamma^\lambda_{\nu\mu}$ and we can see that the electromagnetic field tensor with lower indices, $F_{\mu\nu}$, is metric independent. Of course, one cannot overestimate this fact, because the metric enters the interaction term in the action (14.107) anyway, since we have to contract the indices.

Exercise 30 *(i)* Derive Eq. (14.108) by taking the variational derivative of the action (14.107), or/and by constructing the Lagrange equations for the corresponding Lagrange density. Verify that the electromagnetic tensor $F_{\mu\nu}$ is invariant under the gauge transformation

$$A_\mu \; \rightarrow \; A'_\mu \; = \; A_\mu + \partial_\mu f(x) \tag{14.110}$$

and check that this is also true for the action of the particle (14.107).
(ii) Try to elaborate on the conservation of electric charge in curved space. This is not easy to do, because first the four-dimensional Gauss–Ostrogradsky theorem has to be formulated on a Riemann space.

Exercise 31 Using the previous result for the covariant derivative with torsion, derive the corresponding generalization of (14.109). Then, one can easily verify that this expression is not gauge invariant.

One can note that this fact is important from the physical viewpoint, because it does not allow the introduction of torsion with the same degree of universality as the metric (this means that the metric interacts with all kinds of particles and fields). Since the gauge invariance must be preserved, the standard assumption is that torsion cannot interact directly with photons. Interested reader can see, e.g., the review papers [6, 7] that discuss this issue.

As far as the action of a charged particle can be formulated in a gauge-invariant way, the action of electromagnetic field should have the form of covariant generalization of the action in flat space–time,

$$S_{em} = -\frac{1}{4} \int d^4x \sqrt{|g|} \, F_{\alpha\beta} \, F_{\mu\nu} \, g^{\alpha\mu} g^{\beta\nu} = -\frac{1}{4} \int d^4x \sqrt{|g|} \, F_{\alpha\beta} F^{\alpha\beta}. \tag{14.111}$$

Assuming that the four-vector of the density of the electric current is j^μ, we arrive at the total action of charges and electromagnetic fields in the form

$$S_{total} = S_{\text{free particles}} + S_{em} - \frac{4\pi}{c} \int d^4x \sqrt{|g|} \, A_\mu j^\mu. \tag{14.112}$$

Exercise 32 Derive the equation for electromagnetic potential from the action (14.112) using *(a)* variation with respect to A_μ and *(b)* through the Lagrange equations for the fields. Verify that the equations are gauge invariant and find the covariant form of the conservation law for the four-current.

Solutions. The curved-space Maxwell equations and gauge transformations have the form

$$\nabla_\mu F^{\mu\nu} = \frac{4\pi}{c} \, j^\nu, \qquad A_\mu \; \rightarrow \; A'_\mu = A_\mu + \partial_\mu f(x). \tag{14.113}$$

Exercise 33 Verify whether the action (14.111) is invariant with respect to the global and local conformal transformations,

$$g_{\mu\nu} = \bar{g}_{\mu\nu} \, e^{2\sigma} \,, \quad A_\mu = \bar{A}_\mu \, e^{d_A\sigma} \,, \qquad (14.114)$$

where $\sigma = \lambda = const$ in the global case and $\sigma = \sigma(x)$ in the local version. Try to generalize this result for an arbitrary dimension D. Observe how the *conformal weight* d_A of the field A_μ depends on the dimension D in the global transformation case.

Solution. $d_A = 0$ for $D = 4$.

References

1. C.W. Misner, K.S. Thorn, J.A. Wheeler, *Gravitation* (W.H. Freeman and Co., San Francisco, 1973)
2. R. D'Inverno, *Introducing Einstein's Relativity* (Oxford University Press, Oxford, 1992–1998)
3. A.T. Fomenko, S.P. Novikov, B.A. Dubrovin, *Modern Geometry-Methods and Applications, Part I: The Geometry of Surfaces, Transformation Groups, and Fields* (Springer, Berlin, 1992)
4. S. Weinberg, *Gravitation and Cosmology* (Wiley, Hoboken, 1972)
5. D.F. Carneiro, E.A. Freitas, B. Gonçalves, A.G. de Lima, I.L. Shapiro, On useful conformal tranformations in general relativity. Gravit. Cosmol. **40**, 305 (2004). gr-qc/0412113
6. F.W. Hehl, P. Heide, G.D. Kerlick, J.M. Nester, General relativity with spin and torsion: foundations and prospects. Rev. Mod. Phys. **48**, 393 (1976)
7. I.L. Shapiro, Physical aspects of the space-time torsion. Phys. Repts. **357**, 113 (2002)

Chapter 15
Einstein Equations, Schwarzschild Solution, and Gravitational Waves

15.1 Energy–Momentum Tensor and Einstein Equations

In order to formulate the equations for the metric, the first step is to consider an alternative definition of the energy–momentum tensor of particles and fields. In Chap. 2, we already learned the *canonical* definition of the energy and momentum conservations in classical mechanics and the energy–momentum tensor of particles and fields in special relativity. It happens that in general relativity, things somehow turn out to be simpler—here, we gain an alternative and in fact more economic, *dynamical* definition of the energy–momentum tensor.

15.1.1 Dynamical Definition of Energy–Momentum Tensor

In gravitational theories, we usually call everything that is not gravity, matter. This concerns both massive particles and radiation or electromagnetic fields. Let us give the definition of the dynamical energy–momentum tensor. Let us assume that the action of matter (in the sense described above) is $S_m(g_{\mu\nu}, \Phi)$, where Φ represents all variables, including coordinates of particles and elements of a field, e.g., of the field A_μ, compose matter. Then the dynamical energy–momentum tensor is

$$ T^{\mu\nu} = -\frac{2}{\sqrt{|g|}} \frac{\delta S_m}{\delta g_{\mu\nu}}. \tag{15.1} $$

By construction, (15.1) is a symmetric tensor, and hence one can always raise and lower the indices using the metric. At the same time, one has to be careful with the variation of the metric, which becomes clear from the next exercise.

Exercise 1 Prove that the same dynamical energy–momentum tensor can be derived as (note the change of sign)

© Springer Nature Switzerland AG 2019
I. L. Shapiro, *A Primer in Tensor Analysis and Relativity*, Undergraduate
Lecture Notes in Physics, https://doi.org/10.1007/978-3-030-26895-4_15

$$T_{\mu\nu} = +\frac{2}{\sqrt{|g|}}\frac{\delta S_m}{\delta g^{\mu\nu}}. \tag{15.2}$$

Hint. Use Eq. (14.99).

The canonical and dynamical definitions look so different that it is tempting to ask why the results of these definitions are the same. This is an interesting question, but up to our knowledge the general and comprehensive answer is not known. After a lot of research work in this area (the description of which is beyond the scope of this book), we know two things. First, there is no mathematically complete proof of the equivalence of the two definitions. Second, in all known cases, they are equivalent. One can then ask what is the physical sense of the conservation law, which has to be covariant and therefore, in curved space, should have the form

$$\nabla_\mu T^{\mu\nu} = 0. \tag{15.3}$$

In order to address this question, consider infinitesimal coordinate transformations of coordinates $x^\mu \to x'^\mu = x^\mu + \xi^\mu(x)$. As far as the action of matter is supposed to be a scalar constant (functional), it cannot change under this transformation.

Let us find the transformation rule for the metric. The calculation is similar to what was already done in Chap. 1; hence, we do not need to go into full detail. Keeping only the terms of the first order in ξ and its derivatives, we get

$$\delta g_{\alpha\beta}(x) = g'_{\alpha\beta}(x) - g_{\alpha\beta}(x), \qquad \text{where}$$

$$g'_{\alpha\beta}(x) = g'_{\alpha\beta}(x') - \frac{\partial g'_{\alpha\beta}(x')}{\partial x'^\lambda}\xi^\lambda = g'_{\alpha\beta}(x') - \partial_\lambda g_{\alpha\beta}\,\xi^\lambda \qquad \text{and}$$

$$g'_{\alpha\beta}(x') = \frac{\partial x^\rho}{\partial x'^\alpha}\frac{\partial x^\sigma}{\partial x'^\beta}\,g_{\rho\sigma}(x) = \left(\delta^\rho_\alpha - \partial_\alpha\xi^\rho\right)\left(\delta^\sigma_\beta - \partial_\beta\xi^\sigma\right)g_{\rho\sigma}(x)$$

$$= g_{\alpha\beta} - g_{\rho\beta}\,\partial_\alpha\xi^\rho - g_{\alpha\sigma}\,\partial_\beta\xi^\sigma. \tag{15.4}$$

At the last step, we used the fact that in the linear order in ξ in those terms that are $\mathcal{O}(\xi)$, one can set $x' = x$ and not to care about whether the argument is x or x'. Combining these formulas, and using the relation

$$\partial_\lambda g_{\alpha\beta} = \Gamma^\tau_{\alpha\lambda}g_{\tau\beta} + \Gamma^\tau_{\beta\lambda}g_{\alpha\tau}, \tag{15.5}$$

we arrive at

$$\delta g_{\alpha\beta} = -\nabla_\alpha\xi_\beta - \nabla_\beta\xi_\alpha. \tag{15.6}$$

Exercise 2 Using the same method, calculate δA^μ, δA_ν, $\delta\Psi_{\mu\nu}$ and $\delta\Psi^{\mu\nu}$, where A^μ and $\delta\Psi_{\mu\nu}$ are arbitrary vector and symmetric tensor fields. Discuss the difference between $\delta\Psi_{\mu\nu}$ and the formula (15.6), in light of the metricity condition $\nabla_\lambda g_{\mu\nu} = 0$ for the covariant derivative. Discuss the relation between transformation of co- and contravariant components of the same vector or tensor. Is it possible to obtain one from another without special calculation? Use the answer to obtain $\delta g^{\alpha\beta}$ and verify the result using Eq. (15.6).

Now we are in a position to explore the origin of the conservation law for the energy–momentum tensor. Consider the transformation of the action $S_m(g_{\alpha\beta}, \Phi^i)$, where Φ^i indicates all field variables and coordinates of particles that describe matter. Let us suppose that the general coordinate transformation (also called diffeomorphism) is

$$\delta\Phi^i = R^i_\alpha \xi_\alpha . \tag{15.7}$$

The operators R^i_α are called generators of the transformation. They may be just functions, or are operators that include derivatives, like in (15.6). Taking into account (15.6) and (15.7), we get

$$
\begin{aligned}
\delta S_m &= \int d^4x \sqrt{|g|} \left\{ \frac{1}{\sqrt{|g|}} \frac{\delta S_m}{\delta g_{\mu\nu}} \delta g_{\mu\nu} + \frac{1}{\sqrt{|g|}} \frac{\delta S_m}{\delta \Phi_i} \delta \Phi_i \right\} \\
&= \int d^4x \sqrt{|g|} \left\{ -\frac{1}{2} T^{\mu\nu}(-\nabla_\mu \xi_\nu - \nabla_\nu \xi_\nu) + \frac{1}{\sqrt{|g|}} \frac{\delta S_m}{\delta \Phi_i} R^\alpha_i \xi_\alpha \right\} \\
&= \int d^4x \sqrt{|g|} \left\{ -\nabla_\mu T^{\mu\nu} + \frac{1}{\sqrt{|g|}} \frac{\delta S_m}{\delta \Phi_i} R^\nu_i \right\} \xi_\nu ,
\end{aligned}
\tag{15.8}
$$

where at the last step we integrated by parts in the first term. Since the last term vanishes on-shell (this is a frequently used abbreviation that means "on the mass-shell," or when the variables satisfy their equations of motion) for any $\xi^\nu(x)$, the conservation law

$$\nabla_\mu T^{\mu\nu} = 0 \tag{15.9}$$

follows. This derivation shows a qualitative similarity with the conservation of the canonical energy–momentum tensor, since our consideration was also a consequence of the fundamental symmetries of space–time, meaning the invariance under (15.6) and (15.7).

The last observation is about the integrated form of the conservation law (15.9). We know that in flat space, an integrated form provides the conservation of energy and momentum. However, in the case of curved space, the situation is more complicated. The integral of the $T^{0\nu}$ component over the $3D$ space section of the $4D$ space is a non-covariant procedure that cannot be interpreted in accordance with the equivalence principle, since the last is essentially a local statement. In other words, we cannot define in a covariant (meaning consistent in general relativity) way the notion of energy or momentum of an object that has finite space size. The detailed discussion of this subject goes beyond the framework of this book; however, let us just say that the locality of the conservation law is directly related to the equivalence principle. This feature can not be seen as a shortcoming, but as an intrinsic, fundamental property of the theory.

15.1.2   Energy–Momentum Tensor for Fields, Particle, and Fluid

The energy–momentum tensor of different kinds of matter has fundamental importance in various branches of gravitational physics. Let us consider a few examples, which may be interesting in their own right and also in light of applications to cosmology.

15.1.2.1   Fields

Consider first the case of free electromagnetic radiation. In this case, we know the action (14.111) of the field A^μ, which produce Maxwell equations. Let us divide the calculation into small exercises.

Exercise 3 Staring from the action (14.111), derive the equation for and electromagnetic wave on an arbitrary curved background. The covariant version of Lorentz gauge fixing $\nabla_\mu A^\mu = 0$ can be used.

Solution. The equation in this particular gauge is $\Box A^\mu = 0$, where $\Box = g^{\mu\nu} \nabla_\mu \nabla_\nu$ is a covariant d'Alembert operator.

Exercise 4 Calculate the dynamical energy–momentum tensor of electromagnetic field, and show that its trace is zero. Explain why this is a direct consequence of the result of Exercise 1 in $4D$.

Solution. Starting from Def. (15.1) consider $g_{\mu\nu} \to g'_{\mu\nu} = g_{\mu\nu} + h_{\mu\nu}$ and calculate the variation of the action (14.111) up to the first order in the variation of the metric $\delta g_{\mu\nu} = h_{\mu\nu}$. Using (14.99), (14.100) and (12.43), we arrive at

$$\delta S_{em} = \int d^4x \sqrt{|g|} \left(\frac{1}{2} h\, F_{\alpha\beta} F^{\alpha\beta} - 2 F^{\mu\lambda} F^\nu_{\cdot\lambda} h_{\mu\nu} \right), \qquad (15.10)$$

where $h = g^{\mu\nu} h_{\mu\nu}$. Therefore,

$$T_{\mu\nu} = 4 F_{\mu\lambda} F^\lambda_{\nu\cdot} - g_{\mu\nu} F_{\alpha\beta} F^{\alpha\beta} \qquad (15.11)$$

and it is easy to verify that $T^\mu_\mu = 0$.

The last question of Exercise 4 is quite general, and we can solve it in a general fashion. Consider the action $S(g_{\mu\nu}, \Phi)$, where Φ is a field with conformal weight d_Φ. The last means that under the change of variables

$$g_{\mu\nu} = \bar{g}_{\mu\nu}\, e^{2\sigma}, \qquad \Phi = \bar{\Phi}\, e^{d_\Phi \sigma} \qquad (15.12)$$

the action does not depend on σ,

$$S(g_{\mu\nu}, \Phi) = S(\bar{g}_{\mu\nu}, \bar{\Phi}). \tag{15.13}$$

As we already know from Exercise 1, for the electromagnetic potential in $4D$ the conformal weight is zero, but for other fields it may be different, so it is useful to deal with an arbitrary field Φ with undetermined conformal weight d_Φ. Now, consider the infinitesimal version of the transformation (15.12). Then the variational derivative with respect to σ gives

$$\frac{\delta S}{\delta \sigma} = \frac{\delta g_{\mu\nu}}{\delta \sigma} \frac{\delta S}{\delta g_{\mu\nu}} + \frac{\delta \Phi}{\delta \sigma} \frac{\delta S}{\delta \Phi} = 0,$$

that can be recast into

$$2\bar{g}_{\mu\nu} e^{2\sigma} \frac{\delta S}{\delta g_{\mu\nu}} + d_\Phi \bar{\Phi} e^{d_\Phi \sigma} \frac{\delta S}{\delta \Phi} = 2 g_{\mu\nu} \frac{\delta S}{\delta g_{\mu\nu}} + d_\Phi \Phi \frac{\delta S}{\delta \Phi} = 0$$

or, finally, into

$$T^\mu_\mu = \frac{1}{\sqrt{|g|}} d_\Phi \Phi \frac{\delta S}{\delta \Phi}, \tag{15.14}$$

which is called Noether identity for the local conformal symmetry. We can see that *on-shell*, when the equations of motion for Φ are taken into account, the conformal symmetry implies $T^\mu_\mu = 0$. In the special case of electromagnetic action in $4D$, we have $d_A = 0$ and the identity $T^\mu_\mu = 0$ holds even *off-shell*.

Exercise 5 Calculate again the energy–momentum tensor for the electromagnetic field in curved space using the canonical definition, this time in the presence of the metric. Show that the results of two definitions differ by a total covariant derivative of an antisymmetric tensor, exactly like in the flat space–time, where we have the difference between symmetric and nonsymmetric conserved versions.

Exercise 6 Derive the energy–momentum tensor for the scalar field in curved space using both canonical and dynamical definitions and show that the results coincide. Start with the minimal scalar field with the action

$$S_{\min} = \frac{1}{2} \int d^4 \sqrt{|g|} \left(g^{\mu\nu} \partial_\mu \varphi \partial_\nu \varphi - m^2 \varphi^2 \right). \tag{15.15}$$

Exercise 7 Consider the nonminimal scalar field with the extra term in the action,

$$S_{\text{nonmin}} = S_{\min} + \frac{1}{2} \int d^4 \sqrt{|g|} \, \xi R \varphi^2. \tag{15.16}$$

Here ξ is what is called *nonminimal parameter* of the scalar–curvature interaction. Derive the dynamical equation for the scalar field by taking variation with respect to φ and also by means of the Lagrange equations. After that, repeat the derivation of the canonical energy–momentum tensor for the scalar field. Consider the case of massless conformal theory, with $m^2 = 0$ and $\xi = 1/6$ and show that the trace of the energy–momentum tensor is proportional to the equation of motion for the scalar field, and therefore is zero on-shell.

Hint. The expansion of scalar curvature can be found in the appendix.

Exercise 8 Show that the previously considered scalar field, in the case when $m^2 = 0$ and $\xi = 1/6$, has the action that is invariant under the local conformal transformation,

$$g_{\mu\nu} = \bar{g}_{\mu\nu} e^{2\sigma(x)}, \qquad \varphi = \bar{\varphi} e^{-\sigma(x)}. \tag{15.17}$$

Explain why this invariance guarantees zero trace of the Energy–momentum tensor on-shell.

Exercise 9 The reader may be curious to know whether all theories with $T^\mu_{\ \mu} = 0$ have conformal invariant actions. The answer to this question is negative, and to see this we need an example. Consider this problem for the action

$$S_{\text{surf}} = \int d^4 \sqrt{|g|} \left\{ \xi_1 \Box(\varphi^2) + \frac{1}{2} \xi_2 \nabla_\mu (\varphi \nabla^\mu \varphi) \right\}, \tag{15.18}$$

where $\xi_{1,2}$ are arbitrary constants.

Observation. The terms with coefficients ξ_1 and ξ_2 are identical. Prove this before or after solving the main exercise.

15.1.2.2 Particles

Consider the case of a massive particle. The simplest option is to consider some distribution of such particles with the mass density $\mu(\mathbf{r}, t)$. In order to find the energy–momentum of this mass distribution, we can consider the instantly co-moving inertial reference frame. This means that the observer is moving with the same velocity as the massive particles in the point $\mathbf{r}$ at the instant t. Then we know that the density of the four-momenta of the particles at this space–time point is $T^{0i} = p^i = \mu u^i$, where u^i is the space part of the four-velocity. On the other hand, we know that in the same frame μ is the time component of the four-vector $\frac{\mu}{c} \frac{dx^\alpha}{dt}$. Combining the two things, the formula for the energy–momentum tensor is

$$T^{\mu\nu} = \mu \frac{dx^\mu}{ds} \frac{dx^\nu}{dt} = \mu u^\mu u^\nu \frac{ds}{dt}. \tag{15.19}$$

It is easy to see that the factor $\frac{ds}{dt}$ is not a scalar quantity compensates. However, the product (15.19) is a in $T^{\mu\nu}$ tensor, because $\mu(\mathbf{r}, t)$ is not a tensor quantity too. As a result, there is a compensation between the two factors. After being defined in the special reference frame, the energy–momentum tensor can be transformed to any other frame as usual. The trace of the tensor $T^{\mu\nu}$ can be obtained from Eq. (15.19) and the scalar relation $u^\mu u_\mu = 1$,

$$T^\mu_\mu = \mu \frac{ds}{dt} = \mu \sqrt{1 - \frac{v^2}{c^2}}, \qquad (15.20)$$

where v is the velocity of the particles in an arbitrary frame. In the UV limit $v \to c$, the trace vanishes, which is the same result we already know for the radiation, which can be also seen as the set of waves moving with the speed of light c.

Exercise 10 Derive the formula (15.19) from the action principle for a free particle in the gravitational field. In order to simplify the work, consider an ultrastatic metric of the form $ds^2 = dt^2 - \gamma_{ij}(x)dx^i dx^j$.

15.1.2.3 Fluids

The ideal fluid is a medium for which the stress tensor is isotropic, according to the Pascal law. In the co-moving frame, the energy–momentum tensor is

$$T_{\mu\nu} = \begin{pmatrix} \rho & 0 & 0 & 0 \\ 0 & p & 0 & 0 \\ 0 & 0 & p & 0 \\ 0 & 0 & 0 & p \end{pmatrix}, \qquad (15.21)$$

where $\rho = \mu c^2$ is the energy density and p is the pressure. Assuming that the instantly inertial co-moving metric is Minkowski one, we easily arrive at the trace $T^\mu_\mu = \rho - 3p$, which becomes zero for radiation, when $p = \rho/3$.

In order to construct a covariant version of (15.21), one can perform coordinate transformation. However, there is a simpler way to achieve the same goal. In arbitrary coordinates, $T_{\mu\nu}$ should be a linear combination of the same components, and can depend only on the first derivatives of coordinates, $u_\mu = dx_\mu/ds$. Therefore, the general possible form is

$$T_{\mu\nu} = x u_\mu u_\nu + y g_{\mu\nu}, \qquad (15.22)$$

where the coefficients x and y should linearly depend on ρ and p due to dimensional arguments. Inserting here the values $u_\mu = (1, 0, 0, 0)$ and $g_{\mu\nu} = \eta_{\mu\nu} = \mathrm{diag}\,(1, -1, -1, -1)$ from the comparison with (15.21), one can easily obtain the expression

$$T_{\mu\nu} = (\rho + p)u_\mu u_\nu - p g_{\mu\nu}. \qquad (15.23)$$

The last formula describes any kind of ideal relativistic fluid from so-called pressure-less dust with $p = 0$ up to radiation with $p = \rho/3$. It is important to keep in mind that ρ includes the rest energy of the massive particles, e.g., nmc^2, where n is the concentration of particles. Therefore, many cases of usual matter (e.g., the air in the room when the reader is now) have pressure that is completely negligible compared to the enormous ρ, and the fluid can be safely regarded as pressureless dust.

A more general case of ideal fluid is characterized by the *equation of state* $p = \omega\rho$, and in many cases it is sufficient to regard ω as a constant. However, in the case of adiabatically isolated ideal gas of massive particles, this approximation is not appro-priate. Our intuition hints that when such a gas is very dense, it should be very hot, such that $\omega \approx 1/3$. On the other hand, after a large amount of adiabatic expansion the gas may be close to the dust, with $\omega = 0$. Such a fluid can be described, in the simplest possible case, as an ideal gas of massive particles, with the Maxwell distribution of velocities. This fluid is called the Jüttner model [1] (see also the brief description in the book by Pauli [2]), and was discussed in Chap. 2. The approximated version is the *reduced relativistic gas* (RRG) model, with the equation of state (13.119).

15.1.3   Einstein Equations from the Covariance Arguments

Since we introduced the actions and equations of motion for particles and nongrav-itational fields, it is natural to ask what the equations are for the metric itself. Let us find the answer to this question first from the general covariance arguments and then by constructing the simplest possible action for the dynamical metric. The logic we will follow in this subsection represents a simplified version of the original consid-erations by Einstein, which were also explained in the classical book by Weinberg [3].

Let us remember that the minimal requirements for the gravitational equations are that they must (i) be covariant, meaning a tensor structure; (ii) reduce to the Newton gravitational law in the corresponding limit of nonrelativistic motions for weak and static gravitational field. In principle, any equation that satisfies these two conditions can be considered. But since these conditions admit infinitely many realizations, we can add one extra requirement of (iii) simplicity.

One can assume that the dynamical equations for the metric should be differential equations of the second order, since all other equations we know (Maxwell, Klein–Gordon) are of this type. Moreover, the number of equations should correspond to the number of independent components in the symmetric 4×4 tensor $g_{\mu\nu}$. At present, we know only two tensors that satisfy these conditions, namely, the Ricci tensor $R_{\mu\nu}$ and $g_{\mu\nu}R$, where R is the scalar curvature. Furthermore, the gravitational part (let us call it the left-hand side) of the equation can include the term without the derivative, say $g_{\mu\nu}\Lambda$, where Λ is a constant, called cosmological. Thus, the general *l.h.s.* is $\gamma^{-1}\left(R_{\mu\nu} + bg_{\mu\nu}R\right) + ag_{\mu\nu}\Lambda$, where γ^{-1}, a and b are unknown constants.

The *r.h.s.* of the gravitational equation should be a symmetric tensor that depends on the fundamental characteristics of matter, that is, on the nongravitational fields and particles or fluids. We also know that in the nonrelativistic Newtonian limit the source of the gravitational field is a density of mass distribution. And, finally, we know that for the static particles the mass density is associated with the T_{00} component of the energy–momentum tensor $T_{\mu\nu}$. Since $T_{\mu\nu}$ has exactly the desired tensor structure, we arrive at the natural form of the equation,

$$\gamma^{-1}\left(R_{\mu\nu} + bg_{\mu\nu}R\right) + ag_{\mu\nu}\Lambda = T_{\mu\nu}. \tag{15.24}$$

One can expect that γ^{-1} can be found by comparison with the Newtonian limit and the choice of a is obviously related to the definition for the cosmological constant Λ. It is a standard definition to set $a = -\gamma^{-1}$. The remaining part is to define the coefficient b. For this end, it is useful to remember that the r.h.s. of Eq. (15.24) satisfies the conservation law, $\nabla_\mu T_\nu^\mu = 0$. Therefore, the consistency of equations (15.24) can be achieved only if the l.h.s. also satisfies this condition. Then the Bianchi identity $\nabla_\mu R_\nu^\mu = \frac{1}{2}\nabla_\nu R$ shows that $b = -1/2$. In this way, we arrive at the equations

$$G_{\mu\nu} = R_{\mu\nu} - \frac{1}{2}g_{\mu\nu}R - g_{\mu\nu}\Lambda = \gamma T_{\mu\nu}, \tag{15.25}$$

where the two constants γ and Λ should be defined from experiments and/or observations. After identifying these two parameters, (15.25) will become Einstein equations, which are the main equations of general relativity. The determination of Λ can be achieved only in the framework of cosmology and its value turns out to be very small even at the astrophysical scale, e.g., for the galaxies. In the Newtonian limit associated with the solar system tests, we can safely consider $\Lambda = 0$.

Exercise 11 Prove that trading $T_{\mu\nu} \to T_{\mu\nu} - \zeta T_\alpha^\alpha g_{\mu\nu}$ with $\zeta \neq 1/4$ does not increase the generality of Eq. (15.24). Discuss the peculiarity of the case $\zeta = 1/4$ and find the value of b that makes Eq. (15.24) consistent for $\Lambda = 0$. What changes if $\Lambda \neq 0$?

15.1.4 Einstein Equations from the Einstein–Hilbert Action

Another way to obtain the equations for the gravitational field is to postulate the action of the metric. In order to do so, let us formulate what we expect from such an action. Our experience shows that the metric-dependent quantities that are covariant are non-polynomial in the metric. This can be seen, for instance, from the expansions (14.98) and (14.99). Thus, it would not make much sense to require the action to be polynomial in $g_{\mu\nu}$. At the same time, we can ask for a local functional of the metric, and also require that the equations of motion would have no more than second derivatives. Then, the action also should be at most of the second order in derivatives. Then, we have only two candidate terms for a covariant Lagrangian, namely R and $\Lambda = const$. In this way, we arrive at the action of gravity and matter in the form

$$S_{\text{total}} = S_{\text{grav}} + S_{\text{m}}, \tag{15.26}$$

$$S_{\text{grav}} = -\frac{1}{2\gamma} \int d^4x \sqrt{|g|} \, (R + 2\Lambda). \tag{15.27}$$

The expression (15.27) was first obtained by Hilbert and is usually called Einstein–Hilbert action. In what follows we will see that the choice of the overall coefficient in (15.27) provides perfect correspondence with Eq. (15.25) obtained on the basis of covariance and dimensional arguments.

In order to derive the dynamical equations for the metric, consider its variation in a way that we already anticipated in Exercise 27 of Sect. 14.7.1. The details of the expansion of relevant quantities for the (14.98),

$$g_{\mu\nu} \rightarrow g'_{\mu\nu} = g_{\mu\nu} + h_{\mu\nu}$$

can be found in the appendix of this section. Replacing the first-order terms from (15.31) and (15.42) into the action (15.27), and using the definition of the energy–momentum tensor (15.1), the variation of the total action (15.26) becomes

$$\delta S_{\text{total}} = -\frac{1}{2\gamma} \int d^4x \sqrt{|g|} \left\{ \frac{1}{2} h(R + 2\Lambda) + \nabla_\mu \nabla_\nu h^{\mu\nu} - \Box h - h^{\mu\nu} R_{\mu\nu} \right\}$$
$$- \frac{1}{2} \int d^4x \sqrt{|g|} \, T_{\mu\nu} h^{\mu\nu}. \tag{15.28}$$

Disregarding the irrelevant total derivative terms, this variation gives exactly the equations for the metric (15.25), which were constructed earlier using covariance arguments.

Appendix. Expansions to an Arbitrary Order

Shifting the metric (14.98) into background and perturbation leads to the inverse metric and affine connection expansions, (14.99) and (14.100). Let us remember that these two expansions are non-perturbative in $h_{\mu\nu}$. For deriving the equations of motion, we need only first orders in $h_{\mu\nu}$, but for the sake of generality, it is worthwhile to obtain more general expansions for the quantities of potential interest.

The action (15.27) includes only scalar curvature, but due to our possible interest in more general gravitational theories, let us derive the expansion for Riemann and Ricci tensors.

First of all, it is good to remember that the difference between two affine connections $\delta \Gamma^\lambda_{\mu\nu}$ is a tensor, and hence we are allowed to take its covariant derivatives.

In order to avoid going into full technical detail, let us formulate the two relevant statements as exercises.

Exercise 12 Prove that the variation of the Riemann and Ricci tensors are given by the expressions

$$\delta R^{\lambda}_{.\tau\mu\nu} = \nabla_{\mu}\delta\Gamma^{\lambda}_{\tau\nu} - \nabla_{\nu}\delta\Gamma^{\lambda}_{\tau\mu} + \delta\Gamma^{\rho}_{\tau\nu} \cdot \delta\Gamma^{\lambda}_{\rho\mu} - \delta\Gamma^{\rho}_{\tau\mu} \cdot \delta\Gamma^{\lambda}_{\rho\nu}, \quad (15.29)$$

$$\delta R_{\mu\nu} = \nabla_{\lambda}\delta\Gamma^{\lambda}_{\mu\nu} - \nabla_{\mu}\delta\Gamma^{\lambda}_{\nu\lambda} + \delta\Gamma^{\tau}_{\mu\nu} \cdot \delta\Gamma^{\lambda}_{\tau\lambda} - \delta\Gamma^{\tau}_{\lambda\mu} \cdot \delta\Gamma^{\lambda}_{\lambda\nu}. \quad (15.30)$$

Exercise 13 Using (15.30), (14.99) and (14.100), verify the first-order expansion for the scalar curvature,

$$R' = R + \nabla_{\mu}\nabla_{\nu}h^{\mu\nu} - \Box h - h^{\mu\nu}R_{\mu\nu} + \mathcal{O}(h^2_{..}) \quad (15.31)$$

In this formula, the indices are lowered and raised with the non-perturbed (background) metrics $g_{\mu\nu}$ and $g^{\mu\nu}$. Also, we use notation for the trace $h = h^{\mu}_{\mu}$.

Exercise 14 (i) Using Eqs. (15.29) and (15.30), derive the expansions for Riemann and Ricci tensors up to the second order in $h_{\mu\nu}$. (ii) Make expansions of the same quantities up the first order for $\Phi_{\mu\nu}$ from Eq. (14.101) and for $\gamma_{\mu\nu}$ from (14.102), in the last case on the flat background.

Let us now consider the expansion of the metric determinant and $\sqrt{|g|}$. For this end, it is useful to start from the modified form of the expansions (14.98) and (14.99), with an extra parameter κ,

$$g_{\alpha\beta} \longrightarrow g'_{\alpha\beta} = g_{\alpha\beta} + \kappa h_{\alpha\beta}, \quad (15.32)$$

$$g'^{\alpha\beta} = g^{\alpha\beta} - \kappa h^{\alpha\beta} + \kappa^2 h^{\alpha\lambda} h^{\beta}_{\lambda} - \kappa^3 h^{\alpha\lambda} h^{\tau}_{\lambda} h^{\beta}_{\tau} + \dots . \quad (15.33)$$

Of course, we can set $\kappa = 1$ after the expansion is performed. At the same time, with the new parameter κ one can apply the Tailor series expansion,

$$g' = \det\left(g'_{\mu\nu}\right) = g + \kappa \frac{\partial g'}{\partial\kappa}\bigg|_{\kappa=0} + \frac{\kappa^2}{2} \frac{\partial^2 g'}{\partial\kappa^2}\bigg|_{\kappa=0} + \frac{\kappa^3}{3!} \frac{\partial^3 g'}{\partial\kappa^3}\bigg|_{\kappa=0} + \dots . \quad (15.34)$$

For the first three orders, this gives (with $\kappa = 1$)

$$g' = g\left[1 + h + \frac{1}{2}\left(h^2 - h^2_{\mu\nu}\right) + \frac{1}{3!}\left(h^3 - 3hh^2_{\mu\nu} + 2h^3_{\mu\nu}\right)\right]. \quad (15.35)$$

In this formula, we introduced useful notations

$$h^2_{\mu\nu} = h^{\mu\nu}h_{\mu\nu}, \quad h^3_{\mu\nu} = h^{\nu}_{\mu} h^{\lambda}_{\nu} h^{\mu}_{\lambda}, \quad \text{etc.} \quad (15.36)$$

One can continue the expansion (15.35) to any finite order, but it is simpler to find a non-perturbative solution to any order in $h_{\mu\nu}$. For this end, we start from the well-known relation

$$\Gamma^\lambda_{\alpha\lambda} = \frac{1}{2}\partial_\alpha \log g \tag{15.37}$$

and apply it to the difference between two affine connections, constructed from perturbed and non-perturbed metrics,

$$\Gamma'^\lambda_{\alpha\lambda} - \Gamma^\lambda_{\alpha\lambda} = \frac{1}{2}\partial_\alpha \log\left(\frac{g'}{g}\right). \tag{15.38}$$

According to our previous experience (15.35), the ratio g'/g is a scalar, and therefore the partial derivative in the last formula can be traded to a covariant one.

On the other hand, (14.100) provides the difference between the two connections,

$$\Gamma'^\lambda_{\alpha\beta} - \Gamma^\lambda_{\alpha\beta} = \frac{1}{2}g'^{\lambda\tau}\left(\nabla_\alpha h_{\tau\beta} + \nabla_\beta h_{\tau\alpha} - \nabla_\tau h_{\alpha\beta}\right). \tag{15.39}$$

Taking the trace, we arrive at an alternative form of (15.38),

$$\begin{aligned}
\Gamma'^\lambda_{\alpha\lambda} - \Gamma^\lambda_{\alpha\lambda} &= \frac{1}{2}\left(g^{\lambda\tau} - h^{\lambda\tau} + h^\lambda_\rho h^{\rho\tau} - h^\lambda_\rho h^\rho_\sigma h^{\sigma\tau} + \dots\right)\nabla_\alpha h_{\lambda\tau} \\
&= \frac{1}{2}\nabla_\alpha\left(h - \frac{1}{2}h^2_{\mu\nu} + \frac{1}{3}h^3_{\mu\nu} - \frac{1}{4}h^4_{\mu\nu} + \dots\right).
\end{aligned} \tag{15.40}$$

Comparison of (15.38) and (15.40) gives the following version of the Liouville's formula for a metric determinant:

$$g' = g \times \exp\left\{h - \frac{1}{2}h^2_{\mu\nu} + \frac{1}{3}h^3_{\mu\nu} - \frac{1}{4}h^4_{\mu\nu} + \dots\right\}. \tag{15.41}$$

It is easy to check that the expansion of the exponential into a series gives the result (15.35), but using the general formula (15.41) it is easy to continue the expansion to further orders, and also derive the expansions for arbitrary powers of determinant, $|g'|^d$. In the particular case of $d = 1/2$, we arrive at

$$\sqrt{|g'|} = \sqrt{|g|}\left(1 + \frac{1}{2}h - \frac{1}{4}h^2_{\mu\nu} + \frac{1}{8}h^2 + \frac{1}{6}h^3_{\mu\nu} - \frac{1}{8}hh^2_{\mu\nu} + \frac{1}{48}h^3 + \dots\right). \tag{15.42}$$

Exercise 15 Derive the expansion of the power d of the determinant $|g|^d$, for an arbitrary d in the third order in $h_{\mu\nu}$. Discuss why the expansion does not depend on the dimension of the Riemann space.

Exercise 16 Using the expansions obtained in Exercise 22, calculate the equations of motion for the three four-derivative actions

$$I_1 = \int d^4x \sqrt{-g} \, R^2_{\mu\nu\alpha\beta},$$

$$I_2 = \int d^4x \sqrt{-g} \, R^2_{\mu\nu},$$

$$I_3 = \int d^4x \sqrt{-g} \, R^2. \tag{15.43}$$

Solutions.

$$\frac{1}{\sqrt{-g}} \frac{\delta I_1}{\delta g_{\mu\nu}} = \frac{1}{2} g^{\mu\nu} R^2_{\rho\sigma\alpha\beta} - 2R^{\mu\sigma\alpha\beta} R^\nu_{\sigma\alpha\beta} - 4R^{\mu\alpha\nu\beta} R_{\alpha\beta} + 4R^\mu_\alpha R^{\nu\alpha}$$
$$+ 2\nabla^\mu \nabla^\nu R - 4\Box R^{\mu\nu}, \tag{15.44}$$

$$\frac{1}{\sqrt{-g}} \frac{\delta I_2}{\delta g_{\mu\nu}} = \frac{1}{2} g^{\mu\nu} R^2_{\rho\sigma} - 2R^{\mu\alpha\nu\beta} R_{\alpha\beta} + \nabla^\mu \nabla^\nu R - \frac{1}{2} g^{\mu\nu} \Box R - \Box R^{\mu\nu}, \tag{15.45}$$

$$\frac{1}{\sqrt{-g}} \frac{\delta I_3}{\delta g_{\mu\nu}} = \frac{1}{2} g^{\mu\nu} R^2 + 2\nabla^\mu \nabla^\nu R - 2g^{\mu\nu} \Box R - 2RR^{\mu\nu}. \tag{15.46}$$

Exercise 17 Verify that the following two linear combinations of the terms (15.43) have vanishing traces of the equations of motion,

$$\int d^4x \sqrt{|g|} \, E_4 = I_1 - 4I_2 + I_3 \tag{15.47}$$

and $$\int d^4x \sqrt{|g|} C^2 = I_1 - 2I_2 + \frac{1}{3} I_3. \tag{15.48}$$

Show that the second term is the square of the Weyl tensor, $C^2 = C_{\mu\nu\alpha\beta} C^{\mu\nu\alpha\beta}$ and that the corresponding trace of the equation of motion *must be* zero due to the conformal invariance of the action.

Exercise 18 Consider the first action from the previous exercise, Eq. (15.47), which is called the Gauss–Bonnet topological invariant (or Euler characteristics of the manifold),

$$\int d^4x \sqrt{|g|} E_4 = \int d^4x \sqrt{|g|} \left(R_{\mu\nu\alpha\beta} R^{\mu\nu\alpha\beta} - 4R_{\mu\nu} R^{\mu\nu} + R^2 \right). \tag{15.49}$$

(i) Verify that the integrand can be written as

$$R_{\mu\nu\alpha\beta} R^{\mu\nu\alpha\beta} - 4R_{\mu\nu} R^{\mu\nu} + R^2 = -\frac{1}{4} \varepsilon^{\mu\nu\rho\sigma} \varepsilon^{\lambda\tau\alpha\beta} R_{\mu\nu\alpha\beta} R_{\rho\sigma\lambda\tau}. \tag{15.50}$$

(ii) Use special coordinates in which the affine connection at the given point vanishes, to prove that this integrand is a total derivative.
(iii) Discuss whether the last statement can be formulated in a covariant way.
(iv) Using Eq. (15.50) as an example, try to construct a generalization of the Gauss–

Bonnet action to the space of an arbitrary-even dimension. Prove that the corresponding integrands are total derivatives.

Hint to the point (iii). Prove first that for any vector χ^μ there is an identity

$$\sqrt{|g|}\,\nabla_\mu \chi^\mu = \partial_\mu\left(\sqrt{|g|}\,\chi^\mu\right). \tag{15.51}$$

15.1.5   Newtonian Limit and Calibrating the Constant γ

Let us come back to Eq. (15.25). As it was already mentioned above, the value of the observable cosmological constant Λ is relevant only at the cosmological scale, meaning at the distances typical for the whole universe. Thus, in order to calibrate the remaining constant γ, by using the Newtonian limit we can set $\Lambda = 0$ and deal with the reduced equation

$$R_{\mu\nu} - \frac{1}{2}\,g_{\mu\nu}R = \gamma T_{\mu\nu}\,. \tag{15.52}$$

Taking the trace $-R = \gamma T$, where $T = T_\lambda^\lambda$, one can place it back into Eq. (15.52), raise one of indices and recast this equation in the form

$$R_\mu^\nu = \gamma\left(T_\mu^\nu - \frac{1}{2}\,T\delta_\mu^\nu\right), \tag{15.53}$$

In the Newtonian limit only $T_0^0 = \mu(\mathbf{r})c^2 \neq 0$, therefore $T = T_0^0$ and the relevant part of (15.53) is

$$R_0^0 = \frac{\gamma c^2}{2}\,\mu(\mathbf{r})\,. \tag{15.54}$$

The calculation of R_0^0 is simple, especially (as we shall see right now) because the evaluation of Newtonian limit requires only g_{00} component of the metric. And this component we already know from Eq. (14.97) to be $g_{00} = 1 + 2\varphi/c^2$. As far as the gravitational field is weak, we need to keep only the first order in the Newtonian potential φ and therefore only the first order in the Christoffel symbol $\Gamma_{\mu\nu}^\lambda$. Moreover, since gravity in the Newtonian approximation is static, only space derivatives ∂_i are relevant. Therefore

$$R_{00} = \partial_\lambda \Gamma_{00}^\lambda - \partial_0 \Gamma_{0\lambda}^\lambda + \mathcal{O}(\Gamma \cdot \Gamma) = \partial_i \Gamma_{00}^i\,. \tag{15.55}$$

Next, in the same approximation we can trade $g^{ij} \to -\delta^{ij}$ since the difference is $\mathcal{O}(\varphi^2)$, obtaining (using Euclidean signature in 3D with $\partial^i = \partial_i$ at the last step)

$$\Gamma^i_{00} = \frac{1}{2} g^{i\alpha} \left(2\partial_0 g_{\alpha 0} - \partial_\alpha g_{00} \right) = -\frac{1}{2} g^{ij} \partial_j g_{00}$$

$$= \frac{1}{2} \delta^{ij} \partial_j \left(\frac{2\varphi}{c^2} \right) = \frac{1}{c^2} \partial_i \varphi \tag{15.56}$$

and hence

$$R^0_0 = R_{00} = \frac{1}{c^2} \Delta\varphi. \tag{15.57}$$

Finally, Eq. (15.54) reduces to the Newtonian limit

$$\Delta\varphi = \frac{\gamma c^4}{2} \mu(\mathbf{r}). \tag{15.58}$$

This expression should be compared to the Gaussian law $\Delta\varphi = 4\pi G \mu(\mathbf{r})$. Then the coefficient of the gravitational action is defined to be $\gamma = 8\pi G/c^2$, and the Einstein–Hilbert action with the cosmological constant can be cast into the final form

$$S_{\text{grav}} = -\frac{c^4}{16\pi G} \int d^4x \sqrt{|g|} \left(R + 2\Lambda \right), \tag{15.59}$$

while the Einstein equations with the cosmological constant become

$$G^\nu_\mu = R^\nu_\mu - \frac{1}{2} \delta^\nu_\mu R = \frac{8\pi G}{c^4} T^\nu_\mu + \delta^\nu_\mu \Lambda. \tag{15.60}$$

In what follows we shall mainly use the units with $c = 1$.

According to the existing tradition, we resettled the cosmological constant term into the r.h.s. of the equation. The reason is that this term has the form of exotic matter fluid with the equation of state $p_\Lambda = -\rho_\Lambda$, where the energy density of the fluid is

$$\rho_\Lambda = \frac{\Lambda}{8\pi G}. \tag{15.61}$$

This expression is usually called vacuum energy density, while in reality it is not an energy density of a fluid, but rather the 00-component of the cosmological constant term in the Einstein equations. One of the reasons for this terms mismatch is that in quantum field theory the vacuum is usually associated with the Higgs field of the Minimal Standard Model of particle physics. This field is a complex scalar in the fundamental representation of the $SU(2)$ group, but for the sake of simplicity it is sufficient to consider just a single real scalar field φ. The potential $V(\varphi)$ of such a field enters into the action as

$$S_{scal} = \int d^4x \sqrt{|g|} \left\{ \frac{1}{2} g^{\mu\nu} \partial_\mu \varphi \, \partial_\nu \varphi - V(\varphi) \right\}, \tag{15.62}$$

and is defined as $$V(\varphi) = -\frac{1}{2} m^2 \varphi^2 + \frac{\lambda}{4!} \varphi^4, \tag{15.63}$$

where the mass term is intentionally taken with a wrong sign and the potential is bounded from below due to the positive coupling constant λ. The potential defines the vacuum state, which corresponds to the minimal value $V_m = V(\varphi_0)$, where

$$V'(\varphi_0) = 0 \implies \varphi_0 = \sqrt{\frac{6m^2}{\lambda}}. \tag{15.64}$$

It is easy to see that the value of the potential at the point of minima is

$$V_m = -\frac{3m^4}{2\lambda}. \tag{15.65}$$

Looking at the action (15.62), it is clear that when $\varphi \equiv \varphi_0$, the value of V_m is nothing else but the density ρ_Λ of the cosmological constant term. And since $V(\varphi)$ is regarded as a potential energy term for the scalar, it is somehow natural to identify V_m and therefore ρ_Λ as the energy of the vacuum. This is called *induced* and is denoted as ρ_Λ^{ind}. Numerically, the mass of the Higgs scalar is about 125 GeV, and λ is of the order one. Therefore, $\left| \rho_\Lambda^{ind} \right| \approx 10^8$ GeV^4, while the energy density of the "vacuum energy" which is responsible for the accelerated expansion of the universe (we will discuss this fact in the cosmological section later on) is $\rho_\Lambda^{obs} \propto 10^{-47}$ GeV^4. It is easy to see that the absolute value of V_m is about 55 orders of magnitude larger than the value of the cosmological constant that is required to explain the acceleration of the universe. The problem can be solved by introducing the purely geometric cosmological constant term ρ_Λ^{vac}, which does not depend on the scalar potential. Then the observed value is the sum

$$\rho_\Lambda^{obs} = \rho_\Lambda^{ind} + \rho_\Lambda^{vac}. \tag{15.66}$$

The famous cosmological constant problem is related to the lack of explanation concerning why the two independent contributions ρ_Λ^{ind} and ρ_Λ^{vac} cancel with such a tremendous precision of 55 orders of magnitude. The formation of galaxies, stars, and other observational cosmological data depend on the rate of expansion of the universe, which is defined by the presence and magnitude of the cosmological constant term. The existing estimates (see, e.g., [4]) show that if the mentioned precision were reduced by only two orders of magnitude, the Earth as we know it would not exist and hence we would not be here to discuss this issue. This kind of arguments is called an *antropic approach* and it confirms that the enormous fine-tuning in the elements of the sum (15.66) is vital for our existence. However, thus far, nobody can explain why this fine-tuning takes place.

The understanding of the complicated physical problems related to the Λ-term is possible only in the framework of quantum field theory and is beyond the scope of the present introductory textbook. There are many review papers devoted to this subject, including the very popular [4] and subsequent excellent reviews of different aspects of the problem [5, 6].

15.2 Schwarzschild Solution and Newtonian Limit

The calibration of the constant in the Einstein equations (15.60), which was demonstrated in the previous section, provided the R_0^0-component of the Ricci tensor in the Newtonian limit. It is interesting to derive other components of Ricci or Einstein tensors. It is certainly even more interesting to find an exact solution of Einstein equations in physically interesting situations. The first such solution was found by Karl Schwarzschild in 1916, for the case of a point-like source of mass M in the otherwise empty space. This approximation is also appropriate for the gravitational field created by a nonrotating star of radius R, at distances $r > R$ from its center.

The Schwarzschild solution is based on the assumption of the spherical symmetry of the metric. We shall make calculations in the units with $c = 1$, in the spherical coordinates r, θ, ψ, with the mass M at the central point $r = 0$. The spherical symmetry means that the space–time interval may depend only on the quantities that are invariant under space rotations. Therefore, ds^2 may be a linear combination of the terms dt^2, $d\mathbf{r} \cdot d\mathbf{r}$, $dt\mathbf{r} \cdot d\mathbf{r}$ and $(\mathbf{r} \cdot d\mathbf{r})^2$. In the spherical coordinates

$$d\mathbf{r} \cdot d\mathbf{r} = dr^2, \quad dt\mathbf{r} \cdot d\mathbf{r} = rdrdt, \quad \mathbf{r} \cdot d\mathbf{r} = rdr,$$

hence we can write down the general expression for the interval as

$$ds^2 = F(r)dt^2 - 2rE(r)drdt - D(r)r^2dr^2 - C(r)\left(dr^2 + r^2d\Omega\right), \quad (15.67)$$

where $d\Omega = d\theta^2 + \sin^2\theta d\psi^2$ is the solid angle element. For the sake of simplicity we assume that the functions $F(r)$, $E(r)$, $D(r)$, and $C(r)$ are time-independent; this dependence can be added later on, after some changes of variables. First, one can remove the non-diagonal term in the metric (15.67) by the change of time variable,

$$t' = t + \Phi(r), \quad \text{where} \quad \frac{d\Phi}{dr} = -\frac{rE(r)}{F(r)}. \quad (15.68)$$

Then

$$ds^2 = F(r)dt'^2 - G(r)dr^2 - C(r)\left(dr^2 + r^2d\Omega\right), \quad (15.69)$$

where $G = r^2(D + E^2/F)$. The next change of variables is $r'^2 = C(r)dr^2$, which provides

$$ds^2 = B(r')dt'^2 - A(r')dr'^2 - \left(dr'^2 + r'^2 d\Omega\right), \tag{15.70}$$

where $A(r')dr'^2 = [G(r) + C(r)]dr^2$.

Exercise 19* Consider the most general case, when F, E, D, and C are functions of both r and t, and also F may be zero. We require that the metric is nondegenerate and that its signature is -2. Show that the metric can be still reduced to the same form (15.70).

Warning. This is not an easy problem to solve.

Finally, we can omit the primes and recall the possible time dependence of the functions A and B. Thus, we arrive at the general expression for the interval of the space with spherical symmetry,

$$ds^2 = e^{\nu(r,t)} dt^2 - e^{\lambda(r,t)} dr^2 - \left(dr^2 + r^2 d\Omega\right), \tag{15.71}$$
$$e^{\nu(r,t)} = B(r,t), \qquad e^{\lambda(r,t)} = A(r,t).$$

The nonzero metric elements are

$$g_{tt} = e^{\nu(r,t)}, \quad g_{rr} = -e^{\lambda(r,t)}, \quad g_{\theta\theta} = -r^2, \quad g_{\psi\psi} = -r^2 \sin^2\theta, \tag{15.72}$$
$$g^{tt} = e^{-\nu(r,t)}, \quad g^{rr} = -e^{-\lambda(r,t)}, \quad g^{\theta\theta} = -\frac{1}{r^2}, \quad g^{\psi\psi} = -\frac{1}{r^2 \sin^2\theta}.$$

Since the mass is point-like, the rest of the space has zero density of the energy–momentum tensor, and what we need is to derive the set of equations $G_\alpha^\beta = 0$. After some algebra, one can obtain the nonzero components of the connection,

$$\Gamma_{tt}^t = \frac{\dot{\nu}}{2}, \quad \Gamma_{rr}^r = \frac{\lambda'}{2}, \quad \Gamma_{tt}^r = \frac{\nu'}{2} e^{\nu-\lambda}, \quad \Gamma_{rt}^t = \frac{\nu'}{2}, \quad \Gamma_{rr}^t = \frac{\dot{\lambda}}{2} e^{\lambda-\nu},$$

$$\Gamma_{tr}^r = \frac{\dot{\lambda}}{2}, \quad \Gamma_{\psi\psi}^\theta = -\sin\theta\cos\theta, \quad \Gamma_{\theta\psi}^\psi = \cot\theta, \quad \Gamma_{r\theta}^\theta = \Gamma_{r\psi}^\psi = \frac{1}{r},$$

$$\Gamma_{\theta\theta}^r = -re^{-\lambda}, \quad \Gamma_{\psi\psi}^r = -r\theta e^{-\lambda}\sin^2\theta, \tag{15.73}$$

where we denote the derivative with respect to r by a prime and the one with respect to t by a point.

A little more tedious calculation provides the components of the Ricci tensor and scalar curvature. Let us reproduce the nonzero components to make it easier for the reader to repeat this calculation,

$$R^t_t = e^{-\nu}\left(\frac{\dot\nu\dot\lambda}{4} - \frac{\dot\lambda^2}{4} - \frac{\ddot\lambda}{2}\right) + e^{-\lambda}\left(\frac{\nu''}{2} + \frac{\nu'}{r} + \frac{\nu'^2}{4} - \frac{\nu'\lambda'}{4}\right),$$

$$R^r_r = e^{-\nu}\left(\frac{\dot\nu\dot\lambda}{4} - \frac{\dot\lambda^2}{4} - \frac{\ddot\lambda}{2}\right) + e^{-\lambda}\left(\frac{\nu''}{2} - \frac{\lambda'}{r} + \frac{\nu'^2}{4} - \frac{\nu'\lambda'}{4}\right),$$

$$R^r_t = -\frac{\dot\lambda}{r}e^{-\lambda},$$

$$R^\theta_\theta = R^\psi_\psi = e^{-\lambda}\left(\frac{1}{r^2} + \frac{\nu' - \lambda'}{2r}\right) - \frac{1}{r^2}, \tag{15.74}$$

$$R = e^{-\nu}\left(\frac{\dot\nu\dot\lambda}{2} - \frac{\dot\lambda^2}{2} - \ddot\lambda\right) + e^{-\lambda}\left(\frac{2}{r^2} + \frac{2\nu'}{r} - \frac{2\lambda'}{r} + \nu'' + \frac{\nu'^2}{2} - \frac{\nu'\lambda'}{2}\right) - \frac{2}{r^2}.$$

The nonzero components of the Einstein tensor look much more compact. Finally, equations $G^\nu_\mu = 0$ with the nonvanishing components of the Einstein tensor are

$$G^t_t = \frac{1}{r^2} - e^{-\lambda}\left(\frac{1}{r^2} - \frac{\lambda'}{r}\right), \tag{15.75}$$

$$G^r_r = \frac{1}{r^2} - e^{-\lambda}\left(\frac{1}{r^2} + \frac{\nu'}{r}\right), \tag{15.76}$$

$$G^r_t = -\frac{\dot\lambda}{r}e^{-\lambda}, \tag{15.77}$$

$$G^\theta_\theta = G^\psi_\psi = e^{-\nu}\left(\frac{\ddot\lambda}{2} + \frac{\dot\lambda^2}{4} - \frac{\dot\nu\dot\lambda}{4}\right) + e^{-\lambda}\left(\frac{\lambda' - \nu'}{2r} - \frac{\nu''}{2} + \frac{\nu'\lambda' - \nu'^2}{4}\right). \tag{15.78}$$

Analyzing the last system of equations is not difficult. From Eq. (15.77) follows that $\dot\lambda = 0$, hence $\lambda = \lambda(r)$ and is time-independent. Furthermore, the difference between Eqs. (15.75) and (15.76) provides $\lambda' + \nu' = 0$, so the sum is only time-dependent,

$$\lambda(r) + \nu(r, t) = f(t). \tag{15.79}$$

These two conditions can be satisfied only if $\nu(r, t)$ is a sum of the t-independent and r-independent terms. Since $f(t)$ in (15.79) is an arbitrary function, without loss of generality we can assume that $\nu = \nu(r)$, while the function $f(t)$ can be absorbed into dt^2 in the interval (15.71). Therefore, after this operation we arrive at the relation

$$\lambda(r) + \nu(r) = const. \tag{15.80}$$

Since such a constant can once again be absorbed by a rescaling of time in the interval (15.71), and hence we can assume it to be zero. Later on, we shall see that the time variable defined by the described procedure makes t the standard time variable in Minkowski space at the space infinity $r \to \infty$.

All in all, we learned two things: that the solution we are interested in is static and that $\lambda(r) = -\nu(r)$, such that one has to find only one function of r. Any of Eqs. (15.75) and (15.76) can be written in the most useful form by using the element of the inverse metric of (15.71), $A^{-1} = e^{-\lambda(r)}$. The equation is linear,

$$\frac{dA^{-1}}{dr} + \frac{A^{-1}}{r} = \frac{1}{r} \tag{15.81}$$

and can be easily solved to provide

$$A^{-1} = e^{-\lambda} = e^{\nu} = 1 - \frac{C}{r}. \tag{15.82}$$

An integration constant C can be found from the asymptotic expansion at $r \to \infty$, where the gravitational field is weak, and we meet the Newtonian approximation

$$g_{tt} = 1 - \frac{C}{r} = 1 + 2\varphi, \qquad \varphi = -\frac{GM}{r}. \tag{15.83}$$

Therefore,

$$C = r_g = 2GM, \tag{15.84}$$

which is called *gravitational radius* of the mass M.

The Schwarzschild solution has the form

$$ds^2 = \left(1 - \frac{2GM}{r}\right) dt^2 - \left(1 - \frac{2GM}{r}\right)^{-1} dr^2 - r^2 d\Omega. \tag{15.85}$$

It is easy to see that the metric elements (15.85) are singular at the central point $r = 0$ and on the sphere of the gravitational radius $r = r_g$. One can show that the last singularity is fictitious, in the sense that it depends on the choice of coordinates.[1]

In other words, one can change coordinates such that the singularity at $r = r_g$ disappears. The spherical surface $r = r_g$ is called the event *horizon*. The physical sense of the horizon is that the observer who is far from the mass M will not be able to see the events that occur at $r < r_g$. In order to check this statement, one has to analyze the behavior of free particles in the space with the Schwarzschild metric. We postpone this consideration in the next section.

Another way to verify the statement of coordinate dependence of singularity at the gravitational radius is to calculate the metric invariant, as we suggest in Exercise 22. After making these calculations, one can see that the square of the Riemann tensor is non-singular in all points except $r = 0$.

Let us note that the smallness of the Newton constant

$$G = 6.674 \times 10^{-8} \frac{cm^3}{g \cdot s^2} \tag{15.86}$$

---

[1]The examples of such coordinates were suggested by Finkelstein [7] and Kruskal [8].

makes the gravitational radius of most cosmic bodies be extremely small compared to their real size. For example, for the Earth it is less than 1 cm, and for the Sun 2.95 km. These numbers mean that the gravitational radius is many orders of magnitude smaller than the real radius of these and most other astrophysical objects. The solution (15.85) does not hold inside the massive body, and therefore the Sun and other elements of the solar system do not form a horizon.

The smallest and densest matter objects known in astrophysics are neutron stars. These stars have a typical radius of the order of 10 km and masses from about 1.4 to 3 times of the solar mass. These numbers show that the neutron star is larger than its gravitational radius, but the two quantities are of the same order of magnitude. The density of a neutron star is huge, exceeding that of atomic nuclei, and in this sense such a star cannot be smaller than it is. The stability of a neutron star or of similar smaller objects called white dwarfs is achieved due to the equilibrium between gravitational force and the nuclear forces which do not let it collapse further. However, if the original star has a mass that exceeds the Chandrasekhar limit for white dwarfs or the Tolman–Oppenheimer–Volkoff limit for a neutron star, the nuclear forces on the surface can not compete with the gravitational force. In this case, the star collapses further and the new object forms a horizon. Such an object can be described by the full solution (15.85) if it has zero angular momentum and electric charge. The solution for the rotating black hole is too complicated to be discussed here; it was found in 1963 by Roy Kerr.

At the moment, there are many qualitatively different kinds of observational confirmations of the existence of the black holes. The observed black holes have masses of very different orders of magnitude. For instance, there are some which resulted from the collapse of large stars. On the other hand, there are giant black holes, which can be observed in the center of many galaxies. These huge objects may be responsible for about 5% of the mass of the galaxy, which is equivalent to 10^{10} or more solar masses. The standard hypothesis is that these huge black holes were formed from the initial perturbations at the epoch when the large-scale structure of the universe (the distribution of mass in space) started to be formed.

Black hole physics is an interesting, extensive, and growing field of research. In the last few decades, this theoretical field has become not only theoretical but more and more experimental and observational. Let us mention a few standard references on general relativity and black hole physics [9–11] and existing excellent books on black hole physics [12–14], where the reader can find much more information about these interesting objects.

Exercise 20 Using dimensional arguments, restore the powers of the speed of light c in the Schwarzschild solution (15.85). Furthermore, derive the first-order Newtonian approximation for the elements of the metric at $r \to \infty$.

Exercise 21 Verify that in the presence of the cosmological constant, the modified form of a spherically symmetric vacuum solution is

$$ds^2 = \left(1 - \frac{2GM}{r} - \frac{\Lambda r^2}{3}\right) dt^2 - \left(1 - \frac{2GM}{r} - \frac{\Lambda r^2}{3}\right)^{-1} dr^2 - r^2 d\Omega. \quad (15.87)$$

Exercise 22 Verify that the original Schwarzschild solution (15.85) transforms the angular part of Einstein equations (15.78) into identities. Use the result of the previous exercise for the analog of this equation with the cosmological constant and show that the solution (15.87) also works well.

Exercise 23 For the Ricci-flat metrics, with $R_{\alpha\beta} \equiv 0$, one can check whether it is a coordinate transformation of the flat Minkowski metric by using the square of the Riemann tensor $R_{\mu\nu\alpha\beta}$ or, equivalently, of the Weyl tensor, $C_{\mu\nu\alpha\beta}$. Derive the components of the Riemann tensor for the metric (15.87). Calculate the contraction (square) $R_{\mu\nu\alpha\beta} R^{\mu\nu\alpha\beta}$ and show that it is nonzero even for $\Lambda = 0$. Verify that in the limit $r \to \infty$ the Weyl tensor is vanishing, even for the nonzero Λ.

Observation. Both parts of the exercise may become technically simpler after learning the forthcoming material in Sect. 17.1. The second part can be achieved without real calculations by using the results of Sect. 17.1 and factorization rules of Sect. 14.5.1.

15.2.1 Appendix: Derivation via Conformal Transformation

Consider the practical application of the D-dimensional conformal transformations introduced in Sect. 14.6. We shall also use the factorization theorem of Sect. 14.5.1. The contents of this appendix are based on the paper [15]. A similar derivation using conformal transformation can be found in the book [16].

Let us present an alternative derivation of Einstein equations for the metric with spherical symmetry. The purpose of this appendix is to use the local conformal transformation described in Sect. 14.6. In general, relativity this method is not the most economic one, but in other theories, it can be much faster.

It is useful to start from the slightly more complicated metric

$$ds^2 = g_{\mu\nu}dx^\mu dx^\nu = e^{\nu(r,t)}dt^2 - e^{\lambda(r,t)}dr^2 - e^{2\Phi(r,t)}d\Omega, \qquad (15.88)$$

where $\Phi(r, t)$ is an arbitrary scalar function. The more general metric (15.88) is widely used in the black hole physics (see, e.g., [14]), in particular because it has some advantages in the theories with vacuum quantum corrections to the general relativity. The particular case (15.71) may always be achieved by replacing $\Phi(r, t) = \log r$ into the corresponding expressions.

Our strategy for deriving Einstein equations for the metric (15.88) will be as follows. The first step is the conformal transformation

$$g_{\mu\nu} = \bar{g}_{\mu\nu} e^{2\Phi}. \qquad (15.89)$$

The new metric $\bar{g}_{\mu\nu}$ is diagonal and factorized

$$\bar{g}_{ab} = \text{diag}\,(\bar{g}_{00}, \bar{g}_{11})\,, \qquad \bar{g}_{ij} = \text{diag}\,(\bar{g}_{22}, \bar{g}_{33})\,,$$

where

$$\bar{g}_{00} = e^A, \quad \bar{g}_{11} = -e^B, \quad \bar{g}_{22} = -1, \quad \bar{g}_{22} = -\sin^2\theta \qquad (15.90)$$

and

$$A = A(r,t) = \nu(r,t) - 2\Phi(r,t), \quad B = B(r,t) = \lambda(r,t) - 2\Phi(r,t). \quad (15.91)$$

The relations between the two Ricci tensors and the two scalar curvatures are given by the particular $4D$ form of Eqs. (14.79) and (14.80),

$$R_{\mu\nu} = \bar{R}_{\mu\nu} - \bar{g}_{\mu\nu}(\bar{\nabla}^2\sigma) + 2(\bar{\nabla}_\mu\sigma)(\bar{\nabla}_\nu\sigma) - 2(\bar{\nabla}_\mu\bar{\nabla}_\nu\sigma) - 2\bar{g}_{\mu\nu}(\bar{\nabla}\sigma)^2,$$
$$R = e^{-2\sigma}\left[\bar{R} - 6(\bar{\nabla}^2\sigma) - 6(\bar{\nabla}\sigma)^2\right]. \qquad (15.92)$$

According to the factorization theorem of Sect. 14.5.1, all components of the curvature tensor with the mixed indices vanish and we need to calculate only the curvature tensors and other quantities that emerge in (15.92) for the two-dimensional metrics in (15.90). For the curvatures we have

$$K_{abcd} = \frac{1}{2} K \left(\bar{g}_{ac}\,\bar{g}_{bd} - \bar{g}_{ad}\,\bar{g}_{bc}\right), \qquad a, b, \ldots = t, r, \qquad (15.93)$$

where K_{abcd} is the Riemann tensor for the metric $\bar{g}_{ab}$, K is the corresponding $2D$ scalar curvature. Furthermore,

$$k_{ijkl} = \frac{1}{2} k \left(\bar{g}_{ik}\,\bar{g}_{jl} - \bar{g}_{il}\,\bar{g}_{jk}\right), \qquad i, j, \ldots = \theta, \varphi, \qquad (15.94)$$

where k is the scalar curvature of the $2D$ sphere, $k = -2$.

For the metric $\bar{g}_{ab}$ we obtain, by direct calculation,

$$\bar{\Gamma}^t_{tt} = \frac{\dot{A}}{2}, \quad \bar{\Gamma}^r_{tt} = \frac{A'}{2} e^{A-B}, \quad \bar{\Gamma}^t_{rt} = \frac{A'}{2}$$
$$\bar{\Gamma}^r_{tr} = \frac{\dot{B}}{2}, \quad \bar{\Gamma}^r_{tt} = \frac{\dot{B}}{2} e^{B-A}, \quad \bar{\Gamma}^r_{rr} = \frac{B'}{2}, \qquad (15.95)$$

where the dot stands for d/dt and the prime for the d/dr. After some small algebra, we obtain

$$K = \frac{1}{2} e^{-A}\left(\dot{A}\dot{B} - 2\ddot{B} - \dot{B}^2\right) + \frac{1}{2} e^{-B}\left(2A'' - A'B' + A'^2\right), \qquad (15.96)$$

and check that $K_{tt} = \frac{K}{2} e^A$ and $K_{rr} = -\frac{K}{2} e^B$ according to (15.93). Furthermore

$$\bar{\nabla}_t \bar{\nabla}_t \Phi = \ddot{\Phi} - \frac{1}{2} \dot{A}\dot{\Phi} - \frac{1}{2} A'\Phi' e^{A-B},$$

$$\bar{\nabla}_r \bar{\nabla}_r \Phi = \Phi'' - \frac{1}{2} B'\Phi' - \frac{1}{2} \dot{B}\dot{\Phi} e^{B-A},$$

$$\bar{\nabla}_t \bar{\nabla}_r \Phi = \dot{\Phi}' - \frac{1}{2} A'\dot{\Phi} - \frac{1}{2} \dot{B}\Phi' \tag{15.97}$$

and

$$\bar{\nabla}\Phi^2 = e^{-A}\dot{\Phi}^2 - e^{-B}\Phi'^2,$$

$$\bar{\nabla}^2\Phi = e^{-A}\left(\ddot{\Phi} - \frac{1}{2}\dot{A}\dot{\Phi} + \frac{1}{2}\dot{B}\dot{\Phi}\right) + e^{-B}\left(\frac{1}{2}B'\Phi' - \frac{1}{2}A'\Phi' - \Phi''\right). \tag{15.98}$$

Now we are in a position to calculate R and $R_{\mu\nu}$. Using the second Eq. (15.92), relation $\bar{R} = K + 2$ and formulas (15.98), we arrive at

$$R = -2e^{-2\Phi} + e^{-\nu}\left[(\dot{\nu} - \dot{\lambda})\left(2\dot{\Phi} + \frac{\dot{\lambda}}{2}\right) - \ddot{\lambda} - 4\ddot{\Phi} - 6\dot{\Phi}^2\right]$$

$$+ e^{-\lambda}\left[\nu'' + \left(2\Phi' + \frac{\nu'}{2}\right)(\nu' - \lambda') + 4\Phi'' + 6\Phi'^2\right]. \tag{15.99}$$

Similarly, using the first Eq. (15.92) we obtain the components of the Ricci tensor and finally of the Einstein tensor. For example,

$$R_{\theta\theta} = -2e^{-2\Phi} + e^{-\nu}\left[\dot{\Phi}\left(\frac{\dot{\nu}}{2} - \frac{\dot{\lambda}}{2} - 2\dot{\Phi}\right) - \ddot{\Phi}\right]$$

$$+ e^{-\lambda}\left[\Phi'' + \Phi'\left(\frac{\nu'}{2} - \frac{\lambda'}{2} + 2\Phi'\right)\right] \tag{15.100}$$

leads to

$$G^\theta_\theta = R^\theta_\theta - \frac{1}{2}R\delta^\theta_\theta = e^{-\nu}\left[\ddot{\Phi} + \ddot{\lambda} + \dot{\Phi}^2 + \frac{\dot{\Phi}(\dot{\lambda} - \dot{\nu})}{2} + \frac{\dot{\lambda}(\dot{\lambda} - \dot{\nu} + 2\dot{\Phi})}{4}\right]$$

$$+ e^{-\lambda}\left[\frac{\Phi'(\lambda' - \nu')}{2} - \Phi'' - \Phi'^2 - \frac{\nu''}{2} + \frac{\nu'(\lambda' - \nu')}{4}\right]. \tag{15.101}$$

In the special case $\Phi = \log r$, (15.101) boils down into

$$G^\theta_\theta = e^{-\nu}\left(\frac{\dot{\lambda}^2}{4} - \frac{\dot{\nu}^2}{4} + \frac{\ddot{\lambda}}{2}\right) + e^{-\lambda}\left(\frac{\nu'\lambda'}{4} - \frac{\nu'^2}{4} - \frac{\nu''}{2} + \frac{\lambda' - \nu'}{r}\right), \tag{15.102}$$

in a perfect fit with the previous result [17].

In a similar way, using (15.92) and (15.97) we obtain

$$G_t^r = g^{rr} G_{tr} = g^{rr} R_{tr} - \frac{1}{2} \delta_t^r R = g^{rr} R_{tr}$$

$$= 2e^{-\lambda} \left(\dot{\Phi}' - \frac{\nu' \dot{\Phi}}{2} - \frac{\dot{\lambda} \Phi'}{2} + \Phi' \dot{\Phi} \right), \tag{15.103}$$

$$G_r^r = g^{rr} R_r r - \frac{1}{2} \delta_r^r R = -e^{-\lambda} R_r^r - \frac{1}{2} R$$

$$= e^{-2\varphi} - e^{-\lambda} \left(\nu' \Phi' + \Phi'^2 \right) + e^{-\nu} \left(2\ddot{\Phi} + 3\dot{\Phi}^2 - \dot{\nu} \dot{\Phi} \right), \tag{15.104}$$

$$G_t^t = g^{tt} R_t t - \frac{1}{2} R$$

$$= e^{-2\Phi} + e^{-\nu} \left(\dot{\lambda} \dot{\Phi} + \dot{\Phi}^2 \right) + e^{-\lambda} \left[-2\Phi'' + \lambda' \Phi' - 3\Phi'^2 \right]. \tag{15.105}$$

In the special case $\Phi = \log r$, these expressions become the standard ones, which were obtained in the previous section, namely (15.75)–(15.77). If the derivation is performed directly for the simplest metric (15.71), it is technically simple and not much more difficult than the standard one.

15.3 Motion of Particles in the Schwarzschild Background

Let us consider the motion of massive and massless particles in the gravitational field of the Schwarzschild solution (15.85). The historical importance of these calculations is partially due to the fact that the first tests of general relativity were related to the precession of the perihelion of Mercury and to the prediction of bending of light by the massive bodies, such as Sun. In fact, the Sun is a rotating body and it is not exactly point-like. However, the detailed calculations that take these aspects into account show that the effects of size and rotation for the Sun and many other space objects are small. As a result, the effects of size, rotation, and relativity can be calculated independently, since the double effects would be of the next order of smallness. Thus, we can safely use the approximation of a nonrotating point-like star for the evaluation of relativistic effects in the solar system.

15.3.1 Precession of Perihelion of Planets

In order to make things easier, let us start with the Kepler problem in classical relativistic mechanics and analyze the motion of a test particle with the mass m and conserved energy E in the gravitational field with the potential

$$U(r) = -\frac{\alpha}{r}, \tag{15.106}$$

Fig. 15.1 Effective gravitational potential in the Newtonian case, Eq. (15.108). The point of the minimal is $r_0 = \frac{L^2}{m\alpha}$ and the minimal possible energy of the particle is $U_{\text{eff}}(r_0) = -\frac{m\alpha^2}{2L^2}$, which corresponds to the circular motion

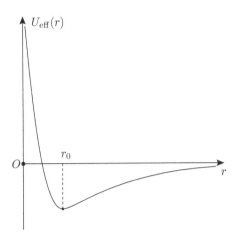

where $\alpha = GM$ and r is the radius of the spherical coordinate system. Elementary analysis shows that the force $\mathbf{F} = -\operatorname{grad} U = -\alpha\hat{\mathbf{r}} r^{-2}$ and the torque $\mathbf{N} = \mathbf{r} \times \mathbf{F} = 0$, such that the angular momentum is conserved, $\mathbf{L} = \mathbf{r} \times \mathbf{p} = const$. This conservation law has the following relevant implications:

(i) The motion is flat, and hence it can be described by $2D$ polar coordinates r and φ.

(ii) The absolute value of the angular momentum is constant, and therefore

$$mr^2\dot{\varphi} = L = const. \tag{15.107}$$

This is equivalent to the Kepler's second law.

(iii) The energy conservation can be formulated as

$$E = \frac{mv^2}{2} + U(r) = \frac{m\dot{r}^2}{2} + \frac{mr^2\dot{\varphi}^2}{2} + U(r) = \frac{m\dot{r}^2}{2} + \frac{mr^2\dot{\varphi}^2}{2} + U(r)$$

$$= \frac{m\dot{r}^2}{2} + U_{\text{eff}}(r), \qquad U_{\text{eff}}(r) = \frac{L^2}{2mr^2} - \frac{\alpha}{r}, \tag{15.108}$$

where at the last step we used conservation of angular momentum (15.107). Let us note that the expression $\frac{L^2}{2mr^2}$ in the effective potential (15.108) is the potential for the force of inertia, which appears in the rotating frame [18]. At this moment the original $3D$ motion was reduced to the $1D$ motion with the effective potential, which is not difficult to analyze (Fig. 15.1).

Obviously, the motions that are compatible with the positivity of the radial part of kinetic energy, $m\dot{r}^2/2$, are those for which $E \geq U_{\text{eff}}(r)$, and the ones that correspond to the finite motions, satisfying the condition $E < 0$. This is the case that we need to analyze for dealing with the precession of the perihelion of Mercury and other planets of the solar system.

As far as we are mainly interested in the trajectory $r(\varphi)$, it makes sense to change the variables according to

$$u(\varphi) = \frac{1}{r} \quad \text{and then} \quad \frac{du}{dt} = \dot{\varphi} \frac{du}{d\varphi} = \frac{Lu^2}{m} u', \quad \text{where} \quad u' = \frac{du}{d\varphi}. \quad (15.109)$$

Then Eq. (15.108) becomes

$$E = \frac{L^2}{2m} \left(u'^2 + u^2 \right) - \alpha u. \quad (15.110)$$

This is the first integral of the oscillator equation with the constant source term,

$$u'' + u = \frac{m\alpha}{L^2}. \quad (15.111)$$

The solution of (15.110) has the form

$$pu = \frac{p}{r} = \frac{p}{e} \cos(\varphi - \varphi_0), \quad p = \frac{L^2}{m\alpha}, \quad e = \sqrt{1 + \frac{2EL^2}{m\alpha^2}}. \quad (15.112)$$

This is the equation of the ellipse in polar coordinates, and hence we got the first of Kepler's laws. The expression (15.112) is a periodic function with the period 2π, which means a closed trajectory without precession of the perihelion of the ellipse.

Exercise 24 Probably, the reader already did at least part of the exercises listed before. But if not, it is worthwhile to do the following: (i) Derive large a and small b semi-axes of the ellipse (15.112) and using (15.107) verify the Kepler's third law, namely, the relation between the time period T of resolution and a,

$$\left(\frac{T}{2\pi} \right)^2 = \frac{m}{\alpha} a^3. \quad (15.113)$$

(ii) Analyze the case of $E = 0$ and $E > 0$, when the eccentricity of the conic session is $e = 1$ and $e > 1$. Show that in the first case the trajectory is a parabola and in the latter it is a hyperbola. For $e > 1$, derive the angle of deviation of the asymptotic motion as a function of the impact parameter. (iii) Repeat the previous problem for the central potential (15.106) with $\alpha < 0$, which corresponds to the repulsion force. (iv) Discover the way of deriving the hyperbolic solution for the cases described in the previous two points without explicit calculations, and instead just using the solution (15.112) for the elliptic curve and mapping to other coordinates and/or parameters α, E, L describing the motion.

Observation. The last part (iv) is not simple to do.

Let us consider the relativistic version by using the Newtonian calculation as a reference. In this section, we use units with $c = 1$. The covariant equation for the particle is (14.88)

$$\frac{d^2 x^\mu}{ds^2} + \Gamma^\mu_{\alpha\beta} \frac{dx^\alpha}{ds} \frac{dx^\beta}{ds} = 0 .$$

The nonzero components of affine connection can be obtained by replacing (15.85) into (15.73), so we arrive at

$$\Gamma^t_{tt} = \Gamma^t_{rr} = \Gamma^r_{tr} = 0 , \qquad \Gamma^r_{rr} = \frac{MG}{r^2} , \qquad \Gamma^t_{rt} = -\frac{MG}{r^2} ,$$

$$\Gamma^r_{tt} = -\frac{MG}{r^2}\left(1 - \frac{2MG}{r}\right) , \quad \Gamma^r_{\theta\theta} = -r\left(1 - \frac{2MG}{r}\right) , \quad \Gamma^r_{\psi\psi} = \Gamma^r_{\theta\theta} \sin^2\theta ,$$

$$\Gamma^\theta_{\psi\psi} = -\sin\theta\cos\theta , \qquad \Gamma^\psi_{\theta\psi} = \cot\theta , \qquad \Gamma^\theta_{r\theta} = \Gamma^\psi_{r\psi} = \frac{1}{r} . \qquad (15.114)$$

Equation (14.88) become as follows:

$$\frac{d^2 t}{ds^2} + v' \frac{dr}{ds}\frac{dt}{ds} = 0 ,$$

$$\frac{d^2 r}{ds^2} + \frac{\lambda'}{2}\left(\frac{dr}{ds}\right)^2 + \frac{v'}{2} e^{v-\lambda}\left(\frac{dt}{ds}\right)^2 - re^{-\lambda}\left(\frac{d\theta}{ds}\right)^2 - re^{-\lambda}\sin^2\theta\left(\frac{d\varphi}{ds}\right)^2 = 0 ,$$

$$\frac{d^2\theta}{ds^2} + \frac{2}{r}\frac{d\theta}{ds}\frac{dr}{ds} - \sin\theta\cos\theta\left(\frac{d\varphi}{ds}\right)^2 = 0 ,$$

$$\frac{d^2\varphi}{ds^2} + \frac{2}{r}\frac{d\varphi}{ds}\frac{dr}{ds} + 2\cot\theta \frac{d\varphi}{ds}\frac{d\theta}{ds} = 0 . \qquad (15.115)$$

Equation (15.115) have full information about the motion of particles in the spherically symmetric Schwarzschild solution of Einstein equations. Due to spherical symmetry, one can expect to reproduce some features of the classical Kepler problem. And in the almost nonrelativistic case the problem can be regarded as a small perturbation over the Kepler solution.

The first observation about Eq. (15.115) is that they can be simplified considerably by a smart choice of coordinates. For instance, choosing the spherical coordinates such that at the initial moment $\theta = \frac{\pi}{2}$, it is easy to check that this relation will hold in the future, since $\frac{d\theta}{ds} = 0$ is a solution. Then the equations simplify to

$$\frac{d^2 t}{ds^2} + v' \frac{dr}{ds}\frac{dt}{ds} = 0 , \qquad (15.116)$$

$$\frac{d^2 r}{ds^2} + \frac{\lambda'}{2}\left(\frac{dr}{ds}\right)^2 + \frac{v'}{2} e^{v-\lambda}\left(\frac{dt}{ds}\right)^2 - re^{-\lambda}\left(\frac{d\theta}{ds}\right)^2 - re^{-\lambda}\left(\frac{d\varphi}{ds}\right)^2 = 0 , \qquad (15.117)$$

$$\frac{d^2\varphi}{ds^2} + \frac{2}{r}\frac{d\varphi}{ds}\frac{dr}{ds} = 0 . \qquad (15.118)$$

The Eqs. (15.116) and (15.118) can be easily integrated by introducing

$$\Phi = \frac{d\varphi}{ds}, \qquad T = \frac{dt}{ds}. \tag{15.119}$$

In terms of these new variables the equations become

$$\frac{dT}{T} = -dv(r), \qquad \frac{d\Phi}{\Phi} = -\frac{2dr}{r}, \tag{15.120}$$

with the solutions of the form

$$\frac{dt}{ds} = T = k\,e^{-v(r)}, \tag{15.121}$$

$$\frac{d\varphi}{ds} = \Phi = \frac{h}{r^2}, \tag{15.122}$$

where we assume $r = r(s)$. It is easy to see that the last equation is the relativistic version of the conservation of angular momentum in the classical case, Eq. (15.107), where h is a relativistic analog of L/m in classical mechanics.

Further consideration can be done on the basis of Eq. (15.117), but it is much simpler to use the equivalent first-order equation, which comes directly from the expression for the interval

$$g_{\mu\nu}\frac{dx^\mu}{ds}\frac{dx^\nu}{ds} = 1 \implies e^v\left(\frac{dt}{ds}\right)^2 - e^\lambda\left(\frac{dr}{ds}\right)^2 - r^2\left(\frac{d\varphi}{ds}\right)^2 = 1, \tag{15.123}$$

where we used $\theta \equiv \pi/2$.

Exercise 25 Derive Eq. (15.117) from (15.123).

Replacing (15.121) into (15.123) and using the Schwarzschild solution

$$e^{-\lambda} = e^v = 1 - \frac{2GM}{r}, \tag{15.124}$$

after some small algebra one can easily obtain the equation

$$\left(\frac{dr}{ds}\right)^2 + r^2\left(\frac{d\varphi}{ds}\right)^2 - \frac{2GM}{r}\left[1 + r^2\left(\frac{d\varphi}{ds}\right)^2\right] = k^2 - 1. \tag{15.125}$$

After using Eq. (15.122), we arrive at the relativistic generalization of Eq. (15.108),

$$\left(\frac{dr}{ds}\right)^2 + \frac{h^2}{r^2} - \frac{2GM}{r}\left(1 + \frac{h^2}{r^2}\right) = k^2 - 1. \tag{15.126}$$

After that we can proceed exactly like in the classical case, looking for the dependence $r(\varphi)$,

$$\frac{dr}{ds} = \frac{dr}{d\varphi}\frac{d\varphi}{ds} = \frac{h}{r^2}\frac{dr}{d\varphi}\,mr^2\dot\varphi = L = const \tag{15.127}$$

and using the variable $u = r^{-1}$. In this way we arrive at the equation

$$h^2\left(\frac{du}{d\varphi}\right)^2 + h^2u^2 - 2GMu\left(1 + h^2u^2\right) = k^2 - 1.   \tag{15.128}$$

Taking the derivative with respect to φ, we arrive at the relativistic generalization of Eq. (15.111),

$$u'' + u = \frac{GM}{h^2} + 3GMu^2.   \tag{15.129}$$

It is easy to see that only the last term in the *r.h.s.* is new compared to the classical analog.

Indeed, (15.129) is the general equation for the motion of a relativistic particle in the central gravitational field of the Schwarzschild metric. It is good to remember that the motion of microscopic particles, such as photons, electrons, protons, or gravitons, can be described by slightly different equations, because these particles possess spin or helicity. Also, the macroscopic astrophysical objects are not exactly spherically symmetric, for example, they may be rotating. However, this equation represents good approximation in many cases, in particular, enabling one to describe the precession of the perihelion of the planets and the light bending by the Sun. Let us consider these two cases.

The motion of the planets is almost classical, and it is not difficult to verify that the second term in the *r.h.s.* of Eq. (15.129) is numerically many orders smaller than the first one. Therefore, let us consider the second term as a small perturbation.

It is useful to make a change of variable and write a dimensionless version of (15.129),

$$u = \frac{GM}{h^2}\, z(\varphi),   \qquad z'' + z = 1 + \gamma z^2,   \tag{15.130}$$

where $\gamma = 3(GM/h)^2$ is a small parameter that can be used for deriving the solution in the form of the perturbation series,

$$z = z_0 + \gamma z_1 + \mathcal{O}(\gamma^2).   \tag{15.131}$$

We shall need only zero and first terms of the series, which can be obtained by placing (15.131) into (15.130). The zero-order term gives exactly the classical mechanics result (15.112),

$$z_0 = 1 + e\cos(\varphi - \varphi_0).   \tag{15.132}$$

One can always make $\varphi_0 = 0$ by properly choosing the coordinate system and me assume this choice in what follows.

With the $\mathcal{O}(\gamma)$ part, one can proceed in two different ways. It is instructive to show them both.

First version. Equation (15.132) is the solution of the harmonic oscillator equation (15.130) with $\gamma = 0$. Let us assume that the presence of nonzero γ will change the position of the minima of the potential and the frequency, but the motion can still be regarded as a harmonic oscillator. The position of the minima of the potential is defined by the equation

$$\frac{dV}{dz} = z - 1 - \gamma z^2 = 0. \tag{15.133}$$

For $\gamma = 0$ the minima is at $z = 1$ and the frequency is

$$\omega_0^2 = \frac{d^2 V}{dz^2} = 1, \tag{15.134}$$

such that the period of oscillations is $\Phi_0 = 2\pi/\omega_0 = 2\pi$. This result exactly corresponds to the basic solution (15.132). In the presence of nonzero γ the point of minima becomes $z = 1 + \gamma$. The new frequency is defined by

$$\omega^2 = \frac{d^2 V}{dz^2} = 1 - 2\gamma \quad \Longrightarrow \quad \omega = \frac{d^2 V}{dz^2} = 1 - \gamma. \tag{15.135}$$

Then the angular period of the oscillations becomes

$$\Phi = \frac{2\pi}{\omega} = 2\pi(1 + \gamma) = 2\pi\left(1 + \frac{3GM}{h^2}\right). \tag{15.136}$$

Second version. The equation for $z_1(\varphi)$ in (15.131) has the form

$$z_1'' + z_1 = z_0^2 = \left(1 + e\cos\varphi\right)^2. \tag{15.137}$$

The solution of this linear nonhomogeneous equation can be easily found. The sum $z_0 + \gamma z_1$ has the form

$$z = 1 + e\cos\varphi + \gamma\left(1 + \frac{e^2}{2} + e\varphi\sin\varphi + \frac{e^2}{6}\cos 2\varphi\right). \tag{15.138}$$

It is clear that the constant term in the parentheses only produces a small change of parameter p of the ellipse. The $\cos 2\varphi$ does not contradict the periodicity with $\Phi_0 = 2\pi$ and therefore the unique relevant term is the one with $\gamma e\varphi\sin\varphi$. In order to understand why this term changes the frequency, one has to remember that all the consideration presented here is valid only in the first order in γ, when

$$\gamma\varphi = \sin\gamma\varphi \quad \text{and} \quad \cos\gamma\varphi = 1.$$

Then

$$e\big(\cos\varphi + \gamma\varphi\sin\varphi\big) = e\cos\big(\varphi - \gamma\varphi\big), \qquad (15.139)$$

which gives exactly the modified frequency (15.135). The Eq. (15.136) means that for the full cycle of radial motion, the angular period is slightly larger than 2π. Therefore, the trajectory is not closed as it is in classical mechanics, indicating a slow precession of the perihelion of the elliptic orbit.

Exercise 26 Using dimensional arguments, restore the powers of c in terms of expression (15.129). Evaluate the numerical values of the potential terms for the planets of the solar system, e.g., for Mercury.

Exercise 27 Consider the motion of a nonrelativistic particle with positive energy $E > 0$ in the central field. First, verify the Classical Mechanics result that the trajectory in the case of potential (15.106) is a hyperbola, and derive the deviation angle as a function of impact parameter. Second, calculate the generalization for the corrected potential that corresponds to Eq. (15.129), and derive the contribution of a small relativistic term to the deviation angle.

15.3.2   Light Bending

In order to consider the simplest approach for the gravitational bending of light, let us come back to the Eq. (15.126),

$$u'' + u = \frac{GM}{h^2} + 3GMu^2$$

and see what changes if we assume the motion with the speed of light c. The quantity h is the relativistic analog of the angular momentum of the particle, such that

$$\frac{1}{h} = u^2\frac{ds}{d\varphi}. \qquad (15.140)$$

For the light $ds = 0$, and therefore $h^{-2} = 0$. This means that the equation of motion becomes much simpler,

$$u'' + u = 3GMu^2. \qquad (15.141)$$

The term in the r.h.s. is the presumably small relativistic correction, and hence it is instructive to start from solving the simpler equation without the term in the r.h.s.. This gives

$$u_0'' + u_0 = 0 \qquad \Longrightarrow \qquad u_0(\varphi) = R^{-1}\sin(\varphi - \varphi_0), \qquad (15.142)$$

which can be written in the conventional form

$$R = r \sin(\varphi - \varphi_0).$$ (15.143)

This is the equation of the straight line, and taking the constant $\varphi_0 = 0$ we arrive at the line which is parallel to the OX-axis, $y = R$. In what follows we assume this orientation of the axes for the zero-order solution. Indeed, the straight line is a natural output in the zero-order approximation , as it shows that in classical mechanics the speed of light is effectively infinite and that the light rays cannot be affected by gravity.

Let us now consider the motion in a weak field of a point-like mass source and regards the r.h.s. of Eq. (15.141) as a small perturbation. Taking $u = u_0 + \Delta u$, in the first order in the perturbation we arrive at the linear equation

$$\Delta u'' + \Delta u = 3MG \, u_0^2(\varphi) = \frac{3MG}{2R^2} (1 - \cos 2\varphi).$$ (15.144)

The solution of this equation is

$$\Delta u = \frac{MG}{R^2} (1 + \cos^2 \varphi)$$ (15.145)

and for the sum we have the first-order solution

$$u = u_0 + \Delta u = \frac{\sin \varphi}{R} + \frac{MG}{R^2} (1 + C \cos \varphi + \cos^2 \varphi).$$ (15.146)

The next task is to solve the last equation in the asymptotic regime, when $r \to \infty$, or $u \to 0$. We assume that the angles slightly differ from the straight line, $\varphi = 0 + \Delta\varphi_1$ and $\varphi = \pi - \Delta\varphi_2$, where positive deviations $\Delta\varphi_{1,2}$ correspond to the attractive action of gravity on a massless particle. We assume that the deviations satisfy the conditions $|\Delta\varphi_{1,2}| \ll 1$. The smallness of the two deviations enable us to consider

$$\sin \Delta\varphi_1 = \Delta\varphi_1, \quad \sin \Delta\varphi_2 = \Delta\varphi_2 \quad \text{and} \quad \sin(\pi - \Delta\varphi_2) = -\Delta\varphi_2. \quad (15.147)$$

Therefore, we arrive at the two equations,

$$\frac{\Delta\varphi_1}{R} + \frac{GM}{R^2} (2 + C) = 0,$$
$$\frac{\Delta\varphi_2}{R} + \frac{GM}{R^2} (2 - C) = 0.$$ (15.148)

Summing up these two solutions, we arrive at the full deviation angle,

$$\Delta\varphi = \Delta\varphi_1 + \Delta\varphi_2 = \frac{4GM}{R}. \tag{15.149}$$

For those stars situated such that the line between them and an observer on Earth passes close to the surface of the Sun, we get the well-known result $\Delta\varphi = 1.75''$, which was historically the first experimental confirmation of general relativity. The discovery occurred in 1919 by the group led by A. Eddington and F. W. Dyson, during a total solar eclipse.

The attraction and bending of light by gravity was originally predicted in the framework of Newtonian gravity, starting from Newton himself. However, as we know from the Chap. 2, the propagation of light cannot be consistently described within Classical Mechanics, and hence the discovery of such bending by Eddington et al. was a real confirmation of general relativity. Nowadays, there is an incredible amount of the new experimental data on gravitational lensing, which is a general version of the effect we have considered here. The effect is important in modern cosmology, where it is regarded as one of the main sources of observational information about the universe.

15.3.3 Red Shift of Light and Radial Motion of Particles

The radial motion of particles in the spherically symmetric background is important from various viewpoints. For instance, the Schwarzschild metric is a very good approximation for describing the gravitational field around many astrophysical objects, such as stars and planets. And in most cases massless particles (photons) and massive particles come to the observer on Earth through an approximately straight line that goes from the star to observer. Therefore, the radial motion is an important example of general motion. On the other hand, radial motion is relatively simple and its consideration enables one to observe many general features of the motion in a static, spherically symmetric metric in the most clear way.

We know that in Classical Mechanics, the particle that goes to the point with greater potential energy is losing its kinetic energy. It is interesting that for the motion in a Schwarzschild background there is a similar effect even for photons. In what follows, we shall partially be guided by an excellent book (it can be called encyclopedia of black holes) [12], but will use only a small piece in the corresponding chapter of this book.

Consider once again the metric (15.85), but this time with restored powers of the speed of light c,

$$ds^2 = \left(1 - \frac{2GM}{c^2 r}\right)c^2 dt^2 - \left(1 - \frac{2GM}{c^2 r}\right)^{-1} dr^2 - r^2 d\Omega, \tag{15.150}$$

$$d\Omega = d\theta^2 + \sin^2\theta^2 d\varphi^2.$$

If we consider the radial motion of a photon, then $ds = 0$ and also $d\theta = d\varphi = 0$, hence we arrive at the derivative

$$\frac{dr}{dt} = \pm\left(1 - \frac{2GM}{c^2 r}\right)c < c. \tag{15.151}$$

The reason for this strange result is that we used the wrong variables to take the derivative. Let us consider this case in some detail. For the sake of simplicity, we ignore the sign flip at the horizon and consider only the outside region with $r > r_g = \frac{2GM}{c^2}$. The reader can explore the situation in the region inside the horizon as an exercise, or just go to specialized books [12, 14].

The time t is the physical time at the space infinite $r \to \infty$, while the "correct" time variable at the point with the radial coordinate r is the proper time τ, defined by

$$d\tau = \sqrt{g_{00}(r)}dt = \sqrt{1 - \frac{2GM}{c^2 r}}\, dt. \tag{15.152}$$

The two time variables coincide very far from the central point $r = 0$, while in the vicinity of the mass they are different. In the same way, the "correct" radial coordinate at the same point is x, which is defined by the relation

$$dx = \sqrt{-g_{11}(r)}dr = \frac{dr}{\sqrt{1 - \frac{2GM}{c^2 r}}}. \tag{15.153}$$

The reason why we called the coordinates τ, x "correct" is that those are exactly the coordinates where the metric is locally Minkowski. It is easy to see that in these coordinates we have the correct value of the derivative $\frac{dx}{d\tau} = c$.

Consider the emitter of the electromagnetic wave with the frequency ν_1 at the point $r = r_1 \geq r_g$. This wave is received by the detector at the point $r = r_2 > r_1$. The question is what is the frequency of the same light ν_2 measured at this point. In fact, the two points do not need to be at the same radial line, so there may be two space points $(r_1, \theta_1, \varphi_1)$ and $(r_2, \theta_2, \varphi_2)$. At the same time, we require that both emitter and detector are in rest, just to separate the effect of the gravitational field intensity from the Doppler effect due to the velocity of the points. The time intervals corresponding to the two frequencies satisfy the relations

$$\frac{\nu_2}{\nu_1} = \frac{\Delta\tau_1}{\Delta\tau_2} \tag{15.154}$$

and at the same time there will be the relation with the time variable that is universal since it is defined at the infinity,

$$\Delta\tau_1 = \sqrt{g_{00}(r_1)}\,\Delta t \quad \text{and} \quad \Delta\tau_2 = \sqrt{g_{00}(r_2)}\,\Delta t. \tag{15.155}$$

The last two relations do not depend on the angular variables θ_1, φ_1 and θ_2, φ_2. Since the metric is static, the expressions depend only on r through the value of $\sqrt{g_{00}}$ at the two given points. In the case of the Schwarzschild Metric, this value characterizes the intensity of the gravitational field and defines the scaling of $\Delta \tau$ with respect to Δt. Another interesting feature of the relations (15.155) is that they are valid only outside the horizon, for $r > r_g = \frac{2GM}{c^2}$. Inside the horizon, the two first terms of the r.h.s. of (15.151) change their signs, and one can say that r plays the role of time and ct plays the role of a space coordinate.[2] For this reason, in what follows we will not discuss the interior of the Schwarzschild solution and concentrate instead on the discussion of the region $r > r_g$.

By combining Eqs. (15.154) and (15.155), we arrive at the general formula for the red shift of the frequency in the static gravitational field

$$\frac{\nu_2}{\nu_1} = \frac{\sqrt{g_{00}(r_1)}}{\sqrt{g_{00}(r_2)}}. \tag{15.156}$$

Let us give the most standard example of the use of the formula (15.156). Consider the emitter at the surface of the Sun and the detector on Earth. In both cases, the gravitational field is relatively weak and one can use the formula (14.97). For the first point

$$g_{00}(R_\odot) = 1 - \frac{2GM_\odot}{c^2 R_\odot}. \tag{15.157}$$

while for the gravitational field on the Earth we can safely approximate $g_{00}(R_\oplus) = 1$. Replacing Eq. (15.157) in Eq. (15.156), it is easy to get the result for the red shift of the spectral lines of the light coming from the Sun,

$$\frac{\Delta \nu}{\nu_\odot} = \frac{\nu_\oplus - \nu_\odot}{\nu_\odot} = -2.12 \times 10^{-6}. \tag{15.158}$$

This formula shows that all spectral lines of light which coming from the Sun are uniformly red shifted compared to the lines emitted by the same types of atoms in a laboratory. This and other similar results successfully passed numerous experimental verifications.

The general result for the redshift (15.156) can be also interpreted in a different way. We know that in Quantum Mechanics the frequency of the quanta of light (photon) is related to its energy $\mathcal{E}$ by the relation $\mathcal{E} = \hbar \nu$, where $\hbar$ is the Planck constant. It is easy to see that the formula (15.156) means that the photon goes from region 1 with a stronger gravitational field to region 2 with a weaker gravitational field. One can say that the photon is losing energy qualitatively similar to Classical

---

[2]Author is especially grateful to V. P. Frolov for explaining this point.

Mechanics, where a particle loses kinetic energy when it moves to the place with a greater potential energy.

From the considerations above follows that (15.156) is a general formula, which does not depend on the metric being spherically symmetric. The proof can be easily extended at least to an arbitrary manifold, which admits the existence of the universal time variable t.

There is one more interesting consequence of the formula (15.156). It is easy to note that for the light emitted from the spherical surface of the horizon $r_1 = r_g$ and received at the greater distance from the center, at $r_2 > r_g$, the frequency $\nu_2 = 0$ independent of the value of initial frequency ν_1. This means that the light emitted on the horizon or inside the horizon cannot arrive at a point lying outside the horizon. For this reason, the object with a horizon, such as the one described by (15.151), can be called a *black hole*. Indeed, the real astrophysical black holes are not invisible; they can be (and many of them are, indeed) detected by the emission of light from charged particles, which are falling to the horizon. Sometimes such energy emission can be extremely intensive.

Exercise 28 Verify the numerical estimate in (15.158).

15.4 Gravitational Waves

The Classical Newtonian gravity is static, and therefore gravitational waves are impossible. Things are different in general relativity, where the metric satisfies dynamical equations and therefore gravitational waves should be expected. In what follows we present the standard derivation of the propagation of gravitational waves, briefly describe their emission, and explain why the detection of gravitational waves is such a difficult task. Since these waves are one of the main predictions of relativistic gravity, the experimental efforts on their detection always attracted a great deal of attention. Finally, quite recently the LIGO collaboration reported on the first two successful events [19], which certainly represents very important evidence of the progress of technology and also in the methods of data treatment. The Nobel prize in Physics for 2017 was awarded for this discovery, and it is expected that the amount of data will grow immensely in the next years, such that gravitational wave astronomy supplements conventional methods for exploring the universe.

The theoretical study of gravitational waves can be divided into two classes. The first one is more complicated and corresponds to the waves on the background of some physically relevant solutions, such as the Schwarzschild, Kerr, or cosmological solutions. The second is much simpler and describes the propagation of gravitational waves on a flat background. Here we consider only the solutions of this class.

Consider the metric (14.90), that is $g_{\mu\nu} = \eta_{\mu\nu} + h_{\mu\nu}$ with $|h_{\mu\nu}| \ll 1$. Our first goal is to construct the linearized version of Einstein equations (15.60) with $\Lambda = 0$. Indeed, the given approximation is supposed to work well for the weak gravitational fields, such as the ones in the LIGO experiments. In this case, the cosmological constant is not going to be relevant and we can set it to zero.

As we already know, the first-order expansion of the affine connection is given by (14.91). After a small calculation, we arrive at the linearized expression for the Ricci tensor,

$$R^{(1)}_{\mu\nu} = \frac{1}{2} \left(\partial_\lambda \partial_\nu h^\lambda_\mu + \partial_\lambda \partial_\mu h^\lambda_\nu - \Box h_{\mu\nu} - \partial_\mu \partial_\nu h \right), \qquad (15.159)$$

where $\Box = \eta^{\mu\nu} \partial_\mu \partial_\nu$, $h = h_{\mu\nu} g^{\mu\nu}$ and all indices are raised and lowered with the flat Minkowski metric. The small perturbations of the metric are described by the equations

$$R^{(1)}_{\mu\nu} = 8\pi G \, S_{\mu\nu} = 8\pi G \left(T_{\mu\nu} - \frac{1}{2} T^\lambda_\lambda g_{\mu\nu} \right), \qquad (15.160)$$

The energy–momentum tensor of matter $T_{\mu\nu}$ can be taken in the lowest-order approximation, that is for the flat metric, and hence satisfies the conservation law $\partial_\mu T^\mu_\nu = 0$.

Exercise 29 Show that the tensor $S_{\mu\nu}$ defined in (15.160) satisfies the identity $\partial_\mu S^\mu_\nu - \frac{1}{2} \partial_\nu S^\lambda_\lambda = 0$, and that the linearized tensor $R^{(1)}_{\mu\nu}$ also satisfies the same condition. The last problem can be solved directly using (15.159) or in the non-perturbative way by means of the third Bianchi identity (14.61).

Equation (15.160) with the nonzero source describes the emission of gravitational waves. The *l.h.s.* of this equation is degenerate because of the gauge invariance, corresponding to the transformation (15.6). In the linearized case, this boils down to

$$h_{\alpha\beta} \rightarrow h'_{\alpha\beta} = h_{\alpha\beta} - \partial_\alpha \xi_\beta - \partial_\beta \xi_\alpha. \qquad (15.161)$$

The gauge symmetry leads to the degeneracy of the main Eq. (15.160), making its solution impossible. As usual, one has to implement the gauge fixing condition. The most useful for us is the *harmonic* of Fock–DeDonder condition,

$$\partial_\alpha h^\alpha_\beta - \frac{1}{2} \partial_\beta h = 0. \qquad (15.162)$$

In order to understand how the relation (15.162) works, let us assume that $h_{\mu\nu}$ satisfies this gauge fixing condition. It is easy to check that under (15.161) Eq. (15.162) transforms as

$$\partial_\alpha h'^\alpha_\beta - \frac{1}{2} \partial_\beta h' = -\Box \xi_\beta. \qquad (15.163)$$

Therefore, the condition (15.162) removes the ambiguity if the gauge parameter ξ_β does not satisfy the equation $\Box \xi_\beta = 0$. The symmetry under the transformation with $\Box \xi_\beta = 0$ remains, and we will see soon what the role is played by this remnant symmetry in the analysis of the propagation of gravitational waves.

In order to explore the propagation of the gravitational waves, consider the same Eq. (15.160) with zero source, $S_{\mu\nu} = 0$. Let us analyze the equation

$$\partial_\lambda \partial_\nu h^\lambda_\mu + \partial_\lambda \partial_\mu h^\lambda_\nu - \Box h_{\mu\nu} - \partial_\mu \partial_\nu h = 0 \qquad (15.164)$$

and see how the gauge transformations (15.161) affect the number of physical degrees of freedom.

The propagation of gravitational waves can be considered in different ways (see, e.g., [20] for a complete introduction). We shall partially follow the classical book [3]. Since (15.164) is a linear homogeneous differential equation, it is useful to make a Fourier transformation of $h_{\mu\nu}$ and consider the individual mode with the momentum k^α,

$$h_{\mu\nu} = e_{\mu\nu}\, e^{ik_\lambda x^\lambda} + e^*_{\mu\nu}\, e^{-ik_\lambda x^\lambda}\,. \qquad (15.165)$$

In what follows, k^λ is regarded as a constant four-vector in flat space. Here $e_{\mu\nu}(k) = e_{\nu\mu}(k)$ is the constant polarization tensor, and the four-momentum of the mode satisfies dispersion relations for the massless fields, $k_\lambda k^\lambda = 0$. The harmonic gauge condition (15.162) provides one more relation,

$$e_{\mu\nu}\, k^\mu = \frac{1}{2}\, e^\mu_\mu\, k_\nu\,. \qquad (15.166)$$

The parameters of the gauge transformations (15.161) can be Fourier transformed too. The mode corresponding to the given four-momentum is

$$\xi_\mu = i\epsilon_\mu\, e^{ik_\lambda x^\lambda} - i\epsilon^*_\mu\, e^{-ik_\lambda x^\lambda}\,. \qquad (15.167)$$

Direct replacement shows that under the gauge transformation (15.161), the polarization operator transforms as

$$e_{\mu\nu} \rightarrow e'_{\mu\nu} = e_{\mu\nu} + k_\mu \epsilon_\nu + k_\nu \epsilon_\mu\,. \qquad (15.168)$$

As we saw in Eq. (15.163), the condition (15.161) does not fix the gauge arbitrariness completely if the gauge parameter satisfies $\Box \xi^\mu = 0$, which follows automatically from (15.167) and $k^\lambda k_\lambda = 0$. Therefore, the relation (15.168) must be taken into account even after the gauge fixing condition (15.162) is imposed. This means that the polarization tensor has less physical components, because the transformation (15.168) does not change the physical contents of the wave.

Let us make the last statement quantitative. Without losing the generality, we can assume that the constant wave vector has only time and Z components,

$$k^\lambda = k,\, 0,\, 0,\, k\,. \qquad (15.169)$$

In this case, the gauge fixing condition (15.162) produces the following relations:

$$e_{31} + e_{01} = e_{32} + e_{02} = 0,$$
$$e_{33} + e_{30} = -e_{00} - e_{30} = \frac{1}{2}(e_{11} + e_{22} + e_{33} - e_{00}), \qquad (15.170)$$

which can be solved in the form

$$e_{01} = -e_{31}, \quad e_{02} = -e_{32}, \quad e_{03} = -\frac{1}{2}(e_{33} + e_{00}). \qquad (15.171)$$

As expected, there remains only six degrees of freedom of the initial ten. Let us now apply the remnant gauge transformation (15.168), making the following change in the polarization tensor:

$$e'_{11} = e_{11}, \quad e'_{12} = e_{12}, \quad e'_{13} = e_{13} + k\epsilon_1,$$
$$e'_{23} = e_{23} + k\epsilon_2, \quad e'_{33} = e_{33} + 2k\epsilon_3, \quad e'_{00} = e_{00} - 2k\epsilon_0. \qquad (15.172)$$

At this moment, we observe that all components of $e_{\mu\nu}$ except the two, e_{11} and e_{12}, can be eliminated by the choice of ϵ_λ. Therefore, only these two components are physical while others are gauge fixing-dependent. This is exactly what should be expected, because the account of physical degrees of freedom means subtracting the number four for the gauge conditions (15.162) and also extra four for the transformations (15.163) from the initial ten components of $h_{\mu\nu}$. The number of physical components of the gravitational wave is two, exactly as in the case of an electromagnetic wave. In both cases, the waves are completely transverse. However, these two similarities do not mean that the two waves are equivalent, because in the gravitational case there are tensor polarizations, different from vector polarizations for the electromagnetic wave.

Another important difference between electromagnetic and gravitational waves is their generation. For the emission of the wave (15.162) is sufficient, because replacing (15.162) into (15.160), after some short algebra we arrive at the equation

$$\Box h_{\alpha\beta} = 8\pi G S_{\alpha\beta}. \qquad (15.173)$$

The last equation is nondegenerate, and its solution can be written, e.g., in the form of retarded potential, exactly as we discussed in Chap. 2 for the electromagnetic case,

$$h_{\alpha\beta}(\mathbf{x}, t) = 4G \int d^3x' \, \frac{S_{\alpha\beta}(\mathbf{x}', t - |\mathbf{x} - \mathbf{x}'|)}{|\mathbf{x} - \mathbf{x}'|}. \qquad (15.174)$$

The solution (15.174) explains why the detection of the gravitational waves is such a difficult problem. Let us compare this solution with the one for the electromagnetic wave. For the electromagnetic wave, the formula is

$$A_\alpha(\mathbf{x}, t) = 4 \int d^3x' \; \frac{j_\alpha(\mathbf{x}', t - |\mathbf{x} - \mathbf{x}'|)}{|\mathbf{x} - \mathbf{x}'|}. \tag{15.175}$$

The main difference between the two cases is that in the electromagnetic theory, the coupling constant is the electric charge, which is dimensionless. In the gravitational case, the coupling constant is G, which is dimensional and very small for most of the gravitational phenomena. In the particle physics units with $c = 1$ and $\hbar = 1$, we have $G = 1/M_P^2$, where the *Planck mass* $M_P \approx 10^{19}$ GeV. This is an incredibly high-mass scale. For example, the energy that can be achieved in the nowadays largest accelerator LHC in CERN is 14×10^3 GeV. The most energetic particles that can be met on our planet are coming from space as the ultrahigh-energy cosmic rays. These particles have typical energies of the order 10^{11} GeV, which is still a huge eight orders of magnitude less than the Planck scale.

Due to the formula (15.174), one can expect that the significant emission of gravitational waves should require the energy–momentum densities $T_{\mu\nu}$, which should be comparable to the Planck densities M_P^4. In fact, we are lucky enough, because no event with such scale of energy emission occurs in space too close to us. The merger of two black holes that produced the gravitational wave detected by LIGO happened more than one billion light-years from the Earth. The energy of the signal weakens proportionally to the square of the distance. And after that, the action of the gravitational wave on the detector is introducing one more factor of $G = 1/M_P^2$. All in all, it is obvious that the detection of such a signal represents the great triumph of technology of precise measurements and also has great scientific importance.

Let us make one more observation concerning the propagation of the gravitational waves. As one can see from Eq. (15.173), the speed of the gravitational wave in flat space is c, exactly like for the electromagnetic wave. However, the real situation can be a little bit different. As we know from the previous sections, the propagation of the massless particle in curved space is different from the one in flat space. For example, the massless particles can change their direction of motions due to the gravitational field. In the zero-order approximation, the motion of photons (quanta of electromagnetic waves) or gravitons (quanta of gravitational waves) follows the geodesic line. However, in the next approximation, there may be small deviations because of the spin (helicity, for the massless particles) of these particles. Spin is the quantum property of the particles, and we do not discuss it in introductory textbook on classical general relativity. But it is good to note that in principle, for the electromagnetic and gravitational waves which were emitted at the same instant of time, but far from us (typically billions of light-years), the instants of their observations can be slightly different. This means that regardless of equal speed of electromagnetic and gravitational waves with respect to the locally Minkowski observers, the average speed of signals passing from one space–time point to another one may be slightly different.

It is worthwhile to mention that quite recently, there was an important first observation of the simultaneous detection of a binary neutron star merger with gravitational waves by LIGO-VIRGO collaboration (events GW170817 for the gravitational and GW170817A for the gamma-rays) and the numerous observations of the same event by conventional optical observations. This remarkable event has shown that the difference between speed of gravitational and electromagnetic waves has such a small upper bound [21] that certain models of modified gravity, which predicted larger difference between these two quantities can be eventually ruled out. This example shows that the science may start to seriously benefit from gravitational waves detection in the study of the universe.

The interested reader can look for special literature such as the standard monograph [20] for much more information about gravitational waves, their propagation, emission, and detection.

Exercise 30 The solution (15.174) is written in the units with $c = 1$. Recover the factors of c in this expression and in the corresponding equation.

Exercise 31 Derive (15.166) from (15.162). Show that the condition $k_\lambda k^\lambda = 0$ is a consequence of the equation $\Box h_{\mu\nu} = 0$, which is (15.164) with the gauge condition (15.162). Discuss Eq. (15.167) and show that the factor i in front of the r.h.s. provides that the given mode of the field ξ_μ is real and provides real contribution into transformation (15.168) of the gravitational perturbation.

Exercise 32 Write the r.h.s. of Eq. (15.159) as $H_{\mu\nu}{}^{\rho\sigma} h_{\rho\sigma}$, where $H_{\mu\nu}{}^{\rho\sigma}$ is a second-order differential operator. Consider the matrix form of this operator in the momentum representation and show that its determinant is zero. This shows the degeneracy of Eq. (15.160), which has been mentioned above.

Hints. This problem may be difficult, but the reader can start from the similar operator in the vector space, $H_\mu^\nu = \delta_\mu^\nu \Box - \partial_\mu \partial^\nu$. In this case, there are two projectors: to the longitudinal and transverse states,

$$\omega_{\mu\nu} = \frac{k_\mu k_\nu}{k^2}, \qquad \theta_{\mu\nu} = \eta_{\mu\nu} - \omega_{\mu\nu}. \tag{15.176}$$

It is sufficient to prove that these projectors satisfy the algebra

$$\omega_{\mu\nu} + \theta_{\mu\nu} = \eta_{\mu\nu}, \quad \theta_\mu^{\ \nu} \theta_\nu^{\ \alpha} = \theta_\mu^{\ \alpha}, \quad \omega_\mu^{\ \nu} \omega_\nu^{\ \alpha} = \omega_\mu^{\ \alpha}, \quad \theta_\mu^{\ \nu} \omega_\nu^{\ \alpha} = 0. \tag{15.177}$$

After that, the proof of degeneracy of the operator H_μ^ν becomes a trivial task.

In the case of gravitational perturbations, the set of projectors (Barnes–Rivers operators in $4D$, which can be generalized to an arbitrary dimension) includes

$$P^{(2)}_{\mu\nu,\,\alpha\beta} = \frac{1}{2}(\theta_{\mu\alpha}\theta_{\nu\beta} - \theta_{\mu\beta}\theta_{\nu\alpha}) - \frac{1}{3}\theta_{\mu\nu}\theta_{\alpha\beta},$$

$$P^{(1)}_{\mu\nu,\,\alpha\beta} = \frac{1}{2}(\theta_{\mu\alpha}\omega_{\nu\beta} + \theta_{\nu\alpha}\omega_{\mu\beta} + \theta_{\mu\beta}\omega_{\nu\alpha} + \theta_{\nu\beta}\omega_{\mu\alpha}),$$

$$P^{(0-s)}_{\mu\nu,\,\alpha\beta} = \frac{1}{3}\theta_{\mu\nu}\theta_{\alpha\beta}, \qquad P^{(0-w)}_{\mu\nu,\,\alpha\beta} = \omega_{\mu\nu}\omega_{\alpha\beta}, \tag{15.178}$$

but the closure of their algebra also requires transfer operators,

$$P^{(ws)}_{\mu\nu,\alpha\beta} = \frac{1}{\sqrt{3}} \theta_{\mu\nu}\omega_{\alpha\beta}, \qquad P^{(sw)}_{\mu\nu,\alpha\beta} = \frac{1}{\sqrt{3}} \omega_{\mu\nu}\theta_{\alpha\beta}. \qquad (15.179)$$

We leave it to the reader to elaborate the algebraic relations similar to (15.177). The result is written as (condensed form taken from Appendix A of the paper [22])

$$P^{i-a} P^{j-b} = \delta^{ij} \delta^{ab} P^{i-a}, \qquad P^{i-ab} P^{j-cd} = \delta^{ij} \delta^{bc} P^{j-a},$$
$$P^{i-a} P^{j-bc} = \delta^{ij} \delta^{ab} P^{i-ac}, \qquad P^{i-ab} P^{j-c} = \delta^{ij} \delta^{bc} P^{i-ac}. \qquad (15.180)$$

After that it becomes an uncomplicated task to prove the degeneracy of the tensor operator in Eq. (15.159). This exercise is especially recommended to readers who plan to work in the area of quantum gravity. It is also instructive to generalize the results to an arbitrary D and discuss the special features of $2D$ and $3D$.

Partial solution. Let us check just one of the relations (15.180), just to show how it works. Consider

$$P^{i-ab} P^{j-cd} = \delta^{ij} \delta^{bc} P^{j-a}$$

for $a = w$, $b = s$, $b = s$, and $b = w$. There is no other options except to assume $i = j = 0$. Then

$$P^{0-ws} P^{0-sw} = \left(P^{0-ws}\right)_{\mu\nu,}{}^{\alpha\beta} \left(P^{0-sw}\right)_{\alpha\beta,\rho\sigma} = \frac{1}{\sqrt{3}} \theta_{\mu\nu} \omega^{\alpha\beta} \times \frac{1}{\sqrt{3}} \omega_{\alpha\beta}\theta_{\rho\sigma}$$
$$= \frac{1}{3} \theta_{\mu\nu}\theta_{\rho\sigma} = \left(P^{0-s}\right)_{\mu\nu,\rho\sigma}. \qquad (15.181)$$

Exercise 33* (i) Derive the bilinear (that means quadratic in $h_{\mu\nu}$) expansion of the actions I_1, I_2 and I_3 from Eq. (15.44). In order to simplify calculations, keep the background metric flat, meaning $g_{\mu\nu} = \eta_{\mu\nu} + h_{\mu\nu}$.
(ii) Using these results, derive the bilinear expansions of the actions from Eqs. (15.47) and (15.48).
(iii) Performing the necessary partial integrations, present the results in all five cases in the form

$$S^{(2)} = \frac{1}{2} \int d^4x \, h_{\mu\nu} H^{\mu\nu,\alpha\beta} h_{\alpha\beta}. \qquad (15.182)$$

(iv) Making a Fourier transformation, write the fourth-order differential operators (all five of them) $H^{\mu\nu,\alpha\beta}$ in the momentum representation, and also write them as linear combinations of the projectors (15.178).

Result. The main result of this big work will be that the expansions for I_3 and the integral of E_4 in (15.47) will have no P^2 mode. This means they do not contribute to the gravitational wave equation.

Exercise 34 By using the results of the previous exercise, show that only one combination (15.48) does contribute to the gravitational wave equation.

Exercise 35 Consider an interesting generalization of Eq. (15.43), namely, the integrals

$$I_1^g = \int d^4x \sqrt{-g}\, R_{\mu\nu\alpha\beta}\, \Phi\left(\frac{\Box}{M^2}\right) R_{\mu\nu\alpha\beta},$$

$$I_2^g = \int d^4x \sqrt{-g}\, R_{\mu\nu}\, \Phi\left(\frac{\Box}{M^2}\right) R^{\mu\nu},$$

$$I_3^g = \int d^4x \sqrt{-g}\, R\, \Phi\left(\frac{\Box}{M^2}\right) R. \tag{15.183}$$

Show that for an arbitrary function $\Phi(x)$, only the combination

$$W^{(g)} = I_1^g - 2I_2^g + \frac{1}{3} I_3^g \tag{15.184}$$

may contribute to the gravitational wave equation on a flat background.

References

1. F. Jüttner, Das Maxwellsche Gesetz der Geschwindigkeitsverteilung in der Relativtheorie. Ann. Phys. **Bd 116**, 145 (1911)
2. W. Pauli, *Theory of Relativity* (Dover, 1981)
3. S. Weinberg, *Gravitation and Cosmology* (Wiley, 1972)
4. S. Weinberg, The cosmological constant problem. Rev. Mod. Phys. **61**, 1 (1989)
5. P.J.E. Peebles, B. Ratra, The cosmological constant and dark energy. Rev. Mod. Phys. **75**, 559 (2003). arXiv:astro-ph/0207347
6. T. Padmanabhan, Cosmological constant: the weight of the vacuum. Phys. Rep. **380**, 235 (2003). arXiv:hep-th/0212290
7. D. Finkelstein, Past-future asymmetry of the gravitational field of a point particle. Phys. Rev. **110**, 965 (1958)
8. M. Kruskal, Maximal extension of Schwarzschild metric. Phys. Rev. **119**, 1743 (1960)
9. S.W. Hawking, G.F.R. Ellis, *The Large Scale Structure of Space–Time* (Cambridge University Press, 1973)
10. C.W. Misner, K.S. Thorn, J.A. Wheeler, *Gravitation* (W.H. Freeman and C, San Francisco, 1973)
11. R. D'Inverno, *Introducing Einstein's Relativity* (Oxford University Press, 1992–1998)
12. V.P. Frolov, I.D. Novikov, *Black Hole Physics—Basic Concepts and New Developments* (Kluwer Academic Publishers, 1989)
13. S. Chandrasekhar, *The Mathematical Theory of Black Holes* (Clarendon Press, Oxford, 1998)
14. V.P. Frolov, A. Zelnikov, *Introduction to Black Hole Physics* (Oxford University Press, 2015)
15. D.F. Carneiro, E.A. Freitas, B. Gonçalves, A.G. de Lima, I.L. Shapiro, On useful conformal tranformations in general relativity. Gravit. Cosmol. **40**, 305 (2004). arXiv:gr-qc/0412113
16. W. Siegel, Fields. arXiv:hep-th/9912205

17. L.D. Landau, E.M. Lifshits, *The Classical Theory of Fields—Course of Theoretical Physics*, vol. 2 (Butterworth-Heinemann, 1987)
18. I.L. Shapiro, G. de Berredo Peixoto, *Lecture Notes on Newtonian Mechanics: Lessons from Modern Concepts* (Springer, 2013)
19. B.P. Abbott et al. (LIGO Scientific and Virgo Collaborations), Observation of gravitational waves from a binary black hole merger. Phys. Rev. Lett. **116**, 061102 (2016). arXiv:1602.03837
20. M. Maggiore, *Gravitational Waves: Volume 1: Theory and Experiments* (Oxford University Press, 2007)
21. B.P. Abbott et al., Gravitational waves and gamma-rays from a binary neutron star merger: GW170817 and GRB 170817A. Astrophys. J. Lett. **848**, L13 (2017)
22. K.S. Stelle, Renormalization of higher derivative quantum gravity. Phys. Rev. D **16**, 953 (1977)

Chapter 16
Basic Elements of Cosmology

16.1 Simplest Cosmological Solutions

According to modern scientific understanding, at the large scale, our universe is an expanding, homogeneous, and isotropic space, approximately 14 billion years (14 bi) old. The size of the universe is defined by the distance which light can cover in this time period, which is quite big. We are learning more about the history of the universe every year. One of the most interesting questions is that how we can explore and understand such an old and gigantic universe, if the region of space where the observations are performed is confined primarily to a small part of our Solar System, which has almost negligible size even compared to the typical distances in our galaxy. The time interval of our observations is also restricted to just a few thousands of years, and of scientific instrumental observations to much less than that. The answer to this question is that there are many qualitatively different observables and methods of observations. It is due to the great improvements in the observational technologies that we can know so much and so well about how the universe is organized and what happened billions of years ago. It is also possible to theoretically describe the formation of galaxies and stars, and eventually explain the inhomogeneity of the picture in the night sky.

Cosmology is nowadays a prosperous area of science. It is sufficient to mention that many of the greatest modern experimental projects belong to observational cosmology and astrophysics. Here we shall discuss the theoretical part and describe only the most basic cosmological solutions. The interested reader can go much further into the subject, e.g., [1–6].

16.1.1 Homogeneous and Isotropic Metric

The starting point in defining the metric to describe the universe at the largest scale is the *cosmological principle*, which states that the universe is homogeneous and

© Springer Nature Switzerland AG 2019

I. L. Shapiro, *A Primer in Tensor Analysis and Relativity*, Undergraduate
Lecture Notes in Physics, https://doi.org/10.1007/978-3-030-26895-4_16

isotropic. Looking at the night sky we would never guess this since at some places there are bright stars and the rest of the sky is dark. The point is that the theory assuming homogeneity and isotropy at large scale is capable of describing the small perturbations of metric and matter density in the universe. The amount of theoretical data which can be obtained in this way is capable to explain the formation of the large-scale structure (meaning the map of the mass distribution at the cosmic scale) of the universe, the spectrum of cosmic microwave radiation, baryon acoustic oscillations, and a number of other relevant observables, which can be verified by a large variety of astronomic data.

The complete mathematical implementation of the cosmological principle can be found elsewhere (see, e.g., [7, 8]); here we shall follow [9] and present arguments that are convincing but not mathematically complete. The $3D$ space section of the homogeneous and isotropic space can be presented as a $3D$ sphere embedded into $4D$ Euclidean space. This construction is completely artificial, mainly because this $4D$ space has nothing to do with the $1 + 3$-dimensional Minkowski space. It is only a mathematical construction which we are using to arrive at the formulas describing a symmetric space. After that, we will make an assumption that all symmetric spaces are of the types we will obtain this way.

The $3D$ sphere of the constant radius a, embedded into $4D$ Euclidean space, is defined by the condition

$$(x_1)^2 + (x_2)^2 + (x_3)^2 + (x_4)^2 = a^2 , \tag{16.1}$$

that immediately gives

$$x_4 = \sqrt{a^2 - x_1^2 - x_2^2 - x_3^2}, \quad \text{and}$$

$$dx_4 = -\frac{x_1 dx_1 + x_2 dx_2 + x_3 dx_3}{\sqrt{a^2 - x_1^2 - x_2^2 - x_3^2}} = -\frac{\mathbf{r} \cdot d\mathbf{r}}{\sqrt{a^2 - r^2}}, \tag{16.2}$$

where $\mathbf{r} = \hat{\mathbf{i}} x_1 + \hat{\mathbf{j}} x_2 + \hat{\mathbf{k}} x_3$ and $r^2 = x_1^2 + x_2^2 + x_3^2$.

As we already know, in spherical coordinates

$$d\mathbf{r} \cdot d\mathbf{r} = dr^2 + r^2 d\Omega , \quad d\Omega = d\theta^2 + \sin^2 \theta d\varphi^2 , \quad \mathbf{r} \cdot d\mathbf{r} = r dr .$$

Therefore, for the line element on the $3D$ sphere we obtain

$$dl^2 = dx_1^2 + dx_2^2 + dx_3^2 + dx_4^2 = dr^2 + \frac{r^2 dr^2}{a^2 - r^2} + r^2 d\Omega + \frac{dr^2}{1 - \frac{r^2}{a^2}} + r^2 d\Omega .$$

$$\tag{16.3}$$

We can observe that the sign of a^2 can be positive or negative. The last case corresponds to the hyperbolic sphere. We will not discuss the geometric sense of this notion and refer the reader to the standard textbook [10].

Exercise 1 Consider the $3D$ space with the metric (16.3) with an arbitrary sign of a^2. Calculate the components of the Ricci tensor R_{ij} and scalar curvature $R = g^{ij} R_{ij}$, and show that the sign of R os the same of a^2.

The metric (16.3) is written in the units where r is dimensional. It proves useful to change the radial variable to r', which is measured in the units of the absolute value of $|a|$, then $r = |a| r'$. Then the sign of a^2 can be taken into account by the coefficient $k = \pm 1$. Also, there is a possibility of a flat space section of the cosmological space–time, which corresponds to the value $k = 0$. Finally, we change the notations $r' \to r$ and obtain the space section of the metric in the form

$$dl^2 = a^2 \left(\frac{dr^2}{1 - kr^2} + r^2 d\Omega \right). \tag{16.4}$$

Now we can let the scale factor be time dependent, $a \to a(t)$, and arrive at the general expression for the homogeneous and isotropic metric,

$$ds^2 = dt^2 - a^2(t) \left[f(r) dr^2 + r^2 d\Omega \right], \tag{16.5}$$

$$\text{where} \quad f(r) = \frac{1}{1 - kr^2}, \quad \text{where} \quad k = 0, \pm 1.$$

The result (16.5) is traditionally called Friedmann–Lemaître–Robertson–Walker metric, but one can also find different combinations of these names. For the sake of brevity, we will call it the FLRW metric.

16.1.2 Derivation of Friedmann–Lemaître Equations

The first cosmological solution was obtained by Alexander Friedmann in 1922, who derived the explicit form of Einstein equations (15.60) with $\Lambda = 0$. In what follows we make all calculations for a nonzero Λ. Such equations are usually called Friedmann–Lemaître's, to stress an important contribution of Georges Lemaître to the progress of cosmology, and, in particular, his recognition of the importance of the cosmological constant term.

In order to derive the Einstein equations, let us consider a more general metric,

$$ds^2 = N^2(t) dt^2 - a^2(t) \left[f(r) dr^2 + r^2 d\Omega \right] = a^2(t) \bar{g}_{\mu\nu} dx^\mu dx^\nu. \tag{16.6}$$

It is clear that the function $N(t)$ can be changed by the change of time variable. Two choices are the most useful. $N(t) = 1$ corresponds to the *physical* or *cosmological time*, while another choice, $N(\eta) = a(\eta)$, corresponds to the *conformal time* η. For a while we shall keep $N(t)$ arbitrary and call the time variable t.

We leave a direct derivation of G^μ_ν as an exercise and will follow the conformal transformation method, in a way similar to Sect. 15.2.1 for the Schwarzschild solution. The metric (16.6) can be written in the form

$$g_{\mu\nu} = a^2 \bar{g}_{\mu\nu}, \qquad a^2 = e^{2\sigma(t)}, \tag{16.7}$$

$$\bar{g}_{\mu\nu} = \mathrm{diag}\left[\frac{N^2(t)}{a^2(t)}, -\gamma_{ij}\right], \quad \gamma_{ij} = \mathrm{diag}\left[f(r), r^2, r^2 \sin^2\theta\right]. \tag{16.8}$$

The space section $\bar{g}_{ij} = -\gamma_{ij}$ of the conformally transformed metric depends only on the space coordinates, and therefore we can apply the factorization theorem of Sect. 14.5.1. In particular, the Ricci tensor $\bar{R}_{\mu\nu}$ may have only space nonzero components, $\bar{R}_{ij}$, while $\bar{R}_{0\mu} \equiv 0$. Then, according to Eq. (14.79), we need only the components of $\bar{\Gamma}^i_{jk}$ to derive $R_{\mu\nu}$. Direct calculation (which we leave to the reader as a simple exercise) gives the nonzero components

$$\bar{\Gamma}^r_{rr} = \frac{f'}{2f}, \quad \bar{\Gamma}^r_{\theta\theta} = -\frac{r}{f}, \quad \bar{\Gamma}^r_{\varphi\varphi} = -\frac{r \sin^2\theta}{f},$$

$$\bar{\Gamma}^\varphi_{r\varphi} = \bar{\Gamma}^\theta_{r\theta} = \frac{1}{r}, \quad \bar{\Gamma}^\theta_{\varphi\varphi} = -\sin\theta\cos\theta, \quad \bar{\Gamma}^\varphi_{\theta\varphi} = \cot\theta. \tag{16.9}$$

Using these components of connection, after some algebra we arrive at the simple relation

$$\bar{R}_{ij} = -2k\bar{g}_{ij} = 2k\gamma_{ij}. \tag{16.10}$$

After direct replacement of (16.10) and (16.7) into (14.79), one can arrive at the following nonzero components of the Ricci tensor and scalar curvature,

$$R_{00} = \frac{3\dot{a}\dot{N}}{a N} - \frac{3\ddot{a}}{a},$$

$$R_{ij} = \gamma_{ij}\left(2k + \frac{a\ddot{a}}{N^2} - \frac{a\dot{a}\dot{N}}{N^3} + \frac{2\dot{a}^2}{N^2}\right),$$

$$R = -\frac{6k}{a^2} - \frac{6\ddot{a}}{aN^2} + \frac{6\dot{a}\dot{N}}{aN^3} - \frac{6\dot{a}^2}{a^2N^2}. \tag{16.11}$$

The most useful form of equations for the FLRW metric is expressed through the Einstein tensor,

$$G^0_0 = \frac{3k}{a^2} + \frac{3\dot{a}^2}{a^2N^2} = 8\pi G\rho + \Lambda,$$

$$G^j_i = \delta^j_i\left(\frac{k}{a^2} + \frac{2\ddot{a}}{aN^2} - \frac{2\dot{a}\dot{N}}{aN^3} + \frac{\dot{a}^2}{a^2N^2}\right) = (-8\pi Gp + \Lambda)\delta^j_i, \tag{16.12}$$

where ρ and p are the sums of energy densities and pressure of all homogeneous and isotropic fluids that fill the universe.

Replacing $N \to 1$ we arrive at the standard form of Friedmann–Lemaître equations in terms of physical time t,

$$\frac{\dot{a}^2}{a^2} + \frac{k}{a^2} = \frac{8\pi G}{3}\rho + \frac{1}{3}\Lambda\,, \tag{16.13}$$

$$\frac{2\ddot{a}}{a} + \frac{\dot{a}^2}{a^2} + \frac{k}{a^2} = -8\pi Gp + \Lambda\,. \tag{16.14}$$

Another useful form of the same equations is related to the Hubble parameter H, which is defined by the relation

$$H = \frac{\dot{a}}{a}. \tag{16.15}$$

The Hubble parameter is important, as it defined the energy scale of a background gravity in the homogeneous and isotropic universe. Another characteristic of the expansion of the universe is the deceleration parameter

$$q = -\frac{\ddot{a}a}{\dot{a}^2}. \tag{16.16}$$

Using the Hubble parameter, the Friedmann–Lemaître equations can be cast into the form

$$H^2 + \frac{k}{a^2} = \frac{8\pi G}{3}\rho + \frac{\Lambda}{3}\,, \tag{16.17}$$

$$2\dot{H} + 3H^2 + \frac{k}{a^2} = -8\pi Gp + \Lambda\,. \tag{16.18}$$

The system of equations (16.13) and (16.14) is incomplete because there are only two equations for at least three variables $a(t)$, $\rho(t)$, $p(t)$. The difference between the number of variables and number of equations increases if there is more than one fluid, as one should naturally expect in a complicated system as our universe. In order to complete the system, let us consider the conservation law, $\nabla_\mu T^\mu_\nu = 0$ for the energy–momentum tensor of the ideal fluid in the co-moving frame, $T^\mu_\nu = \text{diag}\,(\rho, -p, -p, -p)$.

Direct calculation provides the list of nonzero components of affine connection for the FLRW metric with $N = 1$,

$$\Gamma^i_{0k} = H\delta^i_k\,, \qquad \Gamma^0_{ik} = a^2 H\gamma_{ik}\,, \qquad \Gamma^\mu_{0\mu} = 3H\,. \tag{16.19}$$

Using these components of connection, it is easy to arrive at the conservation law (15.3) in the specific case of the FLRW metric,

$$\dot{\rho} + 3H(\rho + p) = 0. \tag{16.20}$$

Now we can conclude that (16.17), (16.18), and (16.20) form a closed system of equations for the three variables $a(t)$, $\rho(t)$ and $p(t)$. In the case of several fluids, the number of equations is supposed to be equal to the enlarged number of variables because of the extra conservation laws.

Exercise 2 Rewrite Eqs. (16.13)–(16.18) and (16.20) in terms of conformal time η. Remember that $a(\eta)d\eta = dt$ or $d\eta = dt/a(t)$. The derivative with respect to η is denoted by prime, e.g.,

$$a' = \frac{da}{d\eta} = \frac{da}{dt}\frac{dt}{d\eta} = a\dot{a}. \tag{16.21}$$

Exercise 3 The main advantage of the form with two functions of time (16.6) is that the variation of the action with respect to $N(\eta)$ gives the 00-component and the variation with respect to $a(\eta)$ gives 11-, 22-, and 33-components of the Einstein equations. Then we can save time by deriving only scalar curvature in the Einstein–Hilbert action, instead of the Einstein tensor in the dynamical equations. Make the corresponding calculation and compare the result with the Friedmann–Lemaître equations obtained in this section.

Exercise 4 Verify the relations (16.9) and (16.10). After this purely calculational exercise the reader can solve a slightly more complicated problem and show that (16.9) can be achieved only for the function $f(r)$ defined in (16.5). Furthermore, try to explain the relation between this result and that the space section of the FLRW metric can be considered as the metric of a $3D$ space embedded in the artificial $4D$ flat Euclidean space.

Hint. Start by analyzing a similar (but simpler) case of the "cosmological" $2D$ metric, assuming that it is embedded into $3D$ flat Euclidean space.

Exercise 5 Verify Eqs. (16.11) and (16.12).

Exercise 6 In order to check the consistency of Friedmann–Lemaître equations, verify that the same conservation law (16.20) holds for the *l.h.s.* of these equations,

$$\rho_g = G_0^0 = \frac{3(\dot{a}^2 + k)}{a^2}, \quad p_g = G_1^1 = G_2^2 = G_3^3 = -\frac{\dot{a}^2 + 2a\ddot{a} + k}{a^2}, \tag{16.22}$$

and also for the cosmological constant term "energy density" and "pressure",

$$\rho_\Lambda = \frac{\Lambda}{8\pi G}, \quad p_\Lambda = -\rho_\Lambda. \tag{16.23}$$

As we have explained earlier, the last expressions can be interpreted as energy density and pressure of a vacuum.

One can use the conservation equation (16.20) to find the dependencies on the scale factor, $\rho(a)$ and $p(a)$. After that, Eqs. (16.13) and (16.17) can be solved. For the cosmological constant case $p_\Lambda = -\rho_\Lambda$, and (16.20) just helps us to verify that

$$\dot{\rho}_\Lambda = 0 \quad \Longrightarrow \quad \rho_\Lambda = const. \tag{16.24}$$

In general, the technical difficulty of the solution depends only on the complexity of the equations of state of the fluids in the given cosmological model.

For the dust matter contents, the same procedure works as follows. The equation of state is $p = 0$ and the conservation equation (16.20) becomes

$$\frac{d\rho}{\rho} = -\frac{3da}{a} \quad \Longrightarrow \quad \rho = \rho_0 \left(\frac{a_0}{a}\right)^3, \tag{16.25}$$

where ρ_0 is the energy density at the initial point $a = a_0$.

For the ideal fluid of radiation, the equation of state is $p = \rho/3$ and we get

$$\frac{d\rho}{\rho + p} = \frac{3}{4}\frac{d\rho}{\rho} = -\frac{3da}{a} \quad \Longrightarrow \quad \rho = \rho_0 \left(\frac{a_0}{a}\right)^4. \tag{16.26}$$

Consider the equation of state $p = \omega\rho$. Then the same procedure gives

$$\rho = \rho_0 \left(\frac{a_0}{a}\right)^{3(1+\omega)}, \tag{16.27}$$

which includes both (16.25) and (16.26) as particular cases.

Exercise 7 Find the dependence on the conformal factor for the reduced relativistic gas model, which was discussed in Chap. 2 as an approximation for the model of ideal gas of relativistic massive particles [11]. The equation of state is [12, 13]

$$p = \frac{1}{3}\rho\left(1 - \frac{\rho_d^2}{\rho^2}\right), \tag{16.28}$$

where ρ_d is the density of mass (in the units $c = 1$).

Solution. $\quad \rho(a) = \left[\rho_1^2 \left(\frac{a_0}{a}\right)^6 + \rho_2^2 \left(\frac{a_0}{a}\right)^8\right]^{1/2}, \tag{16.29}$

where the ratio $\frac{\rho_2}{\rho_1}$ defines the relativistic warmness of the fluid at the scale a_0.

Exercise 8 Show that the conservation law enables one to recover the r.h.s of Eqs. (16.18) from (16.17). Verify that the same can be done for gravitational "energy density" ρ_g and "pressure" p_g from Exercise 6. Repeat the same procedure for ρ_Λ and p_Λ.

Exercise 9 Suppose we know the trace $T^\mu_\mu = T(a, H, \dot{H})$ for a certain fluid. Using the conservation law, recover the energy density $\rho = T^0_0$ and pressure $p = -T^1_1 = -T^2_2 = -T^3_3$. As examples, we suggest considering particular cases of

(i) Dust $T_d = \rho_0 \left(\frac{a_0}{a}\right)^3$.

(ii) Cosmological constant $T_\Lambda = 4\rho_\Lambda$.

(iii) The trace of the *l.h.s.* of Einstein equations in terms of conformal time,

$$T_g = \frac{3\pi G}{4} \frac{a''}{a^3}. \tag{16.30}$$

(iv) Another interesting case is

$$T_\beta = \frac{\beta \bar{F}^2}{a^4}, \qquad \beta \bar{F}^2 = \text{const}, \tag{16.31}$$

that is, the anomalous trace of the energy–momentum tensor in the theory with an external electromagnetic field [14].

(v) And, finally, there are two more complicated examples,

$$T_c = 6\left[-\frac{a''''}{a^5} + 4\frac{a'''a'}{a^6} + 3\left(\frac{a''}{a^3}\right)^2 - 6\frac{a''a'^2}{a^7}\right],$$

$$T_b = -24\left[\left(\frac{a'}{a^2}\right)^4 - \frac{a''a'^2}{a^7}\right], \tag{16.32}$$

which are the expressions that take place for the quantum conformal anomaly [15].[1] In all cases, the prime stands for the derivative with respect to the conformal time η.

Solution. In order to leave for the reader freedom for making independent calculation, let us solve only part of the problem that corresponds to dust, using physical time t. The results for other parts can be found in the references cited above.

We know that for an ideal fluid $T = \rho - 3p$, and that the two quantities satisfy the conservation law (16.20). Assuming that $T = T(a, \dot{a}, \dots)$ is known, the conservation formula boils down to the linear equation,

$$\frac{d\rho}{dt} + 4H\rho - HT = 0. \tag{16.33}$$

This equation is solved by the general expression

$$\rho(t) = \frac{C(t)}{a^4}, \qquad \text{where} \qquad \dot{C} = Ta^3\dot{a}. \tag{16.34}$$

---

[1]The reader can read about the anomaly, e.g., in [15], when this problem was solved. The solution does not require understanding of the quantum aspects of the problem.

In the case of dust (pressureless matter) $T = \rho = \rho_0(a_0/a)^3$, which gives

$$\rho(t) = \frac{C_0}{a^4} + \frac{\rho_0 a_0^3}{a^3} . \tag{16.35}$$

The first term is an integration constant of Eq. (16.33), which appears since for the radiation equation of state, the trace T is zero. Therefore, knowledge of T does not enable one to fix the energy density of radiation. And in all other cases, the radiation energy density is an integration constant for the solution.

Observation An alternative approach (which requires more effort) for solving this problem is to integrate the equation

$$T_\mu^\mu = -\frac{2}{\sqrt{|g|}} g_{\mu\nu} \frac{\delta S_{trace}}{\delta g_{\mu\nu}} , \tag{16.36}$$

and find the action S_{trace}, which generates the trace. Then, taking the derivative with respect to the whole metric $g_{\mu\nu}$, one can obtain the energy–momentum tensor. Obviously, in this case, the conformal functional of the metric will play the role of the integration constant of Eq. (16.36). In this sense, it is correct to call the first term (not only the constant C_0) in Eq. (16.35) the "integration constant". The technique for covariant integration of (16.36) is well known in quantum theories (see, e.g., [16]).

16.2 Cosmological Solutions

Equations. (16.13), (16.14), and (16.20) with $\Lambda = 0$ were used by A. Friedmann in 1922 to obtain the first solution for the expanding universe. This might perhaps be considered one of the most outstanding scientific calculations ever. Even the creator of general relativity Albert Einstein could not imagine that the universe can be non-static and in fact introduced the cosmological constant term to provide a static solution. This attempt was mathematically and physically inconsistent[2] and later on, Einstein expressed regret for introducing the Λ-term. Indeed, the universe does not care about the feelings of physicists, even the greatest ones. And in the last decades, the Λ-term has came back to the scene in all its glory, after analysis of the cosmic microwave background, the mass distribution (large-scale structure) of the universe and especially observations of SN-Ia-type supernovae have shown that the universe is expanding and accelerating. In order to understand how this acceleration is related to the presence of the cosmological constant, let us start by considering an empty universe with $\rho = p = 0$ and $\Lambda > 0$. Also, for the sake of simplicity we consider the flat universe, setting $k = 0$. Then the solution of (16.17) and (16.13) is

---

[2]This historically interesting issue is described in the paper [17].

$$H = H_0 = \sqrt{\frac{\Lambda}{3}} \qquad a(t) = a_0 e^{H_0 t}. \tag{16.37}$$

The deceleration parameter (16.16) is certainly zero because $H = const$. It is easy to check that this solution satisfies the second Friedmann–Lemaître equations (16.13) and (16.14).

The real universe is filled by different kinds of matter, and their appropriate description depends on the epoch of interest. For instance, in the early universe, it makes sense to consider various relativistic fluids with the equation of state that is intermediate between dust and radiation. They can exchange energy such that the model of ideal fluid in equilibrium is not adequate for the situation and it is necessary to use relativistic Boltzmann equations coupled with Einstein equations. The corresponding studies can be found in specialized monographs, but here we intend to give only qualitative presentation of the simplest cosmological solutions. Thus, it proves useful to consider a few simple cases, restricting the analysis to the case of a single ideal fluid of the (16.27) type. Equation (16.13) becomes

$$\dot{a}^2 = \frac{8\pi G}{3} \rho_0 a_o^{3(1+\omega)} a^{-1-3\omega} + \frac{\Lambda}{3} a^2 - k, \tag{16.38}$$

which can be recast in the form

$$\frac{1}{2}\dot{a}^2 + U(a) = 0, \qquad \text{where} \tag{16.39}$$

$$U(a) = -\frac{4\pi G}{3} \rho_0 a_0^{3(1+\omega)} a^{-1-3\omega} - \frac{\Lambda}{6} a^2 + \frac{k}{2}. \tag{16.40}$$

Equation (16.39) has the form of the energy conservation law in Classical Mechanics, with zero total energy of the "particle" and the mass equal to one. Thus, instead of solving this equation (this is perfectly possible but the solution is bulky), we can classify the motions using the requirement $\dot{a}^2 > 0$. One can easily classify the possible motions to finite and infinite depending on the values of k and Λ. Basically, the possible regions of a are those for which $U(a) \leq 0$. Let us illustrate the situation in a few particular cases.

(i) For $\Lambda > 0$ and $k = 0$ or $k = -1$ all terms in the potential (16.40) are negative for all a; therefore, the motion is infinite. This means that the universe may expand infinitely. The sample of the plots of $U(a)$ is given in the left plot in Fig. 16.1. In both cases, we assume that $0 < \omega < \frac{1}{3}$.

(ii) For $k = +1$, the situation may be different depending on the value of Λ. Consider the particular case with $\Lambda = 0$. It is easy to see that there is a return point $a = a_c$, where the expansion of the universe changes to contraction. The illustration is given at the right plot in Fig. 16.1.

The derivation of a_c is not complicated. Consider it for the particular value $\omega = 0$ that is a natural model for the far future. Then Eq. (16.38) becomes

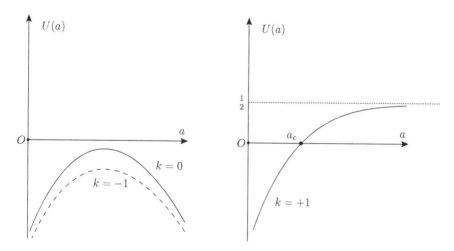

Fig. 16.1 The plots of potentials (16.40) with $\Lambda > 0$ for $k = 0, -1$ on the left and for $k = +1$ with $\Lambda = 0$ on the right. In the last case, a_c is the return point, where the expansion of the universe ends and contraction starts. The region $a > a_c$ is impossible because $U(a) > 0$

$$\dot{a}^2 = \frac{8\pi G M}{3a} - 1, \quad \text{where} \quad M = \rho_0 a_o^3. \tag{16.41}$$

Obviously, in this case, there is a restriction

$$a \leq a_c, \quad a_c = \frac{8\pi G M}{3}, \tag{16.42}$$

such that the universe expands until a reaches the value a_c, after which the contraction phase starts.

A similar behavior occurs in case of $k = 0$ and $\Lambda < 0$, when the equation is

$$\dot{a}^2 = \frac{8\pi G M}{3a} + \frac{\Lambda}{3} a^2 \tag{16.43}$$

and the limit for the scale factor is

$$a \leq a_c, \quad a_c = \frac{8\pi G M}{|\Lambda|}. \tag{16.44}$$

In both cases (16.42) and (16.44), the motion is finite with a single return point.

For the universe with $\Lambda = 0$, one can distinguish flat $k = 0$, open $k = -1$, and closed $k = 1$ universes. As we have just seen, in the presence of the cosmological constant the classification becomes more complicated, and we leave it to the reader as an exercise.

Exercise 10 Consider the case of a small Λ of an arbitrary sign. Derive the return point in this case, by regarding (16.42) as a background non-perturbed value and the term with Λ as a small perturbation.

In the literature, one can find cosmological models with another type of bounce, when the initial contraction at some moment changes to expansion. Such a solution usually requires a gravitational model different from general relativity.

Exercise 11 Is it possible to have such bounce in the universe governed by general relativity, with $k = 0$ and $\Lambda = 0$, filled by several fluids with the equation of state parameters ω_i?

Let us now derive the analytic form of the standard cosmological solutions. Consider the simplest case of the Friedmann equation (16.13) with $\Lambda = 0$, $k = 0$ and with the purely matter or purely radiation contents.

For the pure dust-like matter, we have the equation

$$\frac{\dot{a}^2}{a^2} = \frac{8\pi G \rho_0}{3} \left(\frac{a_0}{a}\right)^3. \tag{16.45}$$

The solution can be easily found and shows time dependence

$$a(t) \sim t^{2/3}, \qquad H(t) \sim t^{-1}. \tag{16.46}$$

According to modern understanding, most of the dust-like content of the universe is the dark matter, which interact with other types of matter and with itself only gravitationally. The existence of dark matter can be explained quite naturally by the existence of generic particle physics models beyond the Minimal Standard Model (MSM) of particle physics. Usually the fact that we do not observe such as extensions of MSM is explained by the huge masses of corresponding "extra" particles. According to the modern approach, called *effective field theory*, this guarantees that these particles decouple (meaning they do not interact, except gravitationally) with other particles (baryonic matter) at energies much lower than their huge masses (in the units with $c = 1$). However, these particles were interacting with baryonic matter and with each other in the early universe, where the energies were much greater. The conditions of equilibrium between these heavy particles and normal baryonic particles at the epoch of decoupling define the relative mass densities of two kinds of particles in the present-day universe.

The experimental determination of these extensions of the MSM is difficult because it may require either experiments with energies many orders of magnitude greater than the ones available in technically possible accelerators, or precision in the measurements of lower energy observables beyond the present-day technologies. Nevertheless, the amount of observational evidence and the variety of the qualitatively different types of observations are such that there is not much doubt concerning the existence of dark matter. Further discussion of this issue can not be given here, so let us only refer to the review on the subject [18] and books [1, 5].

Regarding almost all contents of the universe as dust is a good approximation for the last few billions of years. The reason is that in this "late" epoch, the radiation density was already much smaller than that of one of baryonic and dark matter, and for some part of this time period the cosmological constant density was still much smaller. At the same time, the solution (16.46) tells us that in the earlier period the situation was different because the universe expands. Looking to the past cosmological epochs, radiation density changes as a^{-4}, while the dust density only as a^{-3}. Thus, it is clear that in the past there was a period when radiation dominated over matter and the cosmological constant. Let us consider the solution of the Friedmann equation for the radiation epoch.

In the radiation-dominated universe, the equation is

$$\frac{\dot{a}^2}{a^2} + \ = \ \frac{8\pi G\rho_0}{3}\left(\frac{a_0}{a}\right)^4, \tag{16.47}$$

and the solution shows the behavior

$$a(t) \sim t^{1/2}, \qquad H(t) \sim t^{-1}. \tag{16.48}$$

This is a remarkable result in several respects. Since for a smaller a the dominance of radiation can only increase, in a past radiation-dominated universe one cannot expect qualitative changes in the cosmological equations. Therefore, the solution (16.48) should be assumed valid up to $t = 0$. And at this initial point, the solution (16.48) predicts an infinite Hubble parameter, components of curvature tensor (including scalar invariants such as R^2 and $R^2_{\mu\nu\alpha\beta}$), and the energy density of radiation. This is the initial cosmological singularity, which is qualitatively similar to the one in the black hole solution at $r = 0$.

There are general singularity theorems by Penrose and Hawking (see, e.g., the book [8]), which state that geodesic incompleteness (which is roughly equivalent to the presence of singularities) is typical for the relevant solutions of general relativity. How should we interpret the existence of singularities in the physically relevant solutions of general relativity? The most popular explanation of this is that the singularities indicate the limit of applicability of general relativity. In the regions where curvature becomes strong, the gravitational equations of motion should be complemented by new terms providing the consistency of the theory. The most likely source of these extra terms is related to the quantum theory of gravity or with the semiclassical theory, meaning quantum field theory on a classical metric background. In both cases, the quantum theory predicts relevant quantum corrections to the gravitational action, which become negligible at large (macroscopic) distances, but may be relevant in the vicinity of a singularity.

The problem of singularities in the extended theories of gravity has been extensively studied in the last decades. However, the present-day knowledge of quantum theory does not give definite answers about the form of additional gravitational terms of the quantum origin. The construction of a consistent quantum gravity theory meets serious theoretical difficulties, but this is actually not the worst problem for defining the necessary supplement to the Einstein–Hilbert action (the term used today is *UV completion*, where UV means an ultraviolet, high-energy regime). One can formulate many theoretically interesting extensions of general relativity. The real situation is such that there are many theories that are perhaps able to answer this question, but at present there are no experimental data that would help us to discriminate among them.

Exercise 12 Find an explicit analytical solution for the universe filled by dust in the presence of the nonzero cosmological constant term.

Exercise 13 Derive explicit analytic solutions for the matter-(dust-)dominated universe with $k = \pm 1$ with $\Lambda = 0$. Confirm that the solution with $k = 1$ describes bounded expansion of $a(t)$ with the maximal value (16.42), while in the case of $k = -1$ the growth of $a(t)$ is not restricted from above.

Exercise 14 Consider the $\Lambda = 0$ and $k = 0$ universe filled by the fluid with an intermediate equation of state derived in Eq. (13.119) of Chap. 13,

$$P = \frac{\rho}{3} \cdot \left[1 - \frac{\rho_d^2}{\rho^2} \right],$$

where the density of the rest energy of the particles is $\rho_d = nm \sim 1/a^3(t)$, as it should be for the dust matter contents. The scale dependence of the energy density is given by Eq. (16.29). Replacing this result into the Friedmann equation, establish the functional dependence $a(t)$ in this case. Analyze the initial and late stages of the evolution and show that they correspond to approximately radiation- and dust-dominated universe models.

$$\textbf{Solution.} \quad \left(a^2 + \frac{\rho_2^2}{\rho_1^2} \right)^{3/4} = \sqrt{6\pi G \rho_1} \cdot t. \tag{16.49}$$

16.3 Redshift and Expansion of the Universe

In this section, we introduce the parameterization that is most useful and most frequently used for the description of the dynamics of the universe. The first important element is the Hubble parameter H defined in Eq. (16.15). Its present-day value is defined from the observations to be

$$H_0 = H(t_0) = 100h \, \frac{km}{s \, Mpc}, \quad \text{where} \quad h = 67.77 \pm 1.30, \tag{16.50}$$

where t_0 is the present time. The value of the dimensionless parameter h cited here corresponds to the 2018 data of the Planck collaboration [19] and is slightly different for other observational projects. Let us introduce other useful quantities used for describing the expansion of the universe.

16.3.1 Relative Densities

A useful starting point is the first Friedmann–Lemaître equation (16.17),

$$H^2 = \frac{8\pi G}{3}(\rho + \rho_\Lambda) - \frac{k}{a^2},\tag{16.51}$$

where we used notation (16.23) for the density of the cosmological constant. Remember that the universe is complicated. In particular, this means that the total energy density ρ is a sum of several contributions, such as the energy density of baryons ρ_{BM}, of the dark matter ρ_{DM}, and radiation ρ_r. One can also remember the neutrino part, but it is usually included into ρ_r, since the neutrino masses are small and the equations of state for neutrino are similar to the one of radiation.

It is customary to call the quantity

$$\rho_c = \frac{3H^2}{8\pi G}\tag{16.52}$$

the critical density. The present-day value can be evaluated from (16.50) and the value of G,

$$\rho_c^0 = \rho_c(t_0) = 1.88\, h^2 \times 10^{-29}\,\frac{g}{cm^3}.\tag{16.53}$$

In order to compare this quantity with the particle physics scales, let us rewrite the result in the units of GeV,

$$\rho_c^0 = \frac{3}{8\pi}(M_P \times H_0)^2 \approx \frac{3}{8\pi}\left(10^{19}GeV \times 10^{-42}GeV\right)^2 \approx 10^{-47}GeV^4.\tag{16.54}$$

This quantity can be called the average energy density of the universe *today*, if we regard the cosmological constant density ρ_Λ as energy density. Since the universe is expanding, it is clear that in the past, H was larger and ρ_c was correspondingly larger as well.

In order to parameterize ρ_c at different instants of cosmic time, it is useful to introduce the dimensionless parameters

$$\Omega_i = \frac{\rho_i}{\rho_c}, \quad \text{and} \quad \rho_i = (\rho_\Lambda, \rho_{BM}, \rho_{DM}, \rho_r, \rho_k), \quad \text{where } \rho_k = -\frac{k}{a^2}.\tag{16.55}$$

In the expressions presented above, the indices denote Λ—cosmological constant (also called the energy of vacuum or Dark Energy). BM—baryonic matter, meaning ordinary matter made from protons, neutrons, and electrons. At the cosmological scale, this matter takes the form of stars (also planets, etc) or of the interstellar, intergalactic, etc. gas. DM—dark matter, which we already discussed above. r—radiation, including neutrinos.

Clearly, the sum of all these relative densities is identically one,

$$\Omega_\Lambda + \Omega_{BM} + \Omega_{DM} + \Omega_r + \Omega_k \equiv 1 . \tag{16.56}$$

On the other hand, all Ω_i are time dependent, which means that the relative contents of the universe (sometimes called energy balance) changes with time. This concerns even Ω_Λ, regardless of the fact that $\rho_\Lambda = const$. In this case, the dependence on time comes from the critical density ρ_c. The present-day values are defined from various observational programs, based on the cosmic microwave background (CMB), the large-scale structure (LSS), gravitational lensing, baryon acoustic oscillations (BAO), and others. All these data converge to the approximate values

$$\Omega_\Lambda = 0.7 \quad \Omega_{BM} = 0.05 \quad \Omega_{DM} = 0.25 \quad \Omega_r \approx 10^{-4} \quad \Omega_k \approx 0. \tag{16.57}$$

The next parameter is deceleration q, which is defined by Eq. (16.16). The term "deceleration" is explicit historical evidence that some time ago, most of the cosmologists did not consider that the universe may actually accelerate. In the previous section, we learned that such acceleration is indeed observed. It is most likely related to the positive Λ-term. Hence, for the present universe, the observed value is $q < 0$. One can make a series expansion

$$\frac{a(t)}{a_0} = 1 + H_0(t - t_0) - \frac{q_0}{2} H_0(t - t_0)^2 + \dots , \tag{16.58}$$

where q_0 is the present-day value of q.

Exercise 15 Derive the conditions for the set of Ω_i, which provide $q > 0$, $q = 0$, and $q < 0$ at the given instant of time.

Exercise 16 Consider the universe with the relative densities $\Omega_{matter}^0 = \Omega_\Lambda^0 = 0.5$ at the instant of time t_0, and assume that all matter is dust. Derive the time when the acceleration of the universe starts.

Exercise 17 For the values cited in (16.57), calculate the ratios a_0/a_1 for which $\Omega_r^1 = \Omega_{DM}^1$ and a_0/a_2 for which $\Omega_r^2 = \Omega_{BM}^2$, assuming that both matter components have pressureless equations of state. Which values of Ω_Λ^0 would correspond to zero $q^1 = 0$ and $q^2 = 0$ for each of these cases?

16.3.2 Redshift Parameter

It would be easier to deal with the last exercise of the previous section by using the next useful notion which we will introduce now. It is called the redshift parameter z. The redshift is the most useful way to parameterize the expansion of the universe. There is a direct relation between z and the conformal factor of the metric $a(t)$, but we prefer to arrive at this relation from the definition that better explains the origin of the name of z. The initial definition is

$$z = \frac{\lambda_{\text{obs}} - \lambda_{\text{emit}}}{\lambda_{\text{emit}}}, \tag{16.59}$$

where λ_{emit} and λ_{obs} are the wavelength of the same portion of light (photon) when it was emitted by a distant cosmic object and when it was observed by us. It is easy to rewrite this relation in terms of the frequencies, $\nu = \frac{c}{\lambda}$,

$$z = \frac{\nu_{\text{emit}} - \nu_{\text{obs}}}{\nu_{\text{obs}}}. \tag{16.60}$$

Consider the propagation of light in the FLRW metric,

$$ds^2 = c^2 dt^2 - a^2(t) dl^2,$$
$$dl^2 = \frac{dr^2}{1 - kr^2} - r^2(d\theta^2 + \sin^2\theta d\varphi^2), \tag{16.61}$$

and the light ray that comes to observer at $r = 0$ from the emitter at $r = R$. We shall call the instant of emission of light t_{emit} and the instant of absorbing the same photon by the detector t_{obs}.

Taking the interval of time between the emission of two crests separated by one wavelength, we arrive at another two time instants,

$$t_{\text{emit}} + \frac{\lambda_{\text{emit}}}{c} \quad \text{and} \quad t_{\text{obs}} + \frac{\lambda_{\text{obs}}}{c}. \tag{16.62}$$

For the signal that travels cosmic distances, one can always regard that the trajectory as radial, in the spherical coordinates with the center at the emitter. As far as $d\theta = d\varphi = 0$, and since for the light ray $ds = 0$, from (16.61) follows for the original crest

$$\int_{t_{\text{emit}}}^{t_{\text{obs}}} \frac{cdt}{a(t)} = \int_{R}^{0} \frac{dr}{\sqrt{1 - kr^2}}. \tag{16.63}$$

For the next wave crest, there is

$$\int_{t_{\text{emit}} + \lambda_{\text{emit}}/c}^{t_{\text{obs}} + \lambda_{\text{obs}}/c} \frac{cdt}{a(t)} = \int_{R}^{0} \frac{dr}{\sqrt{1 - kr^2}}. \tag{16.64}$$

Subtracting (16.63) from (16.64), after some algebra we arrive at the relation

$$\int_{t_{\mathrm{emit}}}^{t_{\mathrm{emit}}+\lambda_{\mathrm{emit}}/c} \frac{dt}{a(t)} = \int_{t_{\mathrm{obs}}}^{t_{\mathrm{obs}}+\lambda_{\mathrm{obs}}/c} \frac{dt}{a(t)} . \tag{16.65}$$

Since the universe does not significantly expand during the periods of time $\frac{\lambda_{\mathrm{emit}}}{c}$ or $\frac{\lambda_{\mathrm{obs}}}{c}$, the integrands of the last expressions can be safely regarded as constants, and therefore (16.65) boils down to

$$\frac{\lambda_{\mathrm{emit}}}{a_{\mathrm{emit}}} = \frac{\lambda_{\mathrm{obs}}}{a_{\mathrm{obs}}} \implies \frac{\lambda_{\mathrm{obs}}}{\lambda_{\mathrm{emit}}} = \frac{a_{\mathrm{obs}}}{a_{\mathrm{emit}}} . \tag{16.66}$$

Assuming that $a_{\mathrm{obs}} = a_0$ (usually we set $a_0 = 1$) and $a_{\mathrm{emit}} = a$, one can combine Eqs. (16.66) and (16.59) to arrive at the standard expression for the redshift parameter,

$$1 + z = \frac{a_0}{a} . \tag{16.67}$$

The usefulness of z is based on the fact that (i) it is a direct measure of the relative expansion of the universe; (ii) it is related to the redshift of spectral lines, which can be directly measured by means of spectroscopy.

The entire expansion of the universe can be described in terms of z. This parameter shows how far from us the event is in both space and time. Today $z = 0$, and the redshift of the epoch, when a was half of the present value a_0, is $z = 1$. In the asymptotically far future for the open universe $a \to \infty$ and hence $z \to -1$.

It is instructive to consider an alternative approach to the redshift parameter. Let us start from the action of the free electromagnetic field in curved space (14.111). Let us note that this expression for the action of the four-potential A_μ is the unique possible one, which possesses gauge invariance and is also conformal invariant. For the sake of simplicity, let us consider the flat universe with $k = 0$. In terms of conformal time, one can write the metric in the form

$$g_{\mu\nu} = \eta_{\mu\nu} a^2(\eta) . \tag{16.68}$$

Due to conformal invariance, the action does not depend on the conformal factor of the metric,

$$S_{em} = -\frac{1}{4} \int d^4x \, F_{\alpha\beta} \, F_{\mu\nu} \, \eta^{\alpha\mu} \, \eta^{\beta\nu} . \tag{16.69}$$

Then the basic plane wave solution can be written in the form

$$A_\mu^{(l)} = e_\mu^l \, e^{ik\eta - i\mathbf{k}\cdot\mathbf{x}} , \tag{16.70}$$

where $k = |\mathbf{k}|$ and $e'_\mu = (e'_0, \mathbf{e}')$ with $l = 1, 2$ are polarization space-like four-vectors, which satisfy $\mathbf{e}' \perp \mathbf{k}$.

It is important to remember that $a(\eta)d\eta = dt$, and that η is not a physical time. Thus, k and $|\mathbf{k}|$ are not physical energy and momentum of the wave packet, and one has to make a rescaling of these quantities. The wavelength is $\Delta x = 2\pi/k$ in the $(\eta, \mathbf{x})$ coordinates, but in the physical coordinates $(t, \mathbf{x})$ it is

$$\lambda = a(t)\Delta x = \frac{2\pi a(t)}{k}. \tag{16.71}$$

Similarly, $\Delta \eta$ is a period of the wave in the $(\eta, \mathbf{x})$ coordinates. In the physical time variable t the period is

$$T = a(t)\Delta\eta = \frac{2\pi a(t)}{k} = \lambda, \qquad c = 1. \tag{16.72}$$

Therefore, the physical momentum and frequency of the photon are

$$\mathbf{q} = \frac{\mathbf{k}}{a(t)}, \quad \text{and} \quad \nu = \frac{k}{a(t)}, \quad \text{where} \quad k = |\mathbf{k}|. \tag{16.73}$$

In the expanding universe $\nu = \nu(t)$ decreases, signaling the redshift of the photon frequencies that arrive to us from space. Using (16.73) we arrive at

$$\lambda_{\text{obs}} = \frac{a_0}{a(t)} \lambda_{\text{emit}} = \lambda_{\text{emit}}(1 + z), \tag{16.74}$$

which is equivalent to (16.67).

Exercise 18 Using the Planck formula $\mathcal{E} = \hbar\nu$ for the energy of the individual photon, and Eq. (16.73), discuss the origin of the difference between the scaling of energy densities of dust-like matter $\rho_m \sim a^{-3}$ and radiation $\rho_r \sim a^{-4}$.

Exercise 19 Derive the relation between the derivative with respect to physical time t and with respect to the redshift z. Repeat the same program for conformal time η.
Solution.

$$\dot{X} = -(z + 1)H \frac{dX}{dz}, \quad X' = -a_0 H \frac{dX}{dz}. \tag{16.75}$$

16.3.3 Age of the Universe

Let us end this section by deriving the age of the universe, that is, the time that passed from the initial moment $t = 0$ in the solutions similar to (16.46) and (16.48), but with the cosmological constant included. In fact, since the transition from radiation to matter dominance occurs relatively soon after the "Big Bang" instant $t = 0$, it is possible to consider only matter, without radiation. Let us note that here we ignore the period of extremely fast expansion of the universe, called cosmic inflation. This expansion was so fast that after this period, the universe was almost empty. Then there was a period called *reheating* when the universe was filled by extremely hot matter that can be approximately seen as radiation. The time period of all this, plus the radiation-dominated epoch, was essentially smaller than the matter-dominated period, so we can set $t = 0$ at the moment when matter dominance started, without making a too big mistake.

Starting from

$$\frac{da}{dt} = a(t) H(t) \tag{16.76}$$

we immediately arrive at

$$dt = \frac{da}{a H(a)} \quad \Longrightarrow \quad t_0 = \int_0^1 \frac{da}{a H(a)}, \tag{16.77}$$

where we assumed that today $a_0 = a(t_0) = 1$ and that the initial value of a is zero. In order to see how Eq. (16.77) works, let us consider two particularly relevant cases. (i) Consider the universe dominated by dust-like matter with $k = 0$ and $\Lambda = 0$. Then

$$H^2 = \frac{8\pi G}{3} \rho_0 \left(\frac{a_0}{a}\right)^3 \quad \Longrightarrow \quad H^2 = \frac{H_0^2}{a^3}. \tag{16.78}$$

Replacing the last expression in Eq. (16.77), it is easy to take the integral,

$$t_0 = \frac{1}{H_0} \int_0^1 \sqrt{a}\, da = \frac{3}{2 H_0}. \tag{16.79}$$

Using the value from (16.50) we arrive at the age of the universe $t_0 \approx 9\, bi$, which is not satisfactory. The reason is that there are astrophysical objects older than this. Since nothing can be older than the universe, it is natural to look for another model of cosmic composition.
(ii) Consider the case of $k = 0$ and $\Omega_\Lambda \neq 0$. Repeating the previous considerations, we arrive at

$$H^2 = H_0^2 \left[\Omega_\Lambda^0 + \Omega_m^0 \left(\frac{a_0}{a}\right)^3\right]. \tag{16.80}$$

Replacing the last expression with $a_0 = 1$ in Eq. (16.77), it is easy to obtain the expression

$$t_0 = \frac{1}{H_0} \int_0^1 \frac{a\,da}{\sqrt{\Omega_\Lambda^0 a^4 + \Omega_m^0 a}} . \tag{16.81}$$

The last integral can be taken only numerically. For the values which roughly correspond to the observational data, $\Omega_\Lambda^0 = 0.7$ and $\Omega_m^0 = 0.3$ (remember that Ω_m^0 is a sum of baryonic and dark matter contributions) we get

$$t_0 = \frac{0.964}{H_0} \approx \frac{9.4\,bi}{h} . \tag{16.82}$$

For $h = 0.68$ this gives the age of the universe of approximately 13.8 bi, which is close to the optimized modern estimate.

Taken by itself, the simple calculation presented above does not mean that the positive cosmological constant is favored by observations. However, this is only one of many (and perhaps the simplest) pieces of evidence indicating its presence.

Exercise 20 Rewrite Eqs. (16.80) and (16.82) in terms of the redshift parameter. Construct generalizations of these formulas for nonzero Ω_k and Ω_r.

Exercise 21 Write down the Friedmann equation for the universe filled by reduced relativistic gas, with equation of state (16.28).[3] The solution is

$$H^2 = H_0^2 \left[\Omega_k^0 (1+z)^2 + \frac{\Omega_M^0}{\sqrt{1+b^2}} (1+z)^3 \sqrt{1 + b^2(1+z)^2} + \Omega_\Lambda^0 \right] \tag{16.83}$$

where the warmness parameter b is related to the speed of the particles (it is the same for all particles in this model) as

$$b = \frac{\rho_d^0}{\rho^0} = \frac{\beta}{\sqrt{1 - \beta^2}} . \tag{16.84}$$

The upper bound for the warmness of dark matter today, in the reduced relativistic gas model, is $b \approx 10^{-3}$ [20]. Write the version of (16.81) treating dark matter as reduced relativistic gas, and evaluate the change of the age of the universe, regarding also $\Omega_r^0 = 10^{-4}$ and $\Omega_k^0 = 0$. Compare the effect of warmness of dark matter and the effect of nonzero Ω_r^0.

Hint. Use perturbations in b up to the first order. Since present-day dark matter cannot be too warm (as otherwise it would escape from the galaxies and clusters of galaxies), this is a good approximation.

---

[3]The reader can see the papers [13, 20] for the details of the solution.

References

1. S. Dodelson, *Modern Cosmology* (Academic Press, Cambridge, 2003)
2. V. Mukhanov, *Physical Foundations of Cosmology* (Cambridge University Press, Cambridge, 2005)
3. D.S. Gorbunov, V.A. Rubakov, *Introduction to the Theory of the Early Universe: Vol. I. Hot Big Bang Theory; Vol. II. Cosmological Perturbations and Inflationary Theory* (World Scientific Publishing, Singapore, 2011)
4. S. Weinberg, *Cosmology* (Oxford University Press, Oxford, 2008)
5. Ph.J.E. Peebles, *Principles of Physical Cosmology* (Princeton University Press, Princeton, 1993)
6. P. Peter, J.-Ph Uzan, *Primordial Cosmology* (Oxford University Press, Oxford, 2013)
7. S. Weinberg, *Gravitation and Cosmology* (Wiley, Hoboken, 1972)
8. S.W. Hawking, G.F.R. Ellis, *The large scale structure of space-time* (Cambridge University Press, Cambridge, 1973)
9. L.D. Landau, E.M. Lifshits, *The Classical Theory of Fields - Course of Theoretical Physics*, vol. 2 (Butterworth-Heinemann, Oxford, 1987)
10. A.T. Fomenko, S.P. Novikov, B.A. Dubrovin, *Modern Geometry-Methods and Applications, Part I: The Geometry of Surfaces, Transformation Groups, and Fields* (Springer, Berlin, 1992)
11. F. Jüttner, Das Maxwellsche Gesetz der Geschwindigkeitsverteilung in der Relativtheorie. Ann. der Phys. **Bd 116**, S. 145 (1911)
12. A.D. Sakharov, The initial stage of an expanding universe and the appearance of a nonuniform distribution of matter. Zh. Eksp. Teor. Fiz. **49**(1), 345 [Sov. Phys. JETP **22** (1966) 241]. See also L.P. Grishchuk, Cosmological Sakharov oscillations and quantum mechanics of the early universe. arXiv:1106.5205
13. G. de Berredo-Peixoto, I.L. Shapiro, F. Sobreira, Simple cosmological model with relativistic gas. Mod. Phys. Lett. A **20**, 2723 (2005)
14. A.M. Pelinson, I.L. Shapiro, On the scaling rules for the anomaly-induced effective action of metric and electromagnetic field. Phys. Lett. B **694**, 467 (2011)
15. M.V. Fischetti, J.B. Hartle, B.L. Hu, Quantum effects in the early universe. 1. Influence of trace anomalies on homogeneous, isotropic, classical geometries. Phys. Rev. **D20**, 1757 (1979)
16. I.L. Shapiro, Effective action of vacuum: Semiclassical approach. Class. Quant. Grav. **25**, 103001 (2008). arXiv:0801.0216
17. E. Bianchi, C. Rovelli, Why all these prejudices against a constant? Unpublished. arXiv:1002.3966
18. L. Bergstrom, Non-baryonic dark matter: Observational evidence and detection methods. Rep. Prog. Phys. **63**, 793 (2000). hep-ph/0002126
19. Y. Akrami, and others, *Planck 2018 results. I. Overview and the cosmological legacy of Planck*, Planck Collaboration (2018). arXiv:1807.06205. N. Aghanim, and others, *Planck 2018 results. VI. Cosmological parameters*, Planck Collaboration (2018). arXiv:1807.06209
20. J.C. Fabris, I.L. Shapiro, F. Sobreira, DM particles: how warm they can be? JCAP **02**, 001 (2009). arXiv:0806.1969

Chapter 17
Special Sections

17.1 Metric-Scalar Theories

The theories describing not only a metric but also an additional scalar field (metric-scalar theories) are popular extensions of general relativity. In this section, we consider the conformal transformations in these theories and their mapping into purely metric models with modified Lagrangians. The contents of this section are rather pretty well known and have been discussed in many papers (see, e.g., [1, 2] for reviews). Our exposition will follow the papers [3, 4] and after that [5].

The most general action with no more that second derivative terms in the Lagrangian, describing metric and a single, real scalar field includes arbitrary functions A, B, C of the scalar ϕ.

$$S = \int d^4x \sqrt{-g} \left\{ A(\phi) g^{\mu\nu} \partial_\mu \phi \partial_\nu \phi + B(\phi) R - C(\phi) \right\}. \qquad (17.1)$$

This covers several special cases that will be described below.

17.1.1 Classification of Metric-Scalar Models

As a starting point, we consider the action introduced by O'Hanlon in [6],

$$S = \int d^4x \sqrt{-g'} \left\{ R'\Phi - V(\Phi) \right\}. \qquad (17.2)$$

Here scalar curvature R' corresponds to the metric $g'_{\mu\nu}$ and $g' = \det(g'_{\mu\nu})$. One has to note that the theory (17.2) describes a dynamical scalar Φ, despite there being no kinetic term for Φ in the action. The reason is that the derivatives of Φ emerge in the

© Springer Nature Switzerland AG 2019
I. L. Shapiro, *A Primer in Tensor Analysis and Relativity*, Undergraduate
Lecture Notes in Physics, https://doi.org/10.1007/978-3-030-26895-4_17

equation that corresponds to the variational derivative of (17.2) with respect to the metric.

We can perform simultaneous conformal transformation of the metric and an arbitrary reparameterization of the scalar Φ to the new variables $g_{\mu\nu}$ and ϕ according to

$$g'_{\mu\nu} = g_{\mu\nu}e^{2\sigma(\phi)}, \qquad \Phi = \Phi(\phi), \tag{17.3}$$

where $\sigma(\phi)$ and $\Phi(\phi)$ are arbitrary functions of ϕ. In the next formulas, we use notations where the lower numerical index of a function shows the order of its derivative with respect to ϕ. For instance,

$$B_1 = \frac{dB}{d\phi}, \quad A_2 = \frac{d^2B}{d\phi^2}, \quad \Phi_1 = \frac{d\Phi}{d\phi}, \quad \sigma_1 = \frac{d\sigma}{d\phi}, \quad \text{etc.} \tag{17.4}$$

According to what we learned earlier in Sect. 14.6 (also in Appendix of Sect. 15.2), in the new variables the action becomes

$$S = \int d^4x\sqrt{-g}\, e^{4\sigma(\phi)}\left\{e^{-2\sigma(\phi)}\Phi(\phi)\left[R - 6(\nabla\sigma)^2 - 6\Box\sigma\right] - V(\Phi(\phi))\right\} \tag{17.5}$$

$$= \int d^4x\sqrt{-g}\left\{\Phi(\phi)Re^{2\sigma(\phi)} + 6(\nabla\phi)^2e^{2\sigma(\phi)}[\Phi\sigma_1 + \Phi_1]\sigma_1 - V(\Phi(\phi))e^{4\sigma(\phi)}\right\},$$

where we integrated by parts and omitted total derivatives. It is easy to see that we have transformed the particular action (17.2) to the general form (17.1) with the following identification:

$$A = 6e^{2\sigma(\phi)}[\Phi\sigma_1 + \Phi_1]\sigma_1, \qquad B = \Phi(\phi)e^{2\sigma}, \tag{17.6}$$

or, equivalently,

$$A = 6B_1\sigma_1 - 6B(\sigma_1)^2, \qquad \Phi = Be^{-2\sigma}. \tag{17.7}$$

Replacing (17.7) into (17.6) we obtain the action (17.1) with the identification (17.6) plus

$$C(\phi) = V(\Phi(\phi))\, e^{4\sigma(\phi)}. \tag{17.8}$$

The expressions (17.6) and (17.8) enable one to introduce a useful classification of the metric-scalar models. Consider the special case when (17.2) is just the action of Einstein gravity. For this, we need to set $\Phi = -\kappa^2 = const$. In this case, $\kappa^2 = \frac{1}{16\pi G}$ and $V(-\kappa^2) = -\frac{\Lambda}{8\pi G}$. Since there is no scalar field, such a model has one less degree of freedom compared to the general case. But how can this be seen in the general expression (17.1)?

Let us rewrite the condition $\Phi = -\kappa^2 = const$ in terms of $A(\phi)$, $B(\phi)$, and $C(\phi)$. After some short algebra using (17.6) and (17.8) we obtain

$$2AB - 3(B_1)^2 = -3\,(\Phi_1)^2\,e^{4\sigma(\phi)} = 0, \tag{17.9}$$

$$C(\phi) = \lambda B^2(\phi), \quad \lambda = const. \tag{17.10}$$

It is clear that the model (17.1) with the condition (17.9) qualitatively differs from the general version. In this particular case, the scalar degree of freedom is compensated by the symmetry under conformal transformation of the metric plus reparameterization of the scalar field. This symmetry is a direct generalization of the conformal symmetry, which we already know. The second condition (17.10) is also relevant. If it is not satisfied, then the theory has no classical solutions for the full system of dynamical equations. It is easy to see that the action (17.2) with $\Phi = -\kappa^2$ satisfies both conditions (17.9) and (17.10).

Consider another simple example of the model satisfying (17.9) and (17.10), when $A = \frac{1}{2}$ and $B = \frac{1}{12}\phi^2$. Let us note that the sign of the coefficient A must be positive; otherwise, the square of the time derivative (kinetic term) of the scalar field becomes negative. In this case, the scalar field can be identified as an unphysical ghost, at least if it does not couple with gravity. In the case of a scalar coupled with gravity, the situation may be more complicated, but for the sake of simplicity we take $A = \frac{1}{2}$.

As far as Φ and B are related by the second formula from (17.7), it is impossible to provide that $\Phi = -\kappa^2$, which is dictated by the condition of positive kinetic energy for the tensor mode of the gravitational field. Let us ignore this condition for a while and set $\Phi = +\kappa^2$. Then the first formula of (17.7) tells us that $A = \frac{1}{2}$. Substituting these expressions into (17.8) we see that in the new variables the transformed action has the form

$$S = \int d^4x\sqrt{-g}\left\{\frac{1}{2}\,g^{\mu\nu}\partial_\mu\phi\partial_\nu\phi + \frac{1}{12}\,R\phi^2 - \frac{\lambda}{4!}\phi^4\right\}. \tag{17.11}$$

Thus, we met a contradiction between the positivity of the kinetic term for the tensor mode and the positivity of the kinetic term for the scalar mode. How can this contradiction be resolved? In the new variables, the starting Hilbert–Einstein action with the wrong sign of the kinetic term, meaning Eq. (17.2) with $\Phi = +\kappa^2$, corresponds to the conformally coupled scalar field ϕ. The extra scalar degree of freedom in (17.11) is compensated by the extra symmetry, namely, the local conformal invariance. Both theories are equivalent on the classical level. Thus, the scalar mode is not physical in this theory and the sign of its kinetic term does not matter. The correct choice is certainly $\Phi = -\kappa^2$. Just to complete the story, it is good to remember that in general relativity the gravitational wave on the flat background has two tensor polarizations and no physical scalar mode. Thus, the negative sign of the scalar kinetic term in general relativity should not be taken as a serious physical problem. For instance, this negative sign can be changed to the positive one if we consider the model with torsion [7]. The same can be easily achieved by means of the second scalar field. We leave it to the interested reader to elaborate this possibility.

And so, all the theories (17.1) can be divided into two sets. The actions of the first set satisfy the condition $2AB - 3(B_1)^2 = 0$. These models are conformally equivalent to the general relativity with the cosmological constant if the kinetic term of the scalar field has the wrong (negative) sign. General relativity itself belongs to this class, but it is a special (one can say extreme) case, when the scalar field is frozen and its kinetic term vanishes. In the first two exercises of this section, the reader is invited to prove two statements concerning this class of models.

17.1.2 Conformal Mapping of f(R) Models

For the second kind of metric-scalar models, $2AB - 3(B_1)^2 \neq 0$. Such models are conformally equivalent to the special O'Hanlon action (17.2) with a nonconstant Φ. In what follows we will discuss the transformations between a few special versions of these models and their relations with the models of modified gravity of the important class

$$S_f = \int d^4x \, \sqrt{-g} \, f(R). \tag{17.12}$$

Here $f(R)$ is a differentiable function.

Let us find the metric-scalar dual of the model (17.12) in the form (17.2). The equation of motion that follows from the variation of Φ in (17.2) has the form

$$R = V'(\Phi) = \frac{dV}{d\Phi}. \tag{17.13}$$

Solving (17.13) with respect to Φ and substituting this solution $\Phi = \Phi(R)$ back into (17.2), we arrive at the action (17.12) with

$$\Phi(R) \cdot R - V(\Phi(R)) = f(R). \tag{17.14}$$

This relation means the dynamical equivalence of the two actions. Later on, in this section, we confirm the validity of this procedure through the equations of motion for both metric and Φ.

One can easily find the relation between the functions $V(\Phi)$ and $f(R)$. Taking the derivative with respect to R of the both sides of Eq. (17.14), we arrive at the relation

$$\Phi(R) + R\,\Phi'(R) - V'(\Phi)\,\Phi'(R) = f'(R). \tag{17.15}$$

In this formula $\Phi = \Phi(R)$ or $R = R(\Phi)$. Using (17.13), Eq. (17.15) reduces to the simple relation

$$\Phi = f'(R), \tag{17.16}$$

indicating that the function $R = V'(\Phi)$ is an inverse function of $\Phi = f'(R)$. Thus, the prescription for deriving the potential $V(\Phi)$ for a given $f(R)$ is as follows.

(a) Calculate $\Phi = f'(R)$ and inverting it, obtain the relation $R = V'(\Phi)$. This operation requires $f''(R) \neq 0$, and we will assume that this condition is satisfied in our area of interest.

(b) Integrate over Φ,

$$V(\Phi) = \Omega_0 + \int_0^\Phi R(\Phi) \, d\Phi. \tag{17.17}$$

One comment is in order at this point. Imagine that the function f can be expanded into the power series

$$f(R) = f_0 + f_1 R + f_2 R^2 + \dots . \tag{17.18}$$

An arbitrary integration constant Ω_0 in (17.17) certainly depends on the constant f_0 part of the integrand of Eq. (17.12), which is indeed lost when taking derivative $f'(R)$. On the other hand, when being placed into the covariant action, Ω_0 cannot be regarded as an irrelevant constant because it is multiplied by the metric-dependent factor $\sqrt{-g}$.

In order to fix Ω_0 one can use the following simple considerations. By using (17.14) we arrive at

$$V(\Phi) = R\Phi - f(R), \quad \text{where} \quad \Phi = f'(R). \tag{17.19}$$

As far as (17.17) should be equal to (17.19), one can then fix Ω_0. Later on, we verify the validity of this procedure explicitly for a cosmological dS-like solution.

The prescription given above enables one to establish the potential function $V(\Phi)$ for a given $f(R)$. Let us check the results of the procedure at the level of the equations of motion. Taking variation of Eq. (17.12) with respect to the metric, we get

$$f'\left(R_{\mu\nu} - \frac{1}{2}Rg_{\mu\nu}\right) + \frac{1}{2}g_{\mu\nu}(R f' - f) - \nabla_\mu \nabla_\nu f' + g_{\mu\nu}\Box f = 0. \tag{17.20}$$

Using the same variational derivative for (17.2) we arrive at

$$\Phi\left(R_{\mu\nu} - \frac{1}{2}Rg_{\mu\nu}\right) = -\frac{1}{2}g_{\mu\nu}V(\Phi) + \nabla_\mu \nabla_\nu \Phi - g^{\mu\nu}\Box\Phi = 0. \tag{17.21}$$

The direct replacement shows that the equivalence between (17.20) and (17.21) holds if the relation

$$\frac{f'(R)R - f(R)}{f'(R)} = \frac{V(\Phi)}{\Phi} \qquad (17.22)$$

is satisfied. It is easy to check that the solution of this equation has the form (17.16).

Consider an important example for the procedure described above—the simplest case leading to the linear equation $\Phi = f'(R)$,

$$f(R) = \Omega - \kappa^2 R + \frac{\alpha}{2} R^2. \qquad (17.23)$$

This example is especially interesting because in cosmology it forms the basis of the Starobinsky model of inflation.

Inflation is the period of an extremely fast expansion that almost certainly took place at the very early stage of the universe. There is a lot of observational data supporting the idea of inflation. The quality of this data has grown in the last decades and one can say that nowadays inflation is a part of the standard cosmological model. The theory (17.23) is the basis of the historically first, and up to now most successful from the phenomenological viewpoint model of inflation [8].

Consider the mapping of (17.23) into metric-scalar theory. Using our previous results, we arrive at

$$\Phi(R) = f'(R) = -\kappa^2 + \alpha R \quad \Longrightarrow \quad V'(\Phi) = R = \frac{\Phi + \kappa^2}{\alpha}. \qquad (17.24)$$

Integrating (17.24) we get

$$V(\Phi) = \Omega_0 + \frac{\kappa^2 \Phi}{\alpha} + \frac{\Phi^2}{2\alpha}. \qquad (17.25)$$

Finally, in order to fix the integration constant, one has to substitute $\Phi = -\kappa^2 + \alpha R$ back into (17.25) and compare it to (17.23). This procedure gives

$$\Omega = \Omega_0 - \frac{\kappa^4}{2}. \qquad (17.26)$$

The next step is the conformal mapping of the O'Hanlon model (17.2) with the potential (17.25) into the minimal scalar model

$$S_{min}[g_{\mu\nu}, \varphi] = \int d^4x \sqrt{-g} \left\{ -\kappa^2 R + \frac{1}{2} g^{\mu\nu} \partial_\mu \varphi \partial_\nu \varphi - U(\varphi) \right\}. \qquad (17.27)$$

Our purpose is to find the appropriate conformal transformation and the form of the potential $U(\varphi)$. Using the notation of the metric $g'_{\mu\nu}$ for the model (17.2) with (17.25), we perform the conformal transformation of the metric (17.3). As before, we arrive at Eq. (17.6). One can choose σ such that $\Phi e^{2\sigma} = -\kappa^2$. Then the first term $\Phi e^{2\sigma} R$ coincides with the Einstein–Hilbert term, and the third term $-6\Phi e^{2\sigma} \Box \sigma$

becomes a total derivative, which does not affect the equations of motion. In order to provide the standard form of the kinetic term, we take

$$\varphi = 2\sqrt{3}\,\kappa\,\sigma \quad \text{such that} \quad \Phi = -\kappa^2 \exp\left\{-\frac{\varphi}{\sqrt{3}\,\kappa}\right\}. \tag{17.28}$$

The output for the potential has the form

$$U(\varphi) = e^{4\sigma}\,V(\Phi) = \frac{\kappa^4}{\Phi^2}\,V(\Phi)$$

$$= \Omega + \frac{\kappa^4}{2}\left(1 - \frac{1}{\alpha}\right) + \frac{\kappa^4}{2\alpha}\left(1 - e^{-\frac{\varphi}{\sqrt{(3)}\kappa}}\right)^2. \tag{17.29}$$

where we used (17.26) and (17.28).

Exercise 1 Suggest another example of the function $f(R)$ that enables one to perform the mapping to the O'Hanlon action (17.2). By means of conformal transformation, try to find the analog of the potential (17.29).

17.1.3 Bekenstein's Conformal Duality

As we have learned in the previous subsection, conformal versions of the metric-scalar theories satisfy the conditions (17.9) and (17.10). Therefore, the action of such a model is completely defined by the function $B(\phi)$ and constant λ. This action can be denoted as $S_{B(\phi),\lambda}$. One can change from the theory with $B(\phi)$ to the theory with another function $K(\phi)$ by means of the conformal transformation

$$S_{B(\phi),\lambda}[\bar{g}_{\mu\nu}, \phi] = S_{K(\phi),\lambda}[g_{\mu\nu}, \phi], \tag{17.30}$$

where

$$\bar{g}_{\mu\nu} = \frac{K(\phi)}{B(\phi)}\,g_{\mu\nu}. \tag{17.31}$$

In particular, one can choose $K = const$ and demonstrate the conformal equivalence between general relativity and the models $S_{B(\phi),\lambda}$ with nonconstant B.

Equation (17.30) can be used as a basis for the symmetry called conformal duality. It was discovered by Bekenstein in 1974 [9]. In this subsection, we follow the paper [5], where the result has been obtained in a more economical way.

The conformal duality holds for some of the nonconformal versions of the actions (17.1). Consider the sum of the two conformal actions

$$S_{B(\phi),\lambda}(\bar{g}_{\mu\nu}, \phi) + S_{L(\phi),\tau}(\bar{g}_{\mu\nu}, \phi). \tag{17.32}$$

Here $B(\phi)$ and $L(\phi)$ are some functions, λ and τ are constants. Let us perform the special conformal transformation (17.31). Then (17.32) becomes

$$S_{K(\phi),\lambda}[g_{\mu\nu}, \phi] + S_{M(\phi),\tau}[g_{\mu\nu}, \phi], \quad \text{where} \quad M(\phi) = \frac{L(\phi)K(\phi)}{B(\phi)}. \quad (17.33)$$

An especially interesting example of this transformation is when one of the components in (17.32) corresponds to general relativity, $L = \gamma\kappa^{-2} = const$. For the sake of simplicity, we choose the second function in the form $B(\phi) = \phi\kappa^{-2}$, such that the scalar ϕ is dimensionless. Then the symmetry transformation reads

$$\left[S_{\frac{\phi}{\kappa^2},\lambda} + S_{\frac{\gamma}{\kappa^2},\tau} \right]_{g_{\mu\nu}} = \left[S_{\frac{1}{\phi\kappa^2},\tau} + S_{\frac{1}{\gamma\kappa^2},\lambda} \right]_{\bar{g}_{\mu\nu}} \quad (17.34)$$

and hence for this specific case we face a conformal duality that exchanges

$$\phi \leftrightarrow \phi^{-1}, \quad \gamma \leftrightarrow \gamma^{-1}, \quad \tau \leftrightarrow \lambda, \quad \bar{g}_{\mu\nu} \leftrightarrow g_{\mu\nu} = \phi\gamma\bar{g}_{\mu\nu}. \quad (17.35)$$

The conformal duality links different metrics, different values of scalar field, and different values of coupling constants. In particular, there is a possibility to link the regimes of strong and weak gravitational interactions. Further details can be found in Refs. [5, 9].

Exercise 2 (i) Check the calculations leading to Eq. (17.11) and find the constant λ in terms of V and κ^2. (ii) Prove that the general model that satisfies (17.9) and (17.10) can be obtained from (17.11) by an arbitrary reparameterization of the scalar field $\phi = \phi(\chi)$.

Exercise 3 (i) Answer the following question: is it true that any two models that satisfy the constraints (17.9) and (17.10) can be transformed one into another by the conformal transformation of the metric alone, without reparameterization of the scalar? (ii) Discuss whether is it possible to achieve the same transformation by means of reparameterization of the scalar field only, without conformal transformation of the metric.

Hint. In the second part of the exercise, one has to pay special attention to the extreme case, that is, the theory (17.2) with $\Phi = const$.

Exercise 4 Derive the equations of motion

$$\frac{1}{\sqrt{-g}}\frac{\delta S}{\delta g_{\mu\nu}} \quad \text{and} \quad \frac{1}{\sqrt{-g}}\frac{\delta S}{\delta \Psi} \quad (17.36)$$

for the theory (17.2). Extract the dynamical (with derivatives) equation for Ψ. Explain why in the general model (17.1) the trace of the first equation and the second equation in (17.36) is not linearly dependent. In which kind of theories can one expect linear dependence in this case?

Exercise 5 Verify Eq. (17.20) and check the validity of Eq. (17.22).

Exercise 6 (i) Generalize the conditions of conformal symmetry for the D-dimensional space–time. (ii) Find the form of conformal transformation that compensates an arbitrary reparameterization of the scalar field, $\psi = \psi(\phi)$.

Results:

$$\text{(i)} \ \ A(\phi) = \frac{D-1}{D-2}\frac{B_1{}^2(\phi)}{B(\phi)}, \quad C(\phi) = \lambda \, B^{\frac{D}{D-2}}(\phi), \quad \lambda = const, \quad (17.37)$$

$$\text{(ii)} \ \ e^{(D-2)\sigma(\phi)} = \frac{B(\phi)}{B(\psi(\phi))}. \quad (17.38)$$

Exercise 7 Try to invent more examples of $f(R)$ that admit the explicit mapping to the model (17.2). Making a conformal transformation, find analogs of Eq. (17.29) for these examples.

Exercise 8 Generalize the conformal duality for the space–time of dimension $D \neq 2$. The results can be found in [5]. Discuss the limit $D \to 2$.

Exercise 9 Verify all calculations presented in the appendix.

Appendix

One can perform a simple verification of the described procedure for fixing Ω. For this, let us derive the special de Sitter solution for both theories (17.12) and (17.2) in the case of (17.23). The metric we are interested in is

$$ds^2 = g_{\mu\nu}dx^\mu dx^\nu = dt^2 - a^2(t)\left(\frac{dr^2}{1-kr^2} + r^2\,d\Omega\right),$$

where $a(t) = e^{\sigma(t)}$. Afterward we can fix this function to be $\sigma(t) = H_0 t$. The equation for σ in the theory (17.2) is

$$\frac{1}{\sqrt{-g}}\frac{\delta S_1}{\delta \sigma} = -6e^{-2\sigma}\left(2\Phi k + \Phi'' + 2\sigma''\Phi + 2\Phi'\sigma' + 2\sigma'^2\Phi\right) - 4V(\Phi) = 0. \quad (17.39)$$

Here the prime stands for the derivative with respect to conformal time, e.g.,

$$\sigma' = \frac{d\sigma}{d\eta} = a(t)\frac{d\sigma}{dt},$$

while the derivative with respect to the physical time t is denoted by a dot. In terms of the physical time and for the simplest case $k = 0$, we arrive at the equation

$$- 12H_0^2\, \Phi - 9H_0\, \dot{\Phi} - 3\, \ddot{\Phi} = 2\, V. \tag{17.40}$$

Assuming $\Phi = -\kappa^2 + \alpha\, R$ and taking into account $R = -12\, H_0^2$ for the metric of our interest, we get

$$V = 6H_0^2\kappa^2 - 6\, H_0^2\alpha R,$$

and finally

$$\Omega = 6\, H_0^2\, \kappa^2\,. \tag{17.41}$$

On the other hand, starting from

$$S_f = \int d^4x\, \sqrt{-g}\, f(R) = \int d^4x\, \sqrt{-g}\left[\Omega - \kappa^2\, R + \frac{\alpha}{2}\, R^2\right] \tag{17.42}$$

we obtain

$$\frac{1}{\sqrt{-g}}\, \frac{\delta\, S_f}{\delta\, \sigma} = 4\Omega - 6\kappa^2 e^{-2\sigma}(2\sigma'^2 + 2\sigma'')$$
$$+ 18\alpha e^{-4\sigma}(2\sigma^{(4)} - 12\sigma'^2\sigma'') = 0. \tag{17.43}$$

The straightforward calculation confirms that the solution $\sigma = H_0 t = -\ln(H_0|\eta|)$ corresponds to the relation (17.26).

17.2 Vierbein and Spinor Connection

In the previous section, we learned that general relativity is a covariant theory of the metric field. On the other hand, sometimes it is useful to have variables describing the gravitational field. In particular, for defining the covariant derivative of a fermion, one needs an object called a tetrad. The German name *vierbein* is also frequently used. The covariant description of fermions coupled to gravity is important, in particular, because all elementary particles (electrons and other leptons, quarks) are half-integer particles, meaning they are fermions. The same concerns many of the microscopic composite particles, starting from the proton. For instance, in case we want to describe the motion of high energy and ultrahigh energy cosmic rays in space, the main candidate for the corresponding particles is the proton; its spin and its interaction with gravity have to be taken into account. In this case, we need to consider vierbein and formulate the covariant derivative of fermions. Here we will not resolve this problem completely, but describe the tetrad formalism and explain how the covariant derivative of the tetrad can be formulated. Starting from this point, one can easily learn the formalism of fermions in curved space. The contents of this section are a slightly reduced version of the pedagogical article [10] where consideration on spinors and their covariant derivatives is given in full detail.

The names tetrad or vierbein indicate that the object is four dimensional, but the formalism described below can be easily generalized to the space of any dimension and also does not depend on the signature of the metric. Anyway, for the sake of definiteness we will refer to the four-dimensional space–time with the $M_{1,3}$ signature.

17.2.1 Definition of Tetrad

Let us start by defining the tetrad formalism on the $M_{1,3}$ Riemann space. Locally, at point P one can introduce the flat metric η_{ab}. This means that the vector basis includes four orthonormal vectors $\mathbf{e}_a$, such that $\mathbf{e}_a \cdot \mathbf{e}_b = \eta_{ab}$. Here $\mathbf{e}_a$ are four-dimensional vectors in the tangent space of the manifold at the point P. Furthermore, X^a are local coordinates on $M_{1,3}$, which can be also seen as coordinates in the tangent space in a close vicinity of the point P. With respect to the general coordinates x^μ, we can write $\mathbf{e}_a = e_a^\mu \mathbf{e}_\mu$, where $\mathbf{e}_\mu$ is a corresponding local basis and e_a^μ are transition coefficients from one basis to another. Hence

$$e_a^\mu = \frac{\partial x^\mu}{\partial X^a}, \quad e_\nu^a = \frac{\partial X^a}{\partial x^\nu}$$

and therefore

$$e_a^\mu e^{a\nu} = e_a^\mu e_b^\nu \eta^{ab} = g^{\mu\nu}, \quad e_\mu^a e_{a\nu} = g_{\mu\nu}, \quad e_\mu^a e_b^\mu = \delta_b^a, \quad e_\mu^a e_a^\alpha = \delta_\mu^\alpha. \tag{17.44}$$

We assume that the Greek indices $\mu, \nu, \ldots$ are raised and lowered by the covariant metrics $g_{\mu\nu}$ and $g^{\mu\nu}$, while the indices $a, b, c \ldots$ are raised and lowered by η_{ab} and η^{ab}. The transition from one type or indices to another is done using the tetrad since it is just a change of basis. Taking this into account, in some cases, we admit objects with mixed indices, e.g.,

$$T^a{}_{\cdot\mu} = T^\nu{}_{\cdot\mu}\, e_\nu^a = T^a{}_{\cdot b}\, e_\mu^b. \tag{17.45}$$

Formulas (17.44) show that the descriptions in terms of the metric and in terms of the tetrad are equivalent. The same concerns also the invariant volume of integration. It is easy to show that

$$\det\left(e_\mu^a\right) = \sqrt{|g|}, \quad g = \det g_{\mu\nu}, \tag{17.46}$$

because

$$g = \det\left(e_\mu^a e_\nu^b \eta_{ab}\right) = \left(\det e_\mu^a\right)^2 \cdot \det \eta_{ab}. \tag{17.47}$$

Indeed, the local frame coordinates X^a are not unique, since even for the locally Minkowski metric it is possible to make transformations $\mathbf{e}'_a = \Lambda^b_{a'} e_b$. The metric doesn't change,

$$\mathbf{e}'_a \cdot \mathbf{e}'_c = \Lambda^b_{a'} \Lambda^d_{c'} \mathbf{e}_b \cdot \mathbf{e}_d = \eta_{bd} \Lambda^b_{a'} \Lambda^d_{c'} = \eta_{ac} . \tag{17.48}$$

In the flat space–time case, such frame rotations with constant $\Lambda^b_{a'}$ form a Lorentz group. The situation changes in curved space–time manifolds, because then $\Lambda^a_{b'}$ depends on the point P. For example, we can consider an infinitesimal transformation

$$\Lambda^a_{b'} = \delta^a_{b'} + \Omega^a_{b'}(x) , \tag{17.49}$$

which preserves the form of the tangent-space metric. Then (17.48) gives

$$(\delta^a_{b'} + \Omega^a_{b'})(\delta^d_{c'} + \Omega^d_{c'}) \eta_{ad} = \eta_{bc} . \tag{17.50}$$

It is an easy exercise to show that in this case $\Omega_{ab}(x) = -\Omega_{ba}(x)$. Here and in all similar situations Latin indices are raised and lowered by the Minkowski metric η_{ab}.

Exercise 10 Consider the metric of the form $g_{\mu\nu} = \eta_{\mu\nu} + \kappa h_{\mu\nu}$, where κ is a small parameter. The flat metric with $\kappa = 0$ corresponds to the tetrad $\bar{e}^a_\mu$. Metric-dependent functions can be presented as a power series in κ. (i) Find the expansions, at least up to the order $O(\kappa^2)$, for the two tetrads e^a_μ and e^ν_b.
(ii) Consider a more general expansion $g'_{\mu\nu} = g_{\mu\nu} + h_{\mu\nu}$, with a perturbation over an arbitrary background. Find the nonperturbative expansion (to *all* orders in $h_{\mu\nu}$) for e^a_μ and e^ν_b.

Exercise 11 Present the Schwarzschild solution in terms of the tetrads e^a_μ, e^ν_b, e_{vb} and $e^{\mu a}$. Does the invariant expression for ds^2 change in this case?

17.2.2 Spinor Connection in General Relativity

Let us consider the covariant derivative in tetrad formalism and introduce the notion of spin connection. As we have mentioned earlier, this name appears here because this new object is especially useful for formulating fermionic (spinor) fields on a curved background, but this is beyond the scope of the present book.

In the metric formalism, the covariant derivative of a vector V^μ is

$$\nabla_\nu V^\mu = \partial_\nu V^\mu + \Gamma^\mu_{\lambda\nu} V^\lambda , \tag{17.51}$$

where the affine connection is given by a standard Christoffel symbol expression,

$$\Gamma^\lambda_{\mu\nu} = \frac{1}{2} g^{\mu\tau} \left(\partial_\mu g_{\lambda\nu} + \partial_\nu g_{\lambda\mu} - \partial_\lambda g_{\mu\nu} \right) . \tag{17.52}$$

Our purpose is to construct a version of a covariant derivative for objects, such as a tetrad, which have local Lorentz indices. Obviously, this is a nontrivial task since a tetrad has a Lorentz index, it is not a usual vector or tensor and thus its covariant derivative cannot be conventionally defined. Hence, our constructions will necessarily involve a certain amount of ad hoc assumptions.

It is easy to understand that (different from the flat space–time) in the Lorentz frame X^a the covariant derivative of the same vector cannot just be equal to $\partial_a V^b$ because, otherwise, we arrive at a contradiction. The details will become clear below: let us just say that qualitatively; the reason is that in curved space a shift from the point P means a change from one tangent space to another; therefore, deriving the corresponding difference between two values of the vector requires an additional definition. Let us suppose that the desired covariant derivative is a linear operator and satisfies Leibniz rule. The general expression satisfying these conditions is

$$\nabla_a V^b = \partial_a V^b + \tilde{\omega}^a_{\cdot ca} V^c, \tag{17.53}$$

where $\tilde{\omega}^a_{\cdot ca}$ are unknown coefficients. In the flat space–time $\tilde{\omega}^a_{\cdot ca} = 0$.

We request that the vector components satisfy $V^\mu = e^\mu_a V^a$ and the tensor components satisfy $\nabla_\nu V^\mu = e^a_\nu e^\mu_b \nabla_a V^b$. Then, according to (17.51) and (17.53), we have

$$\nabla_\lambda V^\mu = \partial_\lambda V^\mu + \Gamma^\mu_{\tau\lambda} V^b = e^a_\lambda e^\mu_b \nabla_a V^b = e^a_\lambda e^\mu_b \left(\partial_a V^b + \tilde{\omega}^b_{\cdot ca} V^c \right). \tag{17.54}$$

Let us elaborate the usual partial derivative in the covariant terms,

$$\partial_a V^b = e^\lambda_a \partial_\lambda \left(V^\tau e^b_\tau \right) = e^\lambda_a e^b_\tau \partial_\lambda V^\tau + e^\lambda_a V^\tau \partial_\lambda e^b_\tau.$$

Then we get in (17.54),

$$\nabla_\lambda V^\mu = \delta^\mu_\tau \partial_\lambda V^\tau + V^\tau e^a_\lambda e^\mu_b \partial_a e^b_\tau + e^\mu_b e^c_b e^c_\tau V^\tau \tilde{\omega}^b_{\cdot c\lambda}$$
$$= \partial_\lambda V^\mu + V^\tau (e^\mu_b \partial_\lambda e^b_\tau + e^\mu_b e^c_\tau \tilde{\omega}^b_{\cdot c\lambda}).$$

Therefore, we arrive at the equation for $\tilde{\omega}^b_{\cdot c\lambda}$,

$$\Gamma^\mu_{\tau\lambda} = e^\mu_b \partial_\lambda e^b_\tau + e^\mu_d e^c_\tau \tilde{\omega}^d_{\cdot c\lambda}. \tag{17.55}$$

Multiplying the last equation by $e^a_\mu e^{\tau b}$ we arrive at the following solution:

$$\tilde{\omega}^{ab}_{\cdot\cdot\lambda} = e^a_\mu e^{\tau b} \Gamma^\mu_{\tau\lambda} - e^{\tau b} \partial_\lambda e^a_\tau. \tag{17.56}$$

One of the important features of the last expression is antisymmetry in (a, b), that means $\tilde{\omega}^{ab}_{\cdot\cdot\mu} = -\tilde{\omega}^{ba}_{\cdot\cdot\mu}$. In order to see this, consider the sum

$$\tilde{\omega}^{ab}_{\;\cdot\cdot\mu} + \tilde{\omega}^{ba}_{\;\cdot\cdot\mu} = e^a_\nu e^{\lambda b}\Gamma^\nu_{\lambda\mu} + e^b_\nu e^{\lambda a}\Gamma^\nu_{\lambda\mu} - e^{\lambda b}\partial_\mu e^a_\lambda - e^{\lambda a}\partial_\mu e^b_\lambda$$

$$= \frac{1}{2}\left(\partial_\lambda g_{\nu\mu} + \partial_\mu g_{\nu\lambda} - \partial_\nu g_{\mu\lambda}\right)\cdot\left(e^{a\nu}e^{\lambda b} + e^{a\lambda}e^{\nu b}\right) - e^{b\lambda}\partial_\mu e^a_\lambda - e^{a\lambda}\partial_\mu e^b_\lambda$$

$$= e^{a\nu}e^{\lambda b}e^c_\nu\partial_\mu e_{c\lambda} + e^{a\nu}e^{\lambda b}e_{c\lambda}\partial_\mu e^c_\nu - e^{\lambda b}\partial_\mu e^a_\lambda - e^{a\lambda}\partial_\mu e^b_\lambda = 0. \tag{17.57}$$

Another observation follows from Eqs. (17.51) and (17.53). We can rewrite (17.51) in the form

$$\nabla_\nu(V^a e^\mu_a) = \partial_\mu(V^a e^\mu_a) + \Gamma^\mu_{\lambda\nu}V^a e^\lambda_a \tag{17.58}$$

and construct the "covariant derivative" of the tetrad $\nabla_\nu e^\mu_a$ in a way consistent with the Leibnitz rule and with both (17.58) and (17.53). Then

$$\nabla_\nu V^\mu = \partial_\nu V^\mu + \Gamma^\mu_{\lambda\nu}V^\lambda = \nabla_\nu(V^b e^\mu_b) = e^a_\nu\nabla_a(V^b e^\mu_b) \tag{17.59}$$
$$= e^a_\nu e^\mu_b\nabla_a V^b + e^a_\nu V^b\nabla_a e^\mu_b = e^a_\nu e^\mu_b(\partial_a V^b + \tilde{\omega}^b_{\cdot ca}V^c) + e^a_\nu V^b\nabla_a e^\mu_b.$$

At the same time, we have

$$\partial_\nu V^\mu = e^a_\nu\partial_a(V^b e^\mu_b) = e^a_\nu e^\mu_b\partial_a V^b + V^b e^a_\nu\partial_a e^\mu_b. \tag{17.60}$$

By combining (17.59) and (17.60) one can easily obtain the relation

$$V^\lambda e^c_\lambda\tilde{\omega}^b_{\cdot ca}e^a_\nu e^\mu_b + e^a_\nu e^b V^\lambda\nabla_a e^\mu_b = V^\lambda e^b_\lambda e^a_\nu\partial_a e^\mu_b + V^\lambda\Gamma^\mu_{\lambda\nu}. \tag{17.61}$$

Since this relation should be true for all V^λ, we arrive at

$$e^a_\nu e^b_\lambda\nabla_a e^\mu_b = e^a_\nu e^b_\lambda\partial_a e^\mu_b + \Gamma^\mu_{\lambda\nu} - e^a_\nu e^c_\lambda e^\mu_b\tilde{\omega}^b_{\cdot ca}. \tag{17.62}$$

Multiplying this equation by $e^\nu_d e^\lambda_e$ one can get (changing some indices)

$$\nabla_a e^\mu_b = \partial_a e^\mu_b + \Gamma^\mu_{\lambda\nu}e^\lambda_b e^\nu_a - \tilde{\omega}^{cb}_{\cdot\cdot\nu}e^\nu_a e^\mu_c. \tag{17.63}$$

Since this is a tensor quantity, we can multiply it by e^a_τ and obtain the final result

$$\nabla_\tau e^\mu_b = e^a_\tau\partial_a e^\mu_b + \Gamma^\mu_{\lambda\tau}e^\lambda_b - \tilde{\omega}^c_{\cdot b\tau}e^\mu_c. \tag{17.64}$$

One can show that, also,[1]

$$\nabla_\tau e_{\mu a} = e^b_\tau\partial_b e_{\mu a} - \Gamma^\lambda_{\mu\tau}e^\lambda_a - \tilde{\omega}_a{}^b{}_{\cdot\tau}e_{\mu b}. \tag{17.65}$$

---

[1] This can be done by either repeating the steps leading to (17.64) or by directly using (17.64). We recommend the reader to do both things as a small exercise.

Finally, by direct replacement of (17.56) one can easily check that both covariant derivatives vanish, $\nabla_\tau e_b^\mu = 0$ and $\nabla_\tau e_{\mu a} = 0$. This fact is a direct consequence of the metricity property of the covariant derivative, because

$$\nabla_\tau g_{\mu\nu} = \nabla_\tau \left(e_\mu^a e_\nu^b \eta_{ab}\right) = 2\eta_{ab}\, e_\mu^a \cdot \nabla_\tau e_\nu^b = 0. \tag{17.66}$$

The spinor connection $\omega_\mu^{\;ab}$ is defined as

$$\omega_\mu^{\;ab} = -\frac{1}{2}\tilde{\omega}^{ab}_{\;\;\mu} = \frac{1}{2}(e_\tau^b e^{\lambda a}\Gamma_{\lambda\mu}^\tau - e^{\lambda a}\partial_\mu e_\lambda^b)$$
$$= \frac{1}{4}(e_\tau^b e^{\lambda a} - e_\tau^a e^{\lambda b})\Gamma_{\lambda\mu}^\tau + \frac{1}{4}(e^{\lambda b}\partial_\mu e_\lambda^a - e^{\lambda a}\partial_\mu e_\lambda^b). \tag{17.67}$$

Here we established the relation with Eq. (17.56) and presented both compact and explicitly antisymmetric forms of the connection.

Let us now discuss the relation of the spinor connection to the Riemann tensor. For this, we define the new version of the curvature,

$$R_{\mu\nu}^{\;\;ab} = \partial_\mu\omega_\nu^{\;ab} - \partial_\nu\omega_\mu^{\;ab} + \omega_\mu^{\;ac}\,\omega_{\nu c}^{\;\;b} - \omega_\nu^{\;ac}\,\omega_{\mu c}^{\;\;b}. \tag{17.68}$$

One can show that

$$R_{\mu\nu ab} = R_{\mu\nu\rho\sigma}\, e_a^\rho\, e_b^\sigma. \tag{17.69}$$

The proof consists of a direct replacement of Eq. (17.56) into the expression (17.68) and some algebra which we skip here for brevity and leave as an exercise to the interested reader.

Exercise 12 Verify the formula (17.69) in a way described above. Write down the expressions for the Ricci tensor, scalar curvature, Weyl tensor, and Einstein tensor in terms of the vierbein and spinor connection, without using a metric or $\Gamma_{\mu\nu}^\lambda$.

Exercise 13 The first-order (or Palatini) formalism in general relativity is based on the variables $g_{\rho\sigma}$ and $\Gamma_{\mu\nu}^\lambda$, which are regarded as independent from each other. Consider the corresponding action

$$S_{\mathrm{I}} = -\frac{1}{16\pi G}\int d^D x\sqrt{|g|}\, g^{\mu\nu} R_{\mu\nu}(\Gamma), \tag{17.70}$$

where the Ricci tensor is given by Eq. (14.58).
(i) Consider the dynamical variables $g_{\mu\nu}$ and $\Gamma_{\alpha\beta}^\lambda$ to be independent and take variational derivatives

$$\frac{\delta S_{\mathrm{I}}}{\delta g_{\mu\nu}} \quad \text{and} \quad \frac{\delta S_{\mathrm{I}}}{\delta \Gamma_{\alpha\beta}^\lambda}. \tag{17.71}$$

Demonstrate that the equations of motion in this theory are equivalent to the Einstein equations in general relativity.

(ii) Repeat the procedure with another couple of variables, namely, $\Phi^{\mu\nu} = \sqrt{|g|}\, g^{\mu\nu}$ and $\Gamma^{\lambda}_{\alpha\beta}$. Then the variation derivatives are

$$\frac{\delta S_{\mathrm{I}}}{\delta \Phi^{\mu\nu}} \quad \text{and} \quad \frac{\delta S_{\mathrm{I}}}{\delta \Gamma^{\lambda}_{\alpha\beta}}. \tag{17.72}$$

(iii) Discuss the equivalence between the theory in the two different sets of dynamical variables described in points (i) and (ii). How can the equivalence hold if in the first case the theory is non-polynomial (exactly as in the usual second-order formalism) and in the second case it is polynomial?

(iv) Discuss how the covariance of the theory is preserved in equations (17.72). Why are the covariant Einstein equations equivalent to (17.71), which are apparently noncovariant?

(v) Discuss whether the equivalence between the Palatini formalism and the usual second-order (in derivatives) formalism of general relativity holds in the presence of matter. Which conditions for the matter actions (be it composed of fluids or fields) should be imposed to preserve such an equivalence?

Exercise 14 Consider a modified version of the previous exercise. To this end, rewrite action (17.70) in terms of the vierbein and spinor connection, using the results of Exercise 1. Show that this formulation is completely equivalent to the one based on Eq. (17.71).

Exercise 15 Explore the equivalence of the first- and second-order formalisms from one side and the equivalence between $g^{\alpha\beta} - \Gamma^{\lambda}_{\mu\nu}$ and $e^{\mu}_{c} - \omega^{\lambda}{}_{ab}$ descriptions from another side in the action of modified gravity of the form

$$S_{\mathrm{IF}} = -\frac{1}{16\pi G} \int d^4x \sqrt{|g|}\, F\big(g^{\mu\nu} R_{\mu\nu}(\Gamma)\big), \tag{17.73}$$

where $F(x)$ is an arbitrary function. Show that, in general, neither of these two equivalences holds [11].

17.3 Riemann Normal Coordinates

The Riemann normal coordinates represent a useful tool in the gravitational physics. These coordinates are used not only in classical theory but also for quantum calculations in curved space–time, where they represent a basis for the local momentum representation. We present a simplified and intuitive derivation of the normal coordinates expansion up to the third power in a curvature tensor. In what follows we shall, in general, follow the standard reference on the subject [12]. Also, there are recent publications [13, 14] where one can find more advanced approaches to normal coordinates.

Our general purpose is to construct the expansion of the metric near a given point P, such that the coefficients of the power series expansion in the deviations y^μ are covariant quantities in the point P. We will soon ensure that this condition is satisfied if we impose certain requirements on the symmetrized partial derivatives of the affine connection in the point P.

17.3.1 General Notions

Consider a D-dimensional manifold $\mathcal{M}$ provided by the metric $g_{\mu\nu}$. For the sake of simplicity, we will not introduce torsion or nonmetricity. In what follows we construct special coordinates called normal or Riemann normal. These coordinates correspond to a covariant expansion in the vicinity of a given (arbitrary) point P. All quantities related to this point will be marked by the symbol "$\circ$" on the top. For instance, the coordinates of this point are $\overset{\circ}{x}{}^\mu$, the metric in this point is $g_{\mu\nu}(\overset{\circ}{x}) = \overset{\circ}{g}_{\mu\nu}$, and the affine connection (Christoffel symbol) is $\overset{\circ}{\Gamma}{}^\lambda_{\mu\nu}$.

17.3.2 Affine Connection and Its Transformations

Let us start from arbitrary coordinates and impose the ad hoc restrictions that will provide the desired covariant form of the expansion near the point P. It is known that the partial derivatives of the metric tensor are not tensors, while the covariant derivatives of the metric vanish. As a result, the components of the Christoffel symbol

$$\Gamma^\lambda_{\mu\nu} = \left\{ {\lambda \atop \mu\nu} \right\} = \frac{1}{2} g^{\lambda\tau} \left(\partial_\mu g_{\nu\tau} + \partial_\nu g_{\mu\tau} - \partial_\tau g_{\mu\nu} \right)$$

do not form a tensor. To be precise, when we transform coordinates according to $x'^\mu = x'^\mu(x)$, the components of $\Gamma^\lambda_{\mu\nu}$ transform as (14.7), which we rewrite in the equivalent form as

$$\Gamma'^\lambda_{\alpha\beta} = \frac{\partial x'^{\lambda'}}{\partial x^\tau} \frac{\partial x^\mu}{\partial x^{\alpha'}} \frac{\partial x^\nu}{\partial x^{\beta'}} \Gamma^\tau_{\mu\nu} + \frac{\partial x'^{\lambda'}}{\partial x^\tau} \frac{\partial^2 x^\tau}{\partial x^{\alpha'} \partial x^{\beta'}} . \tag{17.74}$$

Here and below we assume that transformations of coordinates are nondegenerate $x^\mu = x^\mu(x')$, and that the metric is an infinitely differentiable function. Our purpose is to find the transformation of coordinates that provides a desirable form of a series expansion of the metric in the vicinity of the point P. Let us start by proving the following statement.

Theorem *By means of coordinate transformation one can demonstrate that the Christoffel symbol and symmetrizations of its partial derivatives vanish in the point P,*

$$\overset{0}{\Gamma}{}^{\lambda}_{\mu\nu} = 0 \quad and \quad \partial_{(\mu_1\mu_2...\mu_n} \overset{0}{\Gamma}{}^{\lambda}_{\alpha\beta)} = 0. \tag{17.75}$$

In the last formula, we used a useful condensed notation

$$\partial_{\mu_1\mu_2...\mu_n} A = \partial_{\mu_1}\partial_{\mu_2}...\partial_{\mu_n} A = \frac{\partial^n A}{\partial x^{\mu_1}\partial x^{\mu_2}...\partial x^{\mu_n}} \tag{17.76}$$

for the nth order partial derivatives of the quantity A. Similar notation will be also used for the covariant derivative,

$$\nabla_{\mu_1\mu_2...\mu_n} A = \nabla_{\mu_1}\nabla_{\mu_2}...\nabla_{\mu_n} A = A_{;\mu_n\mu_{n-1}...\mu_1}. \tag{17.77}$$

Furthermore, the parentheses in (17.75) mean symmetrization, for instance,

$$\partial_{(\mu}\Gamma^{\lambda}_{\alpha\beta)} = \frac{1}{3}\left(\partial_{\mu}\Gamma^{\lambda}_{\alpha\beta} + \partial_{\alpha}\Gamma^{\lambda}_{\mu\beta} + \partial_{\beta}\Gamma^{\lambda}_{\alpha\mu}\right), \tag{17.78}$$

where we used $\Gamma^{\lambda}_{\alpha\beta} = \Gamma^{\lambda}_{\beta\alpha}$ to simplify the expression.

Proof As previewed before, we introduce shifted coordinates $y^{\mu} = x^{\mu} - \overset{0}{x}{}^{\mu}$, corresponding to the expansion around the point P. Also, let us assume that there are some "initial" coordinates z^{α} with $z^{\alpha}(P) = \overset{0}{z}{}^{\alpha}$, for which conditions (17.75) are not satisfied. Without losing generality one can assume that

$$\overset{0}{g}_{\alpha\beta}(\overset{0}{z}) = \overset{0}{g}_{\alpha\beta}(\overset{0}{x}), \quad \frac{\partial z^{\mu}}{\partial y^{\alpha}}\bigg|_P = \delta^{\mu}_{\alpha}, \quad \frac{\partial y^{\alpha}}{\partial z^{\mu}}\bigg|_P = \delta^{\alpha}_{\mu}. \tag{17.79}$$

Consider the general relation between the two coordinate systems, in the form of an infinite series expansion

$$z^{\mu} = \overset{0}{z}{}^{\mu} + A^{\mu}_{\alpha}y^{\alpha} + \frac{1}{2!}A^{\mu}_{\alpha\beta}y^{\alpha}y^{\beta} + \cdots$$
$$+ \frac{1}{n!}A^{\mu}_{\alpha_1\alpha_2...\alpha_n}y^{\alpha_1}y^{\alpha_2}...y^{\alpha_n} + \cdots. \tag{17.80}$$

Here $A^{\mu}_{\alpha}, \ldots, A^{\mu}_{\alpha_1\alpha_2...\alpha_n}$ are unknown coordinate-independent coefficients that we will choose to satisfy the conditions (17.75). Obviously, $A^{\mu}_{\alpha} = \delta^{\mu}_{\alpha}$ due to (17.79). In order to define the next coefficient, let us apply the transformation rule (17.74),

$$\Gamma^{\lambda}_{\alpha\beta}(y) = \frac{\partial y^{\lambda}}{\partial z^{\tau}}\frac{\partial z^{\mu}}{\partial y^{\alpha}}\frac{\partial z^{\nu}}{\partial y^{\beta}}\Gamma^{\tau}_{\mu\nu}(z) + \frac{\partial y^{\lambda}}{\partial z^{\tau}}\frac{\partial^2 z^{\tau}}{\partial y^{\alpha}\partial y^{\beta}}. \tag{17.81}$$

Using (17.80), at the point P we set $y = 0$ and arrive at the relation

$$\overset{\circ}{\Gamma}{}^{\lambda}_{\alpha\beta} = \delta^{\lambda}_{\tau} \delta^{\mu}_{\alpha} \delta^{\nu}_{\beta} \, \Gamma^{\tau}_{\mu\nu}(\overset{\circ}{z}) + \delta^{\lambda}_{\tau} A^{\tau}_{\alpha\beta} = \Gamma^{\lambda}_{\alpha\beta}(\overset{\circ}{z}) + A^{\lambda}_{\alpha\beta} . \tag{17.82}$$

Therefore, by choosing $A^{\lambda}_{\alpha\beta} = -\Gamma^{\lambda}_{\alpha\beta}(\overset{\circ}{z})$, we provide the first condition in (17.75).

The proof of the second condition of (17.75) is a bit more complicated. Taking the derivative of Eq. (17.81) and then setting $y = 0$, after some small algebra we arrive at the relation

$$\frac{\partial}{\partial y^{\tau}} \Gamma^{\lambda}_{\alpha\beta}(y) \bigg|_{P} = \partial_{\tau} \overset{\circ}{\Gamma}{}^{\lambda}_{\alpha\beta} = \frac{\partial}{\partial z^{\tau}} \Gamma^{\lambda}_{\alpha\beta}(\overset{\circ}{z}) + 2\Gamma^{\lambda}_{\mu\beta}(\overset{\circ}{z}) \, A^{\mu}_{\alpha\tau} - \Gamma^{\nu}_{\alpha\beta}(\overset{\circ}{z}) \, A^{\lambda}_{\nu\tau} \tag{17.83}$$

$$+ A^{\nu}_{\alpha\beta}(\overset{\circ}{z}) \frac{\partial^{2} y^{\lambda}}{\partial z^{\tau} \partial z^{\nu}} \bigg|_{z = \overset{\circ}{z}} + A^{\lambda}_{\alpha\beta\tau} = \partial_{\tau} \Gamma^{\lambda}_{\alpha\beta}(\overset{\circ}{z}) - 2A^{\lambda}_{\mu\beta} A^{\mu}_{\alpha\tau} + A^{\lambda}_{\alpha\beta\tau} .$$

When trying to eliminate the *l.h.s.* of (17.84) by means of the choice of $A^{\lambda}_{\alpha\beta\tau}$ one can immediately see that this is impossible because $A^{\lambda}_{\alpha\beta\tau}$ is symmetric in the lower indices, $A^{\lambda}_{\alpha\beta\tau} = A^{\lambda}_{(\alpha\beta\tau)}$, while the rest of Eq. (17.84) is not symmetric in the same indices, and therefore has more independent components. This output is quite natural because, otherwise, we could eliminate all partial derivatives of the affine connection by the coordinate transformation (17.80) and then the Riemann tensor at the point P would also vanish. However, it works perfectly well for the symmetrization of (17.84) with respect to the three lower indices since in this case (17.84) is a linear nondegenerate equation for $A^{\lambda}_{\alpha\beta\tau}$. Hence, we have just proved (17.75) for the $n = 1$ case. The proof for $n \geq 2$ is essentially the same. Taking further derivatives in the point P, at leach level we arrive at the linear equation for the corresponding absolutely symmetric coefficients $A^{\lambda}_{\alpha_1 \ldots \alpha_k} = A^{\lambda}_{(\alpha_1 \ldots \alpha_k)}$,

$$\partial_{(\mu_1 \ldots \mu_n} \overset{\circ}{\Gamma}{}^{\lambda}_{\alpha\beta)} = , \partial_{(\mu_1 \ldots \mu_n} \Gamma^{\lambda}_{\alpha\beta)}(\overset{\circ}{z}) + (\text{something})_{(\mu_1 \ldots \mu_n \alpha\beta)}$$
$$= A^{\lambda}_{\mu_1 \ldots \mu_n \alpha\beta} . \tag{17.84}$$

As far as a nondegenerate linear equation always has a solution, we proved the second relation (17.75) for an arbitrary n.

The last observation concerning formulas (17.75) concerns their relation to the geodesic equation

$$\frac{d^2 x^{\mu}}{d\tau^2} + \Gamma^{\mu}_{\alpha\beta} \frac{dx^{\alpha}}{d\tau} \frac{dx^{\beta}}{d\tau} = 0 , \tag{17.85}$$

with the initial data

$$\frac{dx^{\alpha}}{d\tau} \bigg|_{x^{\mu} = \overset{\circ}{x}{}^{\mu}} = \xi^{\alpha} . \tag{17.86}$$

The choice of ξ^α defines the metric at the point P and also defines the solution of the geodesic equation. One can assume that τ is an affine parameter along the curve and that $\tau(P) = 0$. Then, since $\overset{\circ}{\Gamma}{}^\lambda_{\alpha\beta} = 0$, the equation for the geodesic line at the point P boils down to

$$\frac{d^2 y^\mu}{d\tau^2} = 0. \tag{17.87}$$

The reader can find much more detail about the geometric aspects of the special coordinates y^μ in the book [12]. For our purposes, it is sufficient to know that the main relations (17.75) can be achieved by a special choice of coordinates.

17.3.3 Expansion of the Metric in Normal Coordinates

Now we are in a position to consider expansions in the Riemann coordinates y^α. For any tensor or a non-tensor quantity $W_{\alpha_1\alpha_2\ldots\alpha_p}$, one can perform the power series expansion in the vicinity of the point P,

$$W_{\alpha_1\alpha_2\ldots\alpha_k}(\overset{\circ}{x} + y) = \overset{\circ}{W}_{\alpha_1\alpha_2\ldots\alpha_k} + \left(\frac{\partial W_{\alpha_1\alpha_2\ldots\alpha_k}}{\partial y^\mu}\right)_P y^\mu + \ldots$$
$$+ \frac{1}{n!} \left(\frac{\partial^n W_{\alpha_1\alpha_2\ldots\alpha_k}}{\partial y^{\mu_1}\ldots\partial y^{\mu_n}}\right)_P y^{\mu_1}\ldots y^{\mu_n} + \ldots. \tag{17.88}$$

Our purpose is to consider such an expansion for the metric. After that one can easily derive the expansion for other quantities, such as the Christoffel symbol and a curvature tensor.

Using condensed notations (17.76), the initial form of the expansion for the metric reads

$$g_{\alpha\beta}(\overset{\circ}{x} + y) = \overset{\circ}{g}_{\alpha\beta} + \partial_\mu \overset{\circ}{g}_{\alpha\beta} \cdot y^\mu + \frac{1}{2!}\partial_{\mu\nu}\overset{\circ}{g}_{\alpha\beta} \cdot y^\mu y^\nu + \ldots$$
$$+ \frac{1}{n!}\partial_{\mu_1\mu_2\ldots\mu_n}\overset{\circ}{g}_{\alpha\beta} \cdot y^{\mu_1} y^{\mu_2}\ldots y^{\mu_n} + \ldots. \tag{17.89}$$

Our purpose is to derive the coefficients of this expansion and show that for the coordinates that satisfy condition (17.75), these coefficients can be expressed via a curvature tensor in the point P, that is, $\overset{\circ}{R}{}^\lambda_{\cdot\alpha\beta\gamma}$ and its covariant derivatives,

$$\nabla_{\mu_1\ldots\mu_k}\overset{\circ}{R}{}^\lambda_{\cdot\alpha\beta\gamma} = \overset{\circ}{R}{}^\lambda_{\cdot\alpha\beta\gamma;\,\mu_k\ldots\mu_1}.$$

17.3.4 Metricity Condition and Its Derivatives

We start from the metricity condition $\nabla_\mu g_{\alpha\beta} = 0$, which gives

$$\partial_\mu g_{\alpha\beta} = g_{\alpha\lambda}\Gamma^\lambda_{\beta\mu} + g_{\lambda\beta}\Gamma^\lambda_{\alpha\mu} = g_{\lambda\alpha}\Gamma^\lambda_{\beta\mu} + (\alpha \leftrightarrow \beta). \qquad (17.90)$$

In the expression, we started to use symmetrization in the pair of indices. This is important because in what follows we will work with the expressions which have *a lot* of indices, some of which will be symmetric. For this reason, it is important to have a useful form of dealing with this situation, which makes the formulas look less cumbersome. Different from other symmetrizations (see below) the one for α and β we will always show explicitly.

Let us start by taking partial derivatives of (17.90) and illustrate at once the way symmetric expressions will be made more compact.

$$\partial_{\mu\nu} g_{\alpha\beta} = \Gamma^\lambda_{\beta(\mu} \cdot \partial_{\nu)} g_{\lambda\alpha} + g_{\lambda\alpha} \cdot \partial_{(\mu}\Gamma^\lambda_{\nu)\beta} + (\alpha \leftrightarrow \beta).$$
$$= \partial_\mu g_{\lambda\alpha} \cdot \Gamma^\lambda_{\beta\nu} + g_{\lambda\alpha} \cdot \partial_\mu \Gamma^\lambda_{\beta\nu} + (\alpha \leftrightarrow \beta). \qquad (17.91)$$

In the last expression, the symmetrization over the pair of indices $(\mu\nu)$ is *assumed*, but not written explicitly. In what follows the same will be consequently done for all indices that stand in the *l.h.s.* of expressions in the partial derivatives of the metric. The next derivative taken with these notations gives

$$\partial_{\mu\nu\rho} g_{\alpha\beta} = \partial_{\mu\nu} g_{\lambda\alpha} \cdot \Gamma^\lambda_{\beta\rho} + 2\partial_\mu g_{\lambda\alpha} \cdot \partial_\nu \Gamma^\lambda_{\beta\rho} + g_{\lambda\alpha} \cdot \partial_{\mu\nu}\Gamma^\lambda_{\beta\rho} + (\alpha \leftrightarrow \beta), \qquad (17.92)$$

where we assume symmetrization over the indices $(\mu\nu\rho)$. Taking more derivatives, with the described symmetrization rule we get

$$\partial_{\mu\nu\rho\sigma} g_{\alpha\beta} = \partial_{\mu\nu\rho} g_{\lambda\alpha} \cdot \Gamma^\lambda_{\beta\sigma} + 3\partial_{\mu\nu} g_{\lambda\alpha} \cdot \partial_\rho \Gamma^\lambda_{\beta\sigma}$$
$$+ 3\partial_\mu g_{\lambda\alpha} \cdot \partial_{\nu\rho} \Gamma^\lambda_{\beta\sigma} + g_{\lambda\alpha} \cdot \partial_{\mu\nu\rho}\Gamma^\lambda_{\beta\sigma} + (\alpha \leftrightarrow \beta), \qquad (17.93)$$

$$\partial_{\mu\nu\rho\sigma\kappa} g_{\alpha\beta} = \partial_{\mu\nu\rho\sigma} g_{\lambda\alpha} \cdot \Gamma^\lambda_{\beta\kappa} + 4\partial_{\mu\nu\rho} g_{\lambda\alpha} \cdot \partial_\sigma \Gamma^\lambda_{\beta\kappa} + 6\partial_{\mu\nu} g_{\lambda\alpha} \cdot \partial_{\rho\sigma} \Gamma^\lambda_{\beta\kappa}$$
$$+ 4\partial_\mu g_{\lambda\alpha} \cdot \partial_{\nu\rho\sigma} \Gamma^\lambda_{\beta\kappa} + g_{\lambda\alpha} \cdot \partial_{\mu\nu\rho\sigma}\Gamma^\lambda_{\beta\kappa} + (\alpha \leftrightarrow \beta), \qquad (17.94)$$

and furthermore

$$\partial_{\mu\nu\rho\sigma\kappa\omega} g_{\alpha\beta} = \partial_{\mu\nu\rho\sigma\kappa} g_{\lambda\alpha} \cdot \Gamma^\lambda_{\beta\omega} + 5\partial_{\mu\nu\rho\sigma} g_{\lambda\alpha} \cdot \partial_\kappa \Gamma^\lambda_{\beta\omega}$$
$$+ 10\partial_{\mu\nu\rho} g_{\lambda\alpha} \cdot \partial_{\sigma\kappa} \Gamma^\lambda_{\beta\omega} + 10\partial_{\mu\nu} g_{\lambda\alpha} \cdot \partial_{\rho\sigma\kappa} \Gamma^\lambda_{\beta\omega} \qquad (17.95)$$
$$+ 5\partial_\mu g_{\lambda\alpha} \cdot \partial_{\nu\rho\sigma\kappa} \Gamma^\lambda_{\beta\omega} + g_{\lambda\alpha} \cdot \partial_{\mu\nu\rho\sigma\kappa}\Gamma^\lambda_{\beta\omega} + (\alpha \leftrightarrow \beta).$$

In all these expressions, we used the symmetrization rule explained before. For instance, in the last expression, the implicit symmetrization is over the indices $(\mu\nu\rho\sigma\kappa\omega)$.

Our intention is to evaluate these derivatives in the point P and replace the result in (17.89). Setting $y^\mu = 0$, expressions (17.90)–(17.96) simplify considerably because $\overset{\circ}{\Gamma}{}^\lambda_{\alpha\beta} = 0$ and therefore $\partial_\mu \overset{\circ}{g}_{\alpha\beta} = 0$.

17.3.5 Expansion of the Metric

It proves useful to derive the symmetrized derivatives of the affine connections first. For this, we start from the definition of the Riemann tensor,

$$R^\lambda_{\cdot\mu\nu\gamma} = \partial_\nu\Gamma^\lambda_{\mu\gamma} - \partial_\gamma\Gamma^\lambda_{\mu\nu} + \Gamma^\tau_{\mu\gamma}\,\Gamma^\lambda_{\tau\nu} - \Gamma^\tau_{\mu\nu}\,\Gamma^\lambda_{\tau\gamma}\,. \tag{17.96}$$

One can note that the last two terms in this expression are irrelevant because $\Gamma^\tau_{\mu\nu}$ and all its symmetrized derivatives vanish at the point P due to (17.75). Therefore, at the point P we can ignore this term. It is important that this does not apply to the second term because the derivative $\partial_\gamma\Gamma^\tau_{\mu\nu}$ and its further symmetrized derivatives do not vanish at P. Using (17.75) we can write

$$\partial_\mu \overset{\circ}{\Gamma}{}^\lambda_{\nu\gamma} + \partial_\nu \overset{\circ}{\Gamma}{}^\lambda_{\mu\gamma} + \partial_\gamma \overset{\circ}{\Gamma}{}^\lambda_{\mu\nu} = 2\partial_\nu \overset{\circ}{\Gamma}{}^\lambda_{\mu\gamma} + \partial_\gamma \overset{\circ}{\Gamma}{}^\lambda_{\mu\nu} = 0\,, \tag{17.97}$$

and therefore

$$\partial_\gamma \overset{\circ}{\Gamma}{}^\lambda_{\mu\nu} = -2\,\partial_\nu \overset{\circ}{\Gamma}{}^\lambda_{\mu\gamma}\,. \tag{17.98}$$

In a similar way, one can derive other useful relations,

$$\begin{aligned}
\partial_{\gamma\rho} \overset{\circ}{\Gamma}{}^\lambda_{\mu\nu} &= -\partial_{\mu\nu} \overset{\circ}{\Gamma}{}^\lambda_{\rho\gamma}\,,\\
\partial_{\gamma\rho\sigma} \overset{\circ}{\Gamma}{}^\lambda_{\mu\nu} &= -\frac{2}{3}\,\partial_{\mu\nu\rho} \overset{\circ}{\Gamma}{}^\lambda_{\sigma\gamma}\,,\\
\partial_{\gamma\rho\sigma\kappa} \overset{\circ}{\Gamma}{}^\lambda_{\mu\nu} &= -\frac{1}{2}\,\partial_{\mu\nu\rho\sigma} \overset{\circ}{\Gamma}{}^\lambda_{\kappa\gamma}\,,\\
\partial_{\gamma\rho\sigma\kappa\omega} \overset{\circ}{\Gamma}{}^\lambda_{\mu\nu} &= -\frac{2}{5}\,\partial_{\mu\nu\rho\sigma\kappa} \overset{\circ}{\Gamma}{}^\lambda_{\omega\gamma}\,.
\end{aligned} \tag{17.99}$$

Using (17.98) and (17.75) at the point P definition (17.96) gives

$$\partial_\mu \overset{\circ}{\Gamma}{}^\lambda_{\nu\gamma} = \frac{1}{3}\,\overset{\circ}{R}{}^\lambda_{\cdot\mu\nu\gamma}\,. \tag{17.100}$$

Taking consequent covariant derivatives of (17.96) and using (17.98) and (17.75) at the point P, after some algebra we obtain

$$\partial_{\mu\nu} \overset{\circ}{\Gamma}{}^{\lambda}_{\rho\gamma} = \frac{1}{2} \overset{\circ}{R}{}^{\lambda}_{\cdot\mu\nu\gamma;\rho} ,$$

$$\partial_{\rho\sigma\mu} \overset{\circ}{\Gamma}{}^{\lambda}_{\nu\gamma} = \frac{3}{5} \overset{\circ}{R}{}^{\lambda}_{\cdot\mu\nu\gamma;\rho\sigma} - \frac{2}{15} \overset{\circ}{R}{}^{\lambda}_{\cdot\mu\nu\tau} \overset{\circ}{R}{}^{\tau}_{\cdot\rho\sigma\gamma} , \qquad (17.101)$$

and furthermore

$$\partial_{\mu\nu\rho\sigma} \overset{\circ}{\Gamma}{}^{\lambda}_{\kappa\gamma} = \frac{2}{3} \overset{\circ}{R}{}^{\lambda}_{\cdot\mu\nu\gamma;\rho\sigma\kappa} - \frac{4}{9} \overset{\circ}{R}{}^{\lambda}_{\cdot\mu\nu\tau} \overset{\circ}{R}{}^{\tau}_{\cdot\rho\sigma\gamma;\kappa} - \frac{2}{9} \overset{\circ}{R}{}^{\lambda}_{\cdot\mu\nu\tau;\kappa} \overset{\circ}{R}{}^{\tau}_{\cdot\rho\sigma\gamma} \qquad (17.102)$$

and

$$\partial_{\mu\nu\rho\sigma\kappa} \overset{\circ}{\Gamma}{}^{\lambda}_{\omega\gamma} = \frac{5}{7} \overset{\circ}{R}{}^{\lambda}_{\cdot\mu\nu\gamma;\rho\sigma\kappa\omega} - \frac{1}{21} \overset{\circ}{R}{}^{\lambda}_{\cdot\mu\nu\tau;\kappa\omega} \cdot \overset{\circ}{R}{}^{\tau}_{\cdot\rho\sigma\beta}$$

$$- \frac{23}{21} \overset{\circ}{R}{}^{\lambda}_{\cdot\mu\nu\tau} \cdot \overset{\circ}{R}{}^{\tau}_{\cdot\rho\sigma\beta;\kappa\omega} - \frac{15}{14} \overset{\circ}{R}{}^{\lambda}_{\cdot\mu\nu\tau;\kappa} \cdot \overset{\circ}{R}{}^{\tau}_{\cdot\rho\sigma\gamma;\omega}$$

$$+ \frac{16}{63} \overset{\circ}{R}{}^{\lambda}_{\cdot\mu\nu\tau} \cdot \overset{\circ}{R}{}^{\tau}_{\cdot\rho\sigma\xi} \cdot \overset{\circ}{R}{}^{\xi}_{\cdot\kappa\omega\gamma} . \qquad (17.103)$$

It is important to note that the derivatives of curvature tensors in these formulas are covariant ones. This means that the derivation of these formulas (we leave it as an exercise), after using (17.99), requires an additional effort to transform partial derivatives into covariant ones.

Now we are in a position to combine the formulas for the partial derivatives of metric (17.90)–(17.96) at the point P with the symmetrized derivatives of affine connection (17.100)–(17.103). It is fairly easy to obtain the following derivatives:

$$\partial_{\mu\nu} \overset{\circ}{g}_{\alpha\beta} = \overset{\circ}{g}_{\lambda\alpha} \cdot \partial_{\mu} \overset{\circ}{\Gamma}{}^{\lambda}_{\beta\nu} + (\alpha \leftrightarrow \beta)$$

$$= \frac{1}{3} \overset{\circ}{R}_{\alpha(\mu\nu)\beta} + (\alpha \leftrightarrow \beta) = \frac{2}{3} \overset{\circ}{R}_{\alpha\mu\nu\beta} \qquad (17.104)$$

and also, after more complicated calculations,

$$\partial_{\mu\nu\rho} \overset{\circ}{g}_{\alpha\beta} = \overset{\circ}{R}_{\alpha\mu\nu\beta;\rho} , \qquad (17.105)$$

$$\partial_{\mu\nu\rho\sigma} \overset{\circ}{g}_{\alpha\beta} = \frac{6}{5} \overset{\circ}{R}_{\mu\alpha\beta\nu;\rho\sigma} + \frac{16}{15} \overset{\circ}{R}_{\alpha\mu\nu\tau} \overset{\circ}{R}{}^{\tau}_{\cdot\rho\sigma\beta} , \qquad (17.106)$$

$$\partial_{\mu\nu\rho\sigma\kappa} \overset{\circ}{g}_{\alpha\beta} = \frac{4}{3} \overset{\circ}{R}_{\alpha\mu\nu\beta;\rho\sigma\kappa}$$

$$+ \frac{8}{3} \left(\overset{\circ}{R}_{\alpha\mu\nu\tau} \overset{\circ}{R}{}^{\tau}_{\cdot\rho\sigma\beta;\kappa} + \overset{\circ}{R}_{\beta\mu\nu\tau} \overset{\circ}{R}{}^{\tau}_{\cdot\rho\sigma\alpha;\kappa} \right) , \qquad (17.107)$$

$$\partial_{\mu\nu\rho\sigma\kappa\omega} \overset{\circ}{g}_{\alpha\beta} = \frac{10}{7} \overset{\circ}{R}_{\alpha\mu\nu\beta;\rho\sigma\kappa\omega} + \frac{16}{7} \overset{\circ}{R}_{\alpha\mu\nu\tau} \overset{\circ}{R}{}^{\tau}_{\cdot\rho\sigma\lambda} \overset{\circ}{R}{}^{\lambda}_{\cdot\kappa\omega\beta}$$

$$
+ \frac{34}{7} \left(\overset{\circ}{R}_{\alpha\mu\nu\tau} \overset{\circ}{R}{}^{\tau}{}_{.\rho\sigma\beta\,;\kappa\omega} + \overset{\circ}{R}_{\beta\mu\nu\tau} \overset{\circ}{R}{}^{\tau}{}_{.\rho\sigma\alpha\,;\kappa\omega} \right)
$$

$$
+ \frac{55}{7} \overset{\circ}{R}_{\alpha\mu\nu\tau\,;\kappa} \overset{\circ}{R}{}^{\tau}{}_{.\rho\sigma\beta\,;\omega}. \tag{17.108}
$$

The result for the metric expansion (17.89) up to the sixth orders in y has the form

$$
\begin{aligned}
g_{\alpha\beta}(y) = {}& \overset{\circ}{g}_{\alpha\beta} + \frac{1}{3} \overset{\circ}{R}_{\alpha\mu\nu\beta} \, y^{\mu} y^{\nu} + \frac{1}{3!} \overset{\circ}{R}_{\alpha\mu\nu\beta\,;\sigma} \, y^{\mu} y^{\nu} y^{\sigma} \\
&+ \left(\frac{1}{20} \overset{\circ}{R}_{\alpha\mu\nu\beta\,;\rho\sigma} + \frac{2}{45} \overset{\circ}{R}_{\alpha\mu\nu\lambda} \overset{\circ}{R}{}^{\lambda}{}_{.\rho\sigma\beta} \right) y^{\mu} y^{\nu} y^{\rho} y^{\sigma} \\
&+ \frac{1}{6!} \left[8 \overset{\circ}{R}_{\alpha\mu\nu\beta\,;\rho\sigma\kappa} + 16 \left(\overset{\circ}{R}_{\alpha\mu\nu\lambda\,;\kappa} \overset{\circ}{R}{}^{\lambda}{}_{.\rho\sigma\beta} + \overset{\circ}{R}_{\alpha\mu\nu\lambda} \overset{\circ}{R}{}^{\lambda}{}_{.\rho\sigma\beta\,;\kappa} \right) \right] y^{\mu} y^{\nu} y^{\rho} y^{\sigma} y^{\kappa} \\
&+ \frac{1}{7!} \left[10 \overset{\circ}{R}_{\alpha\mu\nu\beta\,;\rho\sigma\kappa\omega} + 34 \left(\overset{\circ}{R}_{\alpha\mu\nu\lambda\,;\kappa\omega} \overset{\circ}{R}{}^{\lambda}{}_{.\rho\sigma\beta} + \overset{\circ}{R}_{\alpha\mu\nu\lambda} \overset{\circ}{R}{}^{\lambda}{}_{.\rho\sigma\beta\,;\kappa\omega} \right) \right. \tag{17.109} \\
&\left. + \frac{55}{2} \overset{\circ}{R}_{\alpha\mu\nu\tau\,;\kappa} \overset{\circ}{R}{}^{\tau}{}_{.\rho\sigma\beta\,;\omega} + 16 \overset{\circ}{R}_{\alpha\mu\nu\tau} \overset{\circ}{R}{}^{\tau}{}_{.\rho\sigma\lambda} \overset{\circ}{R}{}^{\lambda}{}_{.\kappa\omega\beta} \right] y^{\mu} y^{\nu} y^{\rho} y^{\sigma} y^{\kappa} y^{\omega} + \dots .
\end{aligned}
$$

The remarkable feature of this expansion is that its coefficients are covariant expressions (curvature tensor and its covariant derivatives) at the point P. This feature will certainly hold in the further orders of expansion, which we do not perform here. One can find the next order terms and more sophisticated approaches to the derivation of these and further coefficients in Refs. [13, 14].

Since the coefficients of this expansion are relatively bulky, it makes sense to introduce condensed notations as follows:

$$
\begin{aligned}
\mathcal{R}_{(2)\alpha\beta} &= \overset{\circ}{R}_{\alpha\mu\nu\beta} \, y^{\mu} y^{\nu}, \quad \mathcal{R}_{(3)\alpha\beta} = \overset{\circ}{R}_{\alpha\mu\nu\beta\,;\kappa} \, y^{\mu} y^{\nu} y^{\kappa}, \dots \\
\mathcal{R}_{(6)\alpha\beta} &= \overset{\circ}{R}_{\alpha\mu\nu\tau\,;\rho\sigma\kappa\omega} \, y^{\mu} y^{\nu} y^{\rho} y^{\sigma} y^{\kappa} y^{\omega}. \tag{17.110}
\end{aligned}
$$

One can easily write (17.109) in the form

$$
g_{\alpha\beta}(y) = \overset{\circ}{g}_{\alpha\beta} + h_{\alpha\beta} \tag{17.111}
$$

and find the inverse metric in the usual form (14.99)

$$
g^{\alpha\beta}(y) = \overset{\circ}{g}{}^{\alpha\beta} - h^{\alpha\beta} + h^{\alpha\lambda} h^{\beta}_{\lambda} - h^{\alpha\lambda} h^{\tau}_{\lambda} h^{\beta}_{\tau} + \dots, \tag{17.112}
$$

where the indices are raised and lowered by means of the metric $\overset{\circ}{g}_{\alpha\beta}$. The expansion for the inverse metric can be written in a more compact form using the compact notation (17.110),

$$g^{\alpha\beta}(y) = \overset{\circ}{g}{}^{\alpha\beta} - \frac{1}{3}\mathcal{R}^{\alpha\beta}_{(2)} - \frac{1}{3!}\mathcal{R}^{\alpha\beta}_{(3)} - \frac{1}{20}\mathcal{R}^{\alpha\beta}_{(4)} + \frac{1}{15}\mathcal{R}^{\alpha\tau}_{(2)}\mathcal{R}^{\beta}_{(2)\tau}$$
$$- \frac{1}{90}\mathcal{R}^{\alpha\beta}_{(5)} + \frac{1}{30}\left[\mathcal{R}^{\alpha\tau}_{(3)}\mathcal{R}^{\beta}_{(2)\tau} + \mathcal{R}^{\alpha\tau}_{(2)}\mathcal{R}^{\beta}_{(3)\tau}\right]$$
$$+ \frac{1}{7!}\left\{-10\mathcal{R}^{\alpha\beta}_{(6)} + 50\left[\mathcal{R}^{\alpha\tau}_{(4)}\mathcal{R}^{\beta}_{(2)\tau} + \mathcal{R}^{\alpha\tau}_{(2)}\mathcal{R}^{\beta}_{(4)\tau}\right]\right.$$
$$\left. + 85\mathcal{R}^{\alpha\tau}_{(3)}\mathcal{R}^{\beta}_{(3)\tau} - \frac{160}{3}\mathcal{R}^{\alpha\tau}_{(2)}\mathcal{R}^{\tau}_{(2)\lambda}\mathcal{R}^{\beta}_{(2)\tau}\right\} + \ldots, \tag{17.113}$$

which also fits with the result of [14]. Finally, the expansion for the Ricci tensor has the form

$$R_{\alpha\beta}(y) = \overset{\circ}{R}_{\alpha\beta} + \overset{\circ}{R}_{\alpha\beta\,;\,\mu}\, y^{\mu}$$
$$- \frac{1}{2}\left[\overset{\circ}{R}_{\alpha\beta\,;\,\mu\nu} + \frac{4}{3}\overset{\circ}{R}{}^{\lambda}{}_{\mu\nu(\alpha}\overset{\circ}{R}_{\beta)\lambda}\right]y^{\mu}y^{\nu} + \ldots. \tag{17.114}$$

It is useful to have the expansion for the metric determinant in an arbitrary power d, that is, $|g(y)|^d$. In order to derive this expansion of normal coordinates, we can start from the general parameterization (15.32) and the corresponding general expressions (15.41) and (15.42). Omitting the technical detail (most of this calculation can indeed be performed by using Wolfram's Mathematica [15]), we present only the final result, which follows directly from expansions for the metric derived in the previous section, Eqs. (15.36) and (15.41),

$$|g(y)|^d = |\overset{\circ}{g}|^d \times F,$$
$$F = 1 + \frac{d}{3}\mathcal{R}^{\alpha}_{(2)\alpha} + \frac{d}{6}\mathcal{R}^{\alpha}_{(3)\alpha} + \frac{d^2}{180}\left\{9d\mathcal{R}^{\alpha}_{(4)\alpha} - 2d\mathcal{R}_{(2)\alpha\beta}\mathcal{R}^{\alpha\beta}_{(2)} + 10[\mathcal{R}^{\alpha}_{(2)\alpha}]^2\right\}$$
$$+ \frac{d^2}{90}\left\{d\mathcal{R}^{\alpha}_{(5)\alpha} - d\mathcal{R}_{(2)\alpha\beta}\mathcal{R}^{\alpha\beta}_{(3)} + 5\mathcal{R}^{\alpha}_{(2)\alpha}\mathcal{R}^{\alpha}_{(3)\alpha}\right\}$$
$$+ \frac{d^3}{18\cdot 7!}\left\{180d^2\mathcal{R}^{\alpha}_{(6)\alpha} + 64d^2\mathcal{R}^{\beta}_{(2)\alpha}\mathcal{R}^{\lambda}_{(2)\beta}\mathcal{R}^{\alpha}_{(2)\lambda} - 288d^2\mathcal{R}_{(2)\alpha\beta}\mathcal{R}^{\alpha\beta}_{(4)}\right.$$
$$- 765d^2\mathcal{R}_{(3)\alpha\beta}\mathcal{R}^{\alpha\beta}_{(3)} - 336d\mathcal{R}_{(2)\alpha\beta}\mathcal{R}^{\alpha\beta}_{(2)}\mathcal{R}^{\lambda}_{(2)\lambda} + 1512d\mathcal{R}^{\alpha}_{(2)\alpha}\mathcal{R}^{\beta}_{(4)\beta}$$
$$\left. + 1260d[\mathcal{R}^{\alpha}_{(3)\alpha}]^2 + 560[\mathcal{R}^{\alpha}_{(2)\alpha}]^3\right\}. \tag{17.115}$$

Exercise 16 Verify the combinatorial expressions (17.99) and subsequent relations (17.101)–(17.103). Pay special attention to the transition from partial to covariant derivatives in Eqs. (17.102) and (17.103).

Exercise 17 Verify the expansion (17.114).

Exercise 18 Derive the first few terms of Eq. (17.115) for the case of $d = 1$ using Eq. (15.35).

Exercise 19 Derive Eq. (17.114) and a similar expansion for $R^{\alpha\beta}(y)$ and for the scalar curvature.

Exercise 20 Derive the expansion for the tetrads e_a^μ, e_ν^b and for the spinor connection ω_μ^{ab} up to the first order in curvature.

Exercise 21 Calculate the expansion for the affine connection up to the first order in curvature. Using this result, obtain the expansion for the d'Alembert operator $\square$ acting on the scalar field.

Solution. The first problem can be easily solved using Eq. (17.100) and the master relation (17.75). As far as

$$\partial_\tau \overset{\circ}{\Gamma}{}^\lambda_{\mu\nu} + 2\partial_\mu \overset{\circ}{\Gamma}{}^\lambda_{\nu\tau} = 0,$$

we get from (17.100)

$$\partial_\tau \overset{\circ}{\Gamma}{}^\lambda_{\mu\nu} = -\frac{2}{3} \overset{\circ}{R}{}^\lambda_{\cdot\mu\nu\tau} \tag{17.116}$$

and therefore

$$\Gamma^\lambda_{\mu\nu} = \overset{\circ}{\Gamma}{}^\lambda_{\mu\nu} - \frac{2}{3} \overset{\circ}{R}{}^\lambda_{\cdot\mu\nu\tau} y^\tau = -\frac{2}{3} \overset{\circ}{R}{}^\lambda_{\cdot\mu\nu\tau} y^\tau. \tag{17.117}$$

Furthermore, for the d'Alembert operator we have

$$\begin{aligned}
\square\varphi &= g^{\mu\nu}\nabla_\mu\partial_\nu\varphi \\
&= \left(\overset{\circ}{g}{}^{\mu\nu} - \frac{1}{3} \overset{\circ}{R}{}^{\mu\nu}_{\alpha\cdot\cdot\beta} y^\alpha y^\beta \right) \left(\partial_\mu\partial_\nu\varphi - \overset{\circ}{\Gamma}{}^\lambda_{\mu\nu}\partial_\lambda\varphi - \partial_\tau\overset{\circ}{\Gamma}{}^\lambda_{\mu\nu}y^\tau\partial_\lambda\varphi \right) + \cdots \\
&= \left(\overset{\circ}{g}{}^{\mu\nu} - \frac{1}{3} \overset{\circ}{R}{}^{\mu\nu}_{\alpha\cdot\cdot\beta} y^\alpha y^\beta \right) \left(\partial_\mu\partial_\nu\varphi + \frac{2}{3} \overset{\circ}{R}{}^\lambda_{\cdot\mu\nu\tau} y^\tau\partial_\lambda\varphi \right) + \cdots \\
&= \left(\partial^2 - \frac{1}{3} \overset{\circ}{R}{}^{\mu\nu}_{\alpha\cdot\cdot\beta} y^\alpha y^\beta \partial_\mu\partial_\nu - \frac{2}{3} \overset{\circ}{R}{}_\tau y^\tau\partial_\lambda \right)\varphi + O\left(\nabla R_{...}, R^2_{...}\right). \tag{17.118}
\end{aligned}$$

The calculation in higher orders can be done in a similar way.

Exercise 22 Rewrite the basic expansion (17.109) using notations (17.110).

References

1. T.P. Sotiriou, V. Faraoni, Rev. Mod. Phys. **82**, 451 (2010). arXiv:0805.1726
2. A. De Felice, S. Tsujikawa, Living Rev. Relativ. **13**, 3 (2010). arXiv:1002.4928
3. I.L. Shapiro, H. Takata, One-loop renormalization of the four-dimensional theory for quantum dilaton gravity. Phys. Rev. **52D**, 2162 (1995). hep-th/9502111; Conformal transformation in gravity. Phys. Lett. B **361**, 31 (1995). hep-th/9504162

4. D.C. Rodrigues, F.de O. Salles, I.L. Shapiro, A.A. Starobinsky, Auxiliary fields representation for modified gravity models. Phys. Rev. D **83**, 084028 (2011). arXiv:1101.5028
5. I.L. Shapiro, On the conformal transformation and duality in gravity. Class. Quantum Gravity **14**, 391 (1997). hep-th/9610129
6. J. O'Hanlon, Intermediate-range gravity: a generally covariant model. Phys. Rev. Lett. **29**, 137 (1972)
7. J.A. Helayel-Neto, A. Penna-Firme, I.L. Shapiro, Conformal symmetry, anomaly and effective action for metric-scalar gravity with torsion. Phys. Lett. B **479**, 411 (2000). gr-qc/9907081
8. A.A. Starobinsky, A new type of isotropic cosmological models without singularity. Phys. Lett. B **91**, 99 (1980)
9. J.D. Bekenstein, Exact solutions of Einstein conformal scalar equations. Ann. Phys. **82**, 535 (1974)
10. I.L. Shapiro, Covariant derivative of fermions and all that. Mens Agitat (Academia Roraimense de Ciências) **11**, 1 (2017). arXiv:1611.02263
11. I.L. Shapiro, Physical aspects of the space-time torsion. Phys. Rep. **357**, 113 (2002)
12. A.Z. Petrov, *Einstein Spaces* (Pergamon Press, Oxford, 1969)
13. U. Muller, C. Schubert, A.E.M. van de Ven, Gen. Relativ. Gravit. **31**, 1759 (1999)
14. A. Hatzinikitas, A note on Riemann normal coordinates. hep-th/0001078; A. Hatzinikitas, R. Portugal, Nucl. Phys. B **613**, 237 (2001). hep-th/0103073
15. Wolfram Research, *Mathematica, Version 9.0* (Champaign, IL, 2012)

Bibliography

1. A.Z. Petrov, *Einstein Spaces* (Pergamon, Oxford, 1969)
2. B.P. Abbott et al., GW170817: observation of gravitational waves from a binary neutron star inspiral. Phys. Rev. Lett. **119**, 161101 (2017). Virgo, LIGO Scientific. arXiv:1710.05832
3. B.F. Schutz, *Geometrical Methods of Mathematical Physics* (Cambridge University Press, Cambridge, 1982)

© Springer Nature Switzerland AG 2019
I. L. Shapiro, *A Primer in Tensor Analysis and Relativity*, Undergraduate
Lecture Notes in Physics, https://doi.org/10.1007/978-3-030-26895-4

Index

© Springer Nature Switzerland AG 2019
I. L. Shapiro, *A Primer in Tensor Analysis and Relativity*, Undergraduate
Lecture Notes in Physics, https://doi.org/10.1007/978-3-030-26895-4

Printed in the United States
By Bookmasters